Lecture Notes in Computer Science 8486

Commenced Publication in 1973
Founding and Former Series Editors:
Gerhard Goos, Juris Hartmanis, and Jan van Leeuwen

Alexander S. Kulikov Sergei O. Kuznetsov
Pavel Pevzner (Eds.)

Combinatorial Pattern Matching

25th Annual Symposium, CPM 2014
Moscow, Russia, June 16-18, 2014
Proceedings

 Springer

Volume Editors

Alexander S. Kulikov
St. Petersburg Department of Steklov Institute of Mathematics
27 Fontanka
St. Petersburg 191023, Russia
E-mail: kulikov@logic.pdmi.ras.ru

Sergei O. Kuznetsov
National Research University Higher School of Economics
3 Bolshoy Trekhsvyatitelskiy pereulok
Moscow 109028, Russia
E-mail: skuznetsov@hse.ru

Pavel Pevzner
University of California at San Diego
9500 Gilman Drive, EBU3b 4236
La Jolla, CA 92093-0404, USA
E-mail: ppevzner@cs.ucsd.edu

ISSN 0302-9743   e-ISSN 1611-3349
ISBN 978-3-319-07565-5   e-ISBN 978-3-319-07566-2
DOI 10.1007/978-3-319-07566-2
Springer Cham Heidelberg New York Dordrecht London

Library of Congress Control Number: 2014939431

LNCS Sublibrary: SL 1 – Theoretical Computer Science and General Issues

Typesetting: Camera-ready by author, data conversion by Scientific Publishing Services, Chennai, India

Printed on acid-free paper

Springer is part of Springer Science+Business Media (www.springer.com)

Preface

This volume contains the papers presented at the 25th Annual Symposium on Combinatorial Pattern Matching held during June 15–17, 2014 in Moscow, Russia. The hosting university was National Research University Higher School of Economics.

There were 54 submissions from 25 countries. Each submission was reviewed by at least three Program Committee members. The committee decided to accept 28 papers, corresponding to an acceptance rate of 54%. We thank the members of the Program Committee and all additional external reviewers for their hard work that resulted in this excellent program. Their names are listed on the following pages. The whole submission and review process was carried out with the invaluable help of the EasyChair conference system.

The year 2014 marks the quarter-of-a-century milestone for the CPM symposium series. CPM 2014 celebrated this event with invited talks by the co-founders of the conference series: Alberto Apostolico (Georgia Institute of Technology, USA, and IASI-CNR, Italy), Maxime Crochemore (King's College London, UK, and Université Paris-Est, France), Zvi Galil (Georgia Institute of Technology, USA), and Udi Manber (Google, USA). In addition, CPM 2014 featured a special keynote lecture by Gene Myers (Max Planck Institute, Germany) that highlighted the contribution of the CPM community to bioinformatics.

The objective of the annual CPM meetings is to provide an international forum for research in combinatorial pattern matching and related applications. It addresses issues of searching and matching strings and more complicated patterns such as trees, regular expressions, graphs, point sets, and arrays. The goal is to derive combinatorial properties of such structures and to exploit these properties in order to achieve superior performance for the corresponding computational problems. The meeting also deals with problems in computational biology, data compression and data mining, coding, information retrieval, natural language processing, and pattern recognition.

The Annual Symposium on Combinatorial Pattern Matching started in 1990, and has since taken place every year. Previous CPM meetings were held in Paris, London (UK), Tucson, Padova, Asilomar, Helsinki, Laguna Beach, Aarhus, Piscataway, Warwick, Montreal, Jerusalem, Fukuoka, Morelia, Istanbul, Jeju Island, Barcelona, London (Ontario, Canada), Pisa, Lille, New York, Palermo, Helsinki, Bad Herrenalb. This years' meeting was the first in Russia. Starting from the third meeting, proceedings of all meetings have been published in the LNCS series, as volumes 644, 684, 807, 937, 1075, 1264, 1448, 1645, 1848, 2089, 2373, 2676, 3109, 3537, 4009, 4580, 5029, 5577, 6129, 6661, 7354, and 7922.

We thank Russian Foundation for Basic Research, National Research University Higher School of Economics, and Yandex for their financial support.

March 2014

Alexander S. Kulikov
Sergei Kuznetsov
Pavel Pevzner

Organization

Program Committee

Max Alekseyev	George Washington University, USA
Amihood Amir	Bar-Ilan University, Israel
Maxim Babenko	Moscow State University, Russia
Martin Farach-Colton	Rutgers University, USA
Paolo Ferragina	University of Pisa, Italy
Johannes Fischer	Technical University of Dortmund, Germany
Dan Gusfield	University of California at Davis, USA
Roman Kolpakov	Moscow State University, Russia
Gregory Kucherov	Université Paris-Est Marne-la-Vallée, France
Alexander S. Kulikov	St. Petersburg Department of Steklov Institute of Mathematics, Russia (Co-chair)
Juha Kärkkäinen	University of Helsinki, Finland
Gad M. Landau	University of Haifa, Israel
Stefano Lonardi	University of California at Riverside, USA
Ian Munro	University of Waterloo, Canada
S. Muthukrishnan	Rutgers University, USA
Gonzalo Navarro	University of Chile, Chile
Kunsoo Park	Seoul National University, South Korea
Pavel Pevzner	University of California San Diego, USA (Co-chair)
Nadia Pisanti	University of Pisa, Italy
Mikhail Roytberg	Higher School of Economics, Russia
Tatiana Starikovskaya	Higher School of Economics, Russia
Jim Storer	Brandeis University, USA
Jens Stoye	University of Bielefeld, Germany
Esko Ukkonen	University of Helsinki, Finland

Steering Committee

Alberto Apostolico	Georgia Institute of Technology, USA, and IASI-CNR, Italy
Maxime Crochemore	Université Paris-Est, France, and Kings College London, UK
Zvi Galil	Georgia Institute of Technology, USA

Organizing Committee

Stepan Artamonov   Moscow State University, Russia
Maxim Babenko   Moscow State University, Russia
Dmitry Ignatov   Higher School of Economics, Russia
Dmitry Ilvovsky   Higher School of Economics, Russia
Alexander S. Kulikov   St. Petersburg Department of Steklov Institute
    of Mathematics, Russia
Sergei Kuznetsov   Higher School of Economics, Russia, (Chair)
Dmitry Morozov   Higher School of Economics, Russia
Kamil Salihov   Moscow State University, Russia
Ruslan Savchenko   Moscow State University, Russia
Tatiana Starikovskaya   Higher School of Economics, Russia

Additional Reviewers

Aganezov, Sergey
Amit, Mika
Antipov, Dmitry
Artamonov, Stepan
Bankevich, Anton
Belazzougui, Djamal
Bowe, Alexander
Braga, Marilia
Butman, Ayelet
Chikhi, Rayan
Cicalese, Ferdinando
Cording, Patrick Hagge
Cunial, Fabio
Fedorov, Sergey
Feijao, Pedro
Fertin, Guillaume
Frid, Yelena
Gagie, Travis
Gawrychowski, Pawel
Giaquinta, Emanuele
Gysel, Rob
He, Meng
Hu, Fei
I, Tomohiro
Inenaga, Shunsuke
Jahn, Katharina
Jiang, Shuai
Kempa, Dominik

Kolesnichenko, Ignat
Kopczynski, Dominik
Labarre, Anthony
Levy, Avivit
Matsieva, Julia
Melsted, Pall
Mirebrahim, Seyed Hamid
Moret, Bernard
Nekrich, Yakov
Nielsen, Jesper Sindahl
Ottaviano, Giuseppe
Paparo, Omer
Pizzi, Cinzia
Polishko, Anton
Prencipe, Giuseppe
Puglisi, Simon
Pérez-Lantero, Pablo
Radoszewski, Jakub
Rogulenko, Sergey
Russo, Luis M. S.
Salikhov, Kamil
Savchenko, Ruslan
Sirotkin, Alexander
Sirén, Jouni
St. John, Katherine
Stevens, Kristian
Tarasov, Pavel
Thankachan, Sharma

Tomescu, Alexandru I.
Vasilevskaya, Maria
Venturini, Rossano
Vildhøj, Hjalte Wedel

Vind, Søren
Välimäki, Niko
Willing, Eyla
Wittler, Roland

Abstracts of Invited Talks

Sequence Comparison in the Time of the Deluge

Alberto Apostolico

Georgia Institute of Technology, USA and IASI-CNR, Italy

It is almost fifty years since the appearance of the famous paper entitled "Binary codes capable of correcting deletions, insertions, and reversals" (1966 English translation, Soviet Physics Doklady) by which Vladimir Levenshtein introduced his eponymous measure of string distance. The notion was since to be re-discovered, variously dithered and more or less efficiently computed in distant domains of application. In particular, it provided the platform for half a century of analysis, alignment, search, taxonomy and phylogeny of molecular sequences. As the size and multiplicity of the sequences produced expand to an unprecedented scale, many elegant techniques of the past no longer work. In molecular taxonomy and phylogeny, for instance, the alignment of whole genomes proves both computationally unbearable and hardly significant. In metagenomics, the elucidation of microbiome compositions under conditions of noisy and largely unidentified reference sequences faces steep barriers during assembly and assignment. In recent studies, classical notions are increasingly being complemented or even supplanted by global similarity measures that refer, implicitly or explicitly, to the composition of sequences in terms of their constituent patterns. Such measures hinge more or less uniformly on an underlying notion of relative compressibility, whereby two sequences are similar if either one can be described using mostly pieces from the other. They can free from chores of alignment and assembly. Their computation poses interesting and variously affordable algorithmic problems. This talk will review some such measures, their computation, and their applications.

Repeats in Strings

Maxime Crochemore

King's College London, UK, and Université Paris-Est, France

Large amounts of text are generated every day in the cyberspace via Web sites, emails, social networks, and other communication networks. These text streams need to be analysed to detect critical events or the monitor business for example. An important characteristics to take into account in this setting is the existence of repetitions in texts. Their study constitutes a fundamental area of combinatorics on words due to major applications to string algorithms, data compression, music analysis, and biological sequences analysis, etc. The talk surveys algorithmic methods used to locate repetitive segments in strings. It discusses the notion of runs that encompasses various types of periodicities considered by different authors, as well as the notion of maximal-exponent factors that captures the most significant repeats occurring in a string. The design and analysis of repeat finders rely on combinatorial properties of words and raise a series of open problems in combinatorics on words.

"Stringology" is 30 Years Old

Zvi Galil

Georgia Institute of Technology, USA

This year we have two anniversaries: This is the 25th CPM. CPM1, the first one, was held in Paris in 1990. But in June 1984 there was a sort of a predecessor of CPM1 which can be considered as CPM0. It was a Nato Workshop on combinatorial algorithms on words in Maratea, Italy. The first paper was my paper: "Open Problems in Stringology". Thus, the term Stringology is exactly 30 years old. The field of stringology is much older. Even though the word stringology cannot be found in any respectable dictionary, Google search yields 23,800 hits; in addition 1640 articles use it, 123 in their title.

In celebration of this 30th anniversary, we will review the status of the 13 open problems of the paper and will introduce some new ones.

How to Think Big

Udi Manber

Google Inc.

"If you have to ask, you'll never know"
Louis Armstrong (answering a question
about the meaning of "swing")

In business, making more impact used to mean making more money. But that changed. In 2010 Twitter's revenues were 40x less than cat litter, but its impact was enormous. Same with Amazon, Google, and Facebook at their beginnings. The key to all of them was that they were doing completely new things. No one thought there was a need to share 140 characters before Twitter, but it turned out to be extremely useful. Very few people thought at the time that shopping on the web, highly relevant search, or communication between friends were important business needs.

When the telephone was introduced to telegraph companies they discounted it. They had a good reason. More than a century later AT&T discounted voice over IP for similar reasons. Yet the 100+ years old AT&T was sold for less than the 5 years old WhatsApp. Things change rather quickly nowadays.

There are some common insights from these stories that can be applied to academic research and I will try to highlight them in this talk.

What's Behind Blast

Gene Myers

Max Planck Institute for Molecular Cell Biology and Genetics

While Blast is one of the most widely used search engines for molecular biology, and while there is a general understanding of how it works, few know the story of how it came about and the theoretical algorithmic result from which it was derived. I will tell the story and explain the theoretical algorithm — an $O(DN^{\mathrm{pow}(D/P)} \log N)$ expected-time algorithm for finding all matches to a query of length P with not more than D differences in a database of length $N \gg P$. Surprisingly, this result, published in early 1994, has to my knowledge not been improved upon to this day.

Table of Contents

On the Efficiency of the Hamming
C-Centerstring Problems

Amihood Amir[1,2,*], Jessica Ficler[1,**], Liam Roditty[1], and Oren Sar Shalom[1]

[1] Department of Computer Science, Bar-Ilan University, Ramat-Gan 52900, Israel
{amir,liamr}@cs.biu.ac.il, {jessica.ficler,oren.sarshalom}@gmail.com
[2] Department of Computer Science, Johns Hopkins University, Baltimore, MD 21218

Abstract. The *Consensus String Problem* is that of finding a string, such that the maximum Hamming distance from it to a given set of strings of the same length is minimized. However, a generalization is necessary for clustering. One needs to consider a partition into a number of sets, each with a distinct centerstring. In this paper we define two natural versions of the consensus problem for c centerstrings. We analyse the hardness and fixed parameter tractability of these problems and provide approximation algorithms.

1 Introduction

The *Hamming distance* between two strings $s_1, s_2 \in |\Sigma|^n$ is defined as the number of positions in which the strings differ and is denoted by $d_H(s_1, s_2)$. The *Consensus String Problem* is that of finding a string, such that the maximum Hamming distance from it to a given set of strings of the same length is minimized. Formally, let $S = \{s_1 \ldots s_k\}$ be a set of k elements in metric space M with metric m, the consensus problem is that of finding an element $\hat{s}$, called a consensus, such that $max\{m(\hat{s}, s_i) \mid i = 1 \ldots k\}$ is smallest. Note that the consensus need not be contained in S. Two metric spaces for which the consensus problem has attracted a lot of work are $\mathbb{R}^\ell$ with the Euclidean distance, and the space of ℓ-length sequences over alphabet Σ and the Hamming distance. Finding the consensus is one popular method for detecting data commonalities of many strings. The consensus problem for strings under the Hamming distance is also known as the *Closest String* Problem. The *Closest String* Problem is $\mathcal{NP}$-complete, even when the characters in the strings are drawn from the binary alphabet [14,8]. The *HRC* and *HRSC* problems deal with finding consensus strings in clusters. Below we define the *HRC* and *HRSC* problems. Denote the *Consensus String Problem* as *1-HRC*, and write *1-HRC(R)* to mean the *radius* of set R. It is the maximum distance between an element of R and the centerstring of R.

---

* Partly supported by NSF grant CCR-09-04581, ISF grant 347/09, and BSF grant 2008217.
** This work is part of Jessica Ficler's M.Sc. thesis.

A.S. Kulikov, S.O. Kuznetsov, and P. Pevzner (Eds.): CPM 2014, LNCS 8486, pp. 1–10, 2014.

Definition 1. *The* Hamming radius c-clustering problem (HRC or c-HRC) *is the following:*
Input: A set of strings S $= \{s_1 \ldots s_k\}$, *each of length* ℓ, *and a constant* d.
Output: A partition of S *into c disjoint sets (clusters)* $\{S_1 \ldots S_c\}$ *such that for each* $S_i \in \{S_1 \ldots S_c\}$: 1-HRC$(S_i) \leq d$, *if such a partition exists.* 1-HRC(S_i) *is the radius of cluster* S_i.

The *Hamming radius local c-clustering problem (HRLC or c-HRLC)* is a special case of *HRC* (defined as *HRC*, with the additional condition that the centers are from the input strings).
Input: A set of strings $S = \{s_1 \ldots s_k\}$, *each of length* ℓ, *and a constant* d.
Output: A partition of S *into c disjoint sets (clusters)* $\{S_1 \ldots S_c\}$ *such that for each* $S_i \in \{S_1 \ldots S_c\}$: $\exists s \in S_i$ *such that* $\forall s' \in S_i : d_H(s', s) \leq d$.
The HRLC problem is clearly polynomial-time computable for the 1 center version, since all that is required is to loop over S, set every $s \in S$ as a potential center and check whether all other strings are within radius d of the center.

Definition 2. *The* Hamming radius c-clustering sum problem (HRSC or c-HRSC) *is:*
Input: A set of strings S $= \{s_1 \ldots s_k\}$, *each of length* ℓ, *and a constant* d.
Output: A partition of S *into c disjoint sets (clusters)* $\{S_1 \ldots S_c\}$ *such that* $\sum_{i=1}^{c}$ 1-HRC$(S_i) \leq d$, *if such a partition exists.*

The Closest String problem has been extensively studied since its introduction by Lanctot, Li, Ma, Wang and Zhang [14] in SODA'99. The problem was studied both in computational biology and theoretical computer science. This extensive research can be roughly divided into three categories: (1) Approximate solutions [15,3,6]. (2) Fixed parameter solutions [11,12,16,15,17,5,1]. (3) Practical solutions [13,12,7,4,2]. For the more general HRC, Gasieniec et al. [9] presented a 2-approximation in $O(ck\ell)$ time based on Gonzalez' farthest-point-clustering algorithm [10]. They also showed that the general *HRC* can not be approximated within any constant factor smaller than two unless $\mathcal{P} = \mathcal{NP}$, and described a $2^{O(cd/\epsilon)}k^{O(c/\epsilon)}\ell^2)$ time algorithm with $(1 + \epsilon)$ approximation.

Clustering is used to solve classification problems in which the elements are partitioned into sets such that all elements in a set are similar in some sense. The focus of this paper is partitioning strings into clusters with a small radius. In web searching, it may be used for classifying data into clusters. Given a number of resources, we can divide our data such that each resource gets a set of similar strings. In biology, it may be used to create collections of similar DNA sequences.

We follow the path of the Consensus String problem. We start (Section 2) by studying the complexity and fixed parameter tractability of the *HRC* and *HRSC* problems. To our knowledge, this paper is the first to prove that these problems are $\mathcal{NP}$-hard for a *fixed constant* number of clusters. However, unlike the Consensus string problem, which is efficiently computable for any fixed parameter, we show the surprising result that this is not the case for the *HRC* and *HRLC* problems. We also provide efficient approximations for these problems.

Table 1. Fixed parameter tractability results

Problem	c fixed	k fixed	d fixed $(d = 1)$	d/ℓ and c fixed	ℓ fixed $(l = 2)$
HRC	$\mathcal{NP}$-hard	polynomial time	$\mathcal{NP}$-hard	polynomial time	?
$HRLC$	polynomial time	polynomial time	?	polynomial time	$\mathcal{NP}$-hard
$HRSC$	$\mathcal{NP}$-hard	polynomial time	?	polynomial time	?

2 $\mathcal{NP}$ Hardness and Parameterized Complexity of HRC, $HRLC$ and $HRSC$

The centerstring problem is fixed parameter tractable for any parameter: k - the number of input strings, ℓ - the length of the strings, and d - the distance from the center (the radius). We investigate fixed parameter tractability in the multi-center case. Table 1 describes the situation under various fixed parameters. The proofs are provided in the rest of this section.

Begin with the $\mathcal{NP}$-hardness of the problem itself. In [9] it was shown that HRC is not approximable to less than constant 2 where c is not a fixed constant. We show that both HRC and $HRSC$ are $\mathcal{NP}$-hard even for 2 centers.

Theorem 1. *2-HRC and 2-HRSC are $\mathcal{NP}$-complete.*

Theorem 2. *HRC,HRLC and HRSC can be solved in polynomial time for fixed k.*

Proof: If $k \le c$ then clearly the input strings can be assigned to c centers where $d = 0$. Otherwise, $c < k$ and there are $c^k \le k^k$ options for partitioning k strings to c sets. For each set, we can find the optimal consensus in polynomial time (because k is fixed). The partition that gives the best result is the optimal solution. $\qquad\square$

Theorem 3. *HRLC can be solved in polynomial time for fixed c.*

Proof: There are $\binom{k}{c}$ options for centers. For every option $C \subseteq S, |C| = c$ we assign each $s \in S$ to its closest center in C. The partition that gives the best result is the optimal solution. $\qquad\square$

Theorem 4. *HRC and HRSC can be solved in polynomial time for fixed d/ℓ and c.*

Proof: Notice that in HRC, $d \le l$, and in $HRSC$, $d \le c\ell$, We conclude that if c and ℓ are fixed, also d is fixed, so it is enough to show polynomial time algorithm for fixed d and c. For every $s_i \in S$ create a set $S_i = \{s | d_H(s, s_i) \le d\}$. S_i contains all strings whose Hamming distance from s_i is $\le d$. Notice that $|S_i| = \sum_{i=1}^{d} \binom{\ell}{i} |\Sigma - 1|^i \le d|\Sigma|^d \ell^d$ and that every $s \in S_i$ is a potential center.

We consider all combinations of c potential centers and for every combination, we go over all the strings in S and assign each string to its closest potential center in the combination. The solution is the combination that gives c sets that gives the best solution for $HRC/HRSC$. The number of combinations we need to check is $\leq \binom{kd|\Sigma|^d \ell^d}{c} \leq (kd)^c(|\Sigma|\ell)^{dc}$.   □

The theorem below is surprising since $1\text{-}HRC$ is tractable for a fixed radius. We see that HRC is not.

Theorem 5. HRC is $\mathcal{NP}$ complete even if the radius is fixed to $d = 1$.

Proof: We show that the decision problem is $\mathcal{NP}$ complete even if the radius is fixed to $d = 1$ and the alphabet is binary, by reduction from *Vertex Cover For Triangle-Free Graphs*. Given (G,t) we construct an input for the HRC problem in the following way:
1. If G has $|V|$ vertices and $|E|$ edges, we set ℓ to $|V|$ and k to $|E|$.
2. For every edge $(u_i, u_j) \in E$ we create a string s ,s is a string filled with zeros except in the indices i, j, there $s[i] = s[j] = 1$.
3. The c parameter (represents the number of sets that the strings will be divided into) is t.
4. The distance parameter d is set to 1.
   We show that: $(G,t) \in VC \Leftrightarrow (S,c,1) \in HRC$:
$(\Rightarrow)$ $(G,t) \in VC$ therefore there are $\{v_{r_1} \cdots v_{r_t}\} \in V$ that cover E. We show a partition of S into c sets $S_{r_1} \cdots S_{r_c}$ such that in every set, the maximum distance from a string to its center is at most 1. Formally, $\forall 1 \leq i \leq c$ there is a string c_{r_i} such that $\forall s \in S_{r_i} : d_H(s, c_{r_i}) \leq 1$. For each $1 \leq i \leq t$, define S_{r_i} to contain $s \in S$ if $s[r_i] = 1$. Notice that if there are $1 \leq i \leq c$ and $1 \leq j \leq c$ such that $(u_i, u_j) \in E$, then there exists a string $s_{ij} \in S$ filled with zeros in all coordinates except i,j, there $s'[i] = s'[j] = 1$. By definition s_{ij} can be assigned to two sets. We put arbitrary s_{ij} in one of them, since we want the sets $S_{r_1}.....S_{r_c}$ to be disjoint. Notice that necessarily each $s \in S$ is assigned to one of the c sets. Each $S_{r_x} \in \{S_{r_1}.....S_{r_c}\}$ contains strings that all of them have symbol 1 in index r_x. Example for set S_{r_x}:

s_{xy} : $0\cdots010\cdots010\cdots0000\cdots00$
s_{xz} : $0\cdots010\cdots000\cdots0100\cdots00$
s_{ux} : $0\cdots010\cdots000\cdots0000\cdots01$
s_{vx} : $0\cdots010\cdots000\cdots0010\cdots00$

(Notice that the string c_{r_x} : $0\cdots010\cdots000\cdots0000\cdots00$ is a center for S_{r_x} with radius 1)
Clearly the string c_{r_x} that is filled with zeros except for index r_x, where the symbol is 1, satisfies that $\forall s \in S_{r_x} : d_H(s, c_{r_x}) \leq 1$, so $(S,c,1) \in HRC$.
$(\Leftarrow)$ $(G,t) \notin VC$ so there are no $\{v_{r_1}.....v_{r_t}\} \in V$ that cover E. It means that we cannot divide the edges to t groups such that in each group there is a vertex that is common to every edge in the group. It follows that for every partition of S to c sets, there is always at least one set S' that has no column i such that $\forall s \in S' : s[i] = 1$. We will show that there is no center with radius 1 for S'. Notice that S' contains at least 2 strings. Consider two cases:

1. There are at least 2 strings in S' that are different in 4 columns (for example 0101 and 1010). For each of these 4 columns, at least one string is different from the center in this column. By the Pigeonhole principle, there is at least one string whose distance from the center is 2.
2. Each string in S' is different in 2 columns from any other string in S'. For each $s_a \in S'$ where s_a has symbol 1 in columns i, j, for every $s_a \neq s_b \in S'$: $s_a[i] = s_b[i]$ and $s_a[j] \neq s_b[j]$, or: $s_a[j] = s_b[j]$ and $s_a[i] \neq s_b[i]$. For example:

   0..010..010..000..000..0
   0..000..010..010..000..0
   0..010..000..010..000..0

   There are exactly 3 i_1, i_2, i_3 columns where the strings disagree. We consider the strings only in these columns, for example 110, 011 and 101. Notice that the existence of these three strings indicates that the edges $(u_{i_1}, u_{i_2}), (u_{i_2}, u_{i_3}), (u_{i_1}, u_{i_3})$ are in E, this contradicts that the graph is triangle-free.

The conclusion is that there is no string c' such that $\forall s \in S' : d_H(s, c') \leq 1$, so $(S, c, 1) \notin HRC$. $\qquad \square$

Theorem 6. *HRLC is $\mathcal{NP}$ complete even if the length is fixed to $l = 2$.*

Proof: We show that the decision problem is $\mathcal{NP}$ complete even if the length is fixed to $\ell = 2$ by reduction from *Mimimum Maximal Matching for Bipartite graphs (MMMB)*. Given (G, t) we construct an input for the *HRLC* problem in the following ,manner:
1. If G has $|V|$ vertices and $|E|$ edges, we set k to $|E|$.
2. For every edge $(u_i, u_j) \in E$ where u_i is in the left side of the bipartite graph and u_j is in the right side, we create the strings $s = (u_i, u_j)$.
3. The number of clusters c is set to t.
4. The distance parameter d is set to 1.
   We need to show that: $(G, t) \in MMMB \Leftrightarrow (S, c, 1) \in HRLC$:
$(\Rightarrow)$ $(G, t) \in MMMB$ therefore there is a group of edges $\{e_{r_1} \cdots e_{r_t}\} \in E$ which is a maximal matching that contains t edges. We show a partition of S into c sets $S_{r_1} ..... S_{r_c}$ such that in every set, the maximum distance from a string to its center is at most 1. For each $1 \leq i \leq c$, define S_{r_i} to contain:
1. The string that was created from e_{r_i} : $(u_{r_i}^1, u_{r_i}^2)$.
2. Every string $s \in S$ such that $s[1] = u_{r_i}^1$ or $s[2] = u_{r_i}^2$.
Clearly, $(u_{r_i}^1, u_{r_i}^2)$ is a center with radius 1 for S_i. Notice that all strings in S are assigned to some S_i. Suppose there is a string $s' = (x, y)$ that is not assigned to any set, hence for all $i \in \{1 \cdots c\} : x \neq u_{r_i}^1$ and $y \neq u_{r_i}^2$, it follows that the edge (x, y) has no common vertex with any e_{r_i}, in contradiction to the fact that the set $\{e_{r_1} \cdots e_{r_t}\}$ is a maximal matching, and hence $(S, c, 1) \in HRLC$.
$(\Rightarrow)$ $(S, c, 1) \in HRLC$ therefore there is a division of S into c sets $S_{r_1} \cdots S_{r_c}$ such that every set has a string $c_{r_i} \in S_{r_i}$ whose distance from any other string is the set is at most 1. Consider two cases:

1. There are no two centers in $c_{r_1} \cdots c_{r_c}$ that have a common symbol : We define e_{r_i} to be the corresponding edge to c_{r_i}. We prove that the edges $\{e_{r_1} \cdots e_{r_t}\}$

are a maximal matching of size $t = c$. Suppose that $c_1 \cdots c_c$ is not a maximal matching, so there is an edge (x, y) such that $\forall e \in \{e_{r_1}.....e_{r_t}\} : e[1] \neq x$ and $e[2] \neq y$. Notice that there is $i \in \{1 \cdots c\}$ such that the string (x, y) is in S_{r_i}. S_{r_i} contains the strings $c_{r_i} = (c_{r_i}^1, c_{r_i}^2)$ and (x, y) at least. All 4 symbols $c_{r_i}^1, c_{r_i}^2, x, y$ are different from each other, so $d_H(c_{r_i}, (x, y)) = 2$, in contradiction to the fact that c_{r_i} is a center for S_{r_i} with a radius at most 1.

2. There are two centers in $c_{r_i}, c_{r_j} \in \{c_{r_1} \cdots c_{r_c}\}$ that have a common symbol : For every two centers c_{r_i}, c_{r_j} that have a common symbol, without loss of generality we assume that $c_{r_i} = (a, b), c_{r_j} = (a, c)$ (notice that $a \neq b$, $a \neq c$, $b \neq c$) have a common symbol at the first coordinate ($c_{r_i}[1] = c_{r_j}[1]$). We show that we can rearrange the strings into at most c sets such that all the center's symbols are different. We do the following:

    (a) S_{r_i} contains strings from form $(a, -)$ or $(-, b)$. We move the strings from the form $(a, -)$ to S_{r_j}. The radius of S_{r_j} remains 1. If S_{r_i} is empty we are done.

    (b) S_{r_i} contains $(-, b)$. For each $(z, b) \in S_{r_i}$, if there is a center $c_{r_m} = (z, -)$, we move (z, b) to S_{r_m}.

    (c) If S_{r_i} still contains strings, we arbitrary choose one of them (w, b) and change c_{r_i} to be (w, b).

We keep going until there are no two centers with common symbol. By (1), $(G, t) \in MMMB$.

3   Approximations for HRC and HRSC

In this section we provide a number of polynomial time approximation algorithms for the c-HRC and c-HRSC problems.

Lemma 1. *Let (S, d) be an input for the c-HRC or c-HRSC problem. Assume $S_1 \cdots S_c$ are a partition of S into c clusters with radius not exceeding d. If $d_H(s_i, s_j) > 2d$ for each $s_i \in S_i$, $s_j \in S_j$ $(i \neq j)$ then the optimal solution can be approximated in polynomial time.*

Proof: Since $\{S_1 \cdots S_c\} \subseteq S$ are the optimal partition of S to c sets then there exist $c_1 \cdots c_c$ such that for every $1 \leq i \leq c$, $s \in S_i$ $d_H(s, c_i) \leq d$. For each $1 \leq i \leq c$ and $s, s' \in S_i$, $d_H(s, s') \leq 2d$ because the distance between two strings from the same cluster can not exceed $2d$. Our algorithm will go over all the $\binom{k}{c}$ options for the group of strings $X = \{x_1 \cdots x_c\}$ with size c. For each such group we construct a partition $S_1 \cdots S_c$ of S in the following manner: S_i is the set of all $s \in S$ for which $d_H(s, x_i) \leq 2d$. Notice that every string $s \in S_i$ can not be contained in any other set S_j, because we are given that for each $1 \leq i \leq c, 1 \leq j \leq c$ such that $i \neq j$, $\forall s_i \in S_i, \forall s_j \in S_j$, $d_H(s_i, s_j) > 2d$. Now use an approximation for finding the consensus strings of each S_i. Choose a result where there is a partition to c clusters with radius d. Clearly the time is polynomial. The approximation ratio is the same as for the *1-HRC* problem. □

Lemma 2. *Let (S, d) be an input for the c-HRC or c-HRSC problem. Assume $S_1 \cdots S_c$ are a partition of S into c clusters, and let $c_1 \cdots c_c$ be the consensus* *strings of the clusters, resp. If for each $1 \leq i \leq c, 1 \leq j \leq c$ $(i \neq j)$, $d_H(c_i, c_j) > 4d$ then the solution for the problem can be approximated in polynomial time.*

Proof: Since $c_1 \cdots c_c$ are the consensus strings of $S_1 \cdots S_c$ then for every $s \in S_i$ $d_H(s, c_i) \leq d$. We will show that for every $s_m \in S_m$ and for every $s_n \in S_n$ $(m \neq n)$: $d_H(s_m, s_n) > 2d$. Let (s_m, s_n) be an arbitrary pair from $S_m \times S_n$. Let $I_c = \{i \mid c_m[i] \neq c_n[i]\}$ be a set of indices where the two centers do not have the same symbol. Similarly let $I_m = \{i \mid c_m[i] \neq s_m[i]\}$ and $I_n = \{i \mid c_n[i] \neq s_n[i]\}$. s_m and s_n are closest when $I_m, I_n \subseteq I_c$ and $I_m \cap I_n = \emptyset$. We are interested in the size of the set $I_s = \{i \mid s_m[i] \neq s_n[i]\}$. Notice that $|I_s| \geq |I_c| - |I_m| - |I_n|$. Clearly $|I_c| > 4d$, $|I_m| \leq d$, and $|I_n| \leq d$. Therefore $|I_s| \geq |I_c| - |I_m| - |I_n| > 4d - d - d = 2d$.

Example:

(Notice that $d_H(s_m, s_n) \geq 2d + x > 2d$)

The distance between each pair from different clusters is $> 2d$, so by Lemma1, we can approximate c-HRC and c-$HRSC$ in polynomial time. □

The algorithm below is a linear-time algorithm that gives a 4-approximation for the *2-HRSC* problem.

Algorithm 4APPX for *2-HRSC*

1. Choose an arbitrary string from S, denote it by v_1. Then search the most distant string from v_1, denote it by v_2
2. Put v_1, v_2 in different sets, v_1 in set $\hat{A}$ and v_2 in set $\hat{B}$
3. For each $v \in S$ such that $v \neq v_1, v_2$, if $d_H(v, v_1) \geq d_H(v, v_2)$ put v in set $\hat{A}$, otherwise put v in set $\hat{B}$.
4. Let $\hat{r_1}$ be the maximum distance between any $v \in \hat{A}$ to v_1 and $\hat{r_2}$ be the maximum distance between any $v \in \hat{B}$ to v_2
5. return $\hat{r_1} + \hat{r_2}$

Claim: Algorithm 4APPX gives a 4-approximation for the *2-HRSC* problem.
Proof: Let A, B to be an optimal partition. Denote the consensus strings of sets A, B by c_a, c_b, resp, and the optimal solution by $r_1 + r_2$ when r_1 is the radius of set A and r_2 is the radius of set B. Consider two cases:

1. **At the optimal solution, v_1 and v_2 are on the same set** - Without loss of generality, we say that v_1 and v_2 are both in set A. Given a string v we consider three cases:

(a) In the optimum, $v \in A$ (in the same set with v_1 and v_2) - that means that $d(v,c_a) \leq r_1$. By our algorithm, v is in a set with v_1 or v_2. By the triangle inequality it holds that $d(v,v_1) \leq 2 \cdot r_1$ and that $d(v,v_2) \leq 2 \cdot r_1$, so v can contribute at most $2 \cdot r_1$ to the group to which the algorithm assigns it.

(b) In the optimum, $v \in B$ and our algorithm assign it to group $\hat{A}$ - v_2 is the most distant string from v_1, so it holds that $d(v_1,v) \leq 2 \cdot r_1$. So v can contribute at most $2 \cdot r_1$ to group $\hat{A}$.

(c) In the optimum, $v \in B$ and our algorithm assign it to group $\hat{B}$ - we showed in the previous sub-case that $d(v_1,v) \leq 2 \cdot r_1$. Out algorithm assign v to group B, that means that $d(v_2,v) \leq d(v_1,v) \leq 2 \cdot r_1$. So v can contribute at most $2 \cdot r_1$ to group $\hat{B}$.

v can contribute at most $2 \cdot r_1$ for group $\hat{A}$ or group $\hat{B}$, so $\hat{r_1} + \hat{r_2} \leq 2 \cdot r_1 + 2 \cdot r_1 \leq 4 \cdot r_1 \leq 4 \cdot r_1 + 4 \cdot r_2$.

2. **At the optimal solution, v_1 and v_2 are not on the same set** - We consider the 4 types of strings:

(a) In the optimum, $v \in A$ and $d_H(v,v_1) \leq d_H(v,v_2)$.

(b) In the optimum, $v \in B$ and $d_H(v,v_2) \leq d_H(v,v_1)$.

(c) In the optimum, $v \in A$ and $d_H(v,v_1) > d_H(v,v_2)$.

(d) In the optimum, $v \in B$ and $d_H(v,v_2) > d_H(v,v_1)$.

In the first case, notice that $d(v,v_1) \leq 2 \cdot r_1$. Symmetrically, in the second case, $d(v,v_2) \leq 2 \cdot r_2$. So in the first two cases, v will contribute twice the optimum for the group it assign to. We need to analyze the last two cases. We will analyze case (c), case (d) is symmetric. For string v that in the optimal solution is in set A but whose distance to v_1 is greater than to v_2, the algorithm will associate v to set $\hat{B}$. In the optimum v and v_1 are in the same group so by the triangle inequality it holds that $d(v,v_1) \leq 2 \cdot r_1$. Our algorithm assign v to be in the same set as v_2 because $d(v,v_2) \leq d(v,v_1) \leq 2 \cdot r_1$. Therefore v will contribute at most $2 \cdot r_1$ to group $\hat{B}$. Symmetrically, in case (d), v will contribute at most $2 \cdot r_2$ to group $\hat{A}$. In total, $\hat{r_1} \leq max(2 \cdot r_1, 2 \cdot r_2)$ and $\hat{r_2} \leq max(2 \cdot r_1, 2 \cdot r_2)$ so the algorithm approximation is $\hat{r_1} + \hat{r_2} \leq 4 \cdot max(r_1, r_2) \leq 4 \cdot (r_1 + r_2)$. $\square$

We can get a better approximation for the *2-HRSC* problem if we sacrifice linearity.

Theorem 7. *There is a polynomial time algorithm that finds an approximation for the 2-HRSC problem whose ratio is 3 times the ratio of 1-HRC approximation.*

Proof: Assume that $A \subseteq S$ and $B \subseteq S$ ($A \cap B = \emptyset$) are an optimal partition of S into 2 sets. There exist c_a, c_b such that for every $s \in A$ $d_H(s, c_a) \leq d_a$ and for every $s \in B$ $d_H(s, c_b) \leq d_b$. Let $s_a \in A$, $s_b \in B$ be the closest such pair. We construct two sets $\hat{A}$ and $\hat{B}$ as follows. First we put s_a in A and s_b in B, now for each $s \in S$, we assign s to set $\hat{A}$ if $d_H(s, s_a) \leq d_H(s, s_b)$ and $\hat{B}$ otherwise. To prove the approximation, we consider 4 cases: (1) In the optimum, $s \in A$ and $d_H(s, s_a) \leq d(s, s_b)$. The algorithm assign s to $\hat{A}$. (2) In the optimum, $s \in B$ and

$d_H(s, s_b) \leq d(s, s_a)$. The algorithm assign s to $\hat{B}$. (3) In the optimum, $s \in A$ and $d_H(s, s_a) > d(s, s_b)$. The algorithm assign s to $\hat{B}$. (4) In the optimum, $s \in B$ and $d_H(s, s_b) > d(s, s_a)$. The algorithm assign s to $\hat{A}$.

We examine the distance of the strings in set $\hat{X}$ ($X \in \{A, B\}$) from the center c_x, notice that for s in case 1, $d_H(s, c_a) \leq d_a$ and for s in case 2, $d_H(s, c_b) \leq d_b$. We analyze case 3, case 4 is symmetric. For string s that in the optimal solution is in set A and $d_H(s, s_b) \leq d_H(s, s_a) \leq 2d_a$, the algorithm will associate s to set $\hat{B}$. According to the triangle inequality $d_H(s, c_b) \leq d_H(s_b, c_b) + d_H(s_b, s) \leq d_b + 2d_a$. The total maximal distance of $s \in \hat{B}$ to c_b is $\leq d_b + 2d_a$ and the maximal distance of $s \in \hat{A}$ to c_a is $\leq d_a + 2d_b$. Now assume that the 1-HRC approximation algorithm gives $nOPT + m$, it follows that if we use this algorithm for $\hat{A}$ and $\hat{B}$ we will get $\leq n(d_b + 2d_a) + m$ and $\leq n(d_a + 2d_b) + m$, so the total solution is $\leq (n(d_b + 2d_a) + m) + (n(d_a + 2d_b) + m) \leq 3n(d_a + d_b) + 2m \leq 3(n(d_a + d_b) + m)$. $\square$

References

1. Amir, A., Landau, G.M., Na, J.C., Park, H., Park, K., Sim, J.S.: Consensus optimizing both distance sum and radius. In: Karlgren, J., Tarhio, J., Hyyrö, H. (eds.) SPIRE 2009. LNCS, vol. 5721, pp. 234–242. Springer, Heidelberg (2009)
2. Amir, A., Paryenty, H., Roditty, L.: Approximations and partial solutions for the consensus sequence problem. In: Grossi, R., Sebastiani, F., Silvestri, F. (eds.) SPIRE 2011. LNCS, vol. 7024, pp. 168–173. Springer, Heidelberg (2011)
3. Andoni, A., Indyk, P., Patrascu, M.: On the optimality of the dimensionality reduction method. In: Proc. 47th IEEE Symposium on the Foundation of Computer Science (FOCS), pp. 449–458 (2006)
4. Verbin, E., Yu, W.: Data structure lower bounds on random access to grammar-compressed strings. In: Fischer, J., Sanders, P. (eds.) CPM 2013. LNCS, vol. 7922, pp. 247–258. Springer, Heidelberg (2013)
5. Boucher, C., Brown, D.G., Durocher, S.: On the structure of small motif recognition instances. In: Amir, A., Turpin, A., Moffat, A. (eds.) SPIRE 2008. LNCS, vol. 5280, pp. 269–281. Springer, Heidelberg (2008)
6. Boucher, C., Wilkie, K.: Why Large CLOSEST STRING instances are easy to solve in practice. In: Chavez, E., Lonardi, S. (eds.) SPIRE 2010. LNCS, vol. 6393, pp. 106–117. Springer, Heidelberg (2010)
7. Chimani, M., Woste, M., Böcker, S.: A closer look at the closest string and closest substring problem. In: Proc. 13th Workshop on Algorithm Engineering and Experiments (ALENEX), pp. 13–24 (2011)
8. Frances, M., Litman, A.: On covering problems of codes. Theory of Computing Systems 30(2), 113–119 (1997)
9. Gasieniec, L., Jansson, J., Lingas, A.: Approximation algorithms for hamming clustering problems. Journal of Discrete Algorithms 2(2), 289–301 (2004)
10. Gonzalez, T.F.: Clustering to minimize the maximum intercluster distance. Theoretical Computer Science 38, 293–306 (1985)
11. Gramm, J., Niedermeier, R., Rossmanith, P.: Exact solutions for closest string and related problems. In: Eades, P., Takaoka, T. (eds.) ISAAC 2001. LNCS, vol. 2223, pp. 441–453. Springer, Heidelberg (2001)
12. Gramm, J., Niedermeier, R., Rossmanith, P.: Fixed-parameter algorithms for closest string and related problems. Algorithmica 37(1), 25–42 (2003)

13. Hufsky, F., Kuchenbecker, L., Jahn, K., Stoye, J., Böcker, S.: Swiftly computing center strings. In: Moulton, V., Singh, M. (eds.) WABI 2010. LNCS, vol. 6293, pp. 325–336. Springer, Heidelberg (2010)
14. Lanctot, K., Li, M., Ma, B., Wang, S., Zhang, L.: Distinguishing string selection problems. Information and Computation 185(1), 41–55 (2003)
15. Ma, B., Sun, X.: More efficient algorithms for closest string and substring problems. SIAM J. Computing 39(4), 1432–1443 (2009)
16. Stojanovic, N., Berman, P., Gumucio, D., Hardison, R., Miller, W.: A linear-time algorithm for the 1-mismatch problem. In: Rau-Chaplin, A., Dehne, F., Sack, J.-R., Tamassia, R. (eds.) WADS 1997. LNCS, vol. 1272, pp. 126–135. Springer, Heidelberg (1997)
17. Sze, S.-H., Lu, S., Chen, J.: Integrating sample-driven and pattern-driven approaches in motif finding. In: Jonassen, I., Kim, J. (eds.) WABI 2004. LNCS (LNBI), vol. 3240, pp. 438–449. Springer, Heidelberg (2004)

Dictionary Matching with One Gap[*]

Amihood Amir[1,2,**], Avivit Levy[3], Ely Porat[1], and B. Riva Shalom[3]

[1] Department of Computer Science, Bar-Ilan University, Ramat-Gan 52900, Israel
{amir,porately}@cs.biu.ac.il
[2] Department of Computer Science, Johns Hopkins University, Baltimore, MD 21218
[3] Department of Software Engineering, Shenkar College, Ramat-Gan 52526, Israel
{avivitlevy,rivash}@shenkar.ac.il

Abstract. The dictionary matching with gaps problem is to preprocess a dictionary D of d gapped patterns $P_1, \ldots, P_d$ over alphabet Σ,
where each gapped pattern P_i is a sequence of subpatterns separated
by bounded sequences of don't cares. Then, given a query text T of
length n over alphabet Σ, the goal is to output all locations in T in
which a pattern $P_i \in D$, $1 \leq i \leq d$, ends. There is a renewed current interest in the gapped matching problem stemming from cyber security. In this paper we solve the problem where all patterns in the
dictionary have one gap with at least α and at most β don't cares,
where α and β are given parameters. Specifically, we show that the
dictionary matching with a single gap problem can be solved in ei-
ther $O(d \log d + |D|)$ time and $O(d \log^\varepsilon d + |D|)$ space, and query time
$O(n(\beta - \alpha) \log \log d \log^2 \min\{d, \log |D|\} + occ)$, where occ is the number
of patterns found, or preprocessing time: $O(d^2 \cdot ovr + |D|)$, where ovr is
the maximal number of subpatterns including each other as a prefix or
as a suffix, space: $O(d^2 + |D|)$, and query time $O(n(\beta - \alpha) + occ)$, where
occ is the number of patterns found. As far as we know, this is the best
solution for this setting of the problem, where many overlaps may exist
in the dictionary.

1 Introduction

Pattern matching has been historically one of the key areas of computer science.
It contributed many important algorithms and data structures that made their
way to textbooks, but its strength is that it has been contributing to applied
areas, from text searching and web searching, through computational biology
and to cyber security. One of the important variants of pattern matching is
pattern matching with variable length gaps. The problem is formally defined
below.

---

[*] This research was supported by the Kabarnit Cyber consortium funded by the Chief
Scientist in the Israeli Ministry of Economy under the Magnet Program.
[**] Partly supported by NSF grant CCR-09-04581, ISF grant 347/09, and BSF grant
2008217.

A.S. Kulikov, S.O. Kuznetsov, and P. Pevzner (Eds.): CPM 2014, LNCS 8486, pp. 11–20, 2014.

Definition 1. Gapped Pattern

A Gapped Pattern is a pattern P of the form p_1 $\{\alpha_1, \beta_1\}$ p_2 $\{\alpha_2, \beta_2\}$ ... $\{\alpha_{k-1}, \beta_{k-1}\}$ p_k,
where each subpattern p_j, $1 \leq j \leq k$ is a string over alphabet Σ, and $\{\alpha_j, \beta_j\}$ refers to a sequence of at least α_j and at most β_j don't cares between the subpatterns P_j and P_{j+1}.

Definition 2. The Gapped Pattern Matching Problem:

Input: A text T of length n, and a gapped pattern P over alphabet Σ
Output: All locations in T, where the pattern P ends.

The problem arose a few decades ago by real needs in computational biology applications [20,12,22,14]. For example, the PROSITE database [14] supports queries for proteins specified by gaps.

The problem has been well researched and many solutions proposed. The first type of solutions [7,19,23] consider the problem as a special case of regular expression. The best time achieved using this method is $O(n(B|\Sigma| + m))$, where n is the text length, $B = \sum_{i=1}^{k} \beta_i$ (the sum of the upper bounds of the gaps), and $m = \sum_{i=1}^{k} p_i$ (the length of the non-gapped part of the pattern).

Naturally, a direct solution of the gapped pattern matching problem should have a better time complexity. Indeed, such a solution exists [8] whose time is essentially $O(nk)$.

A further improvement [18,24,6] analyses the time as a function of *socc*, which is the number of times all segments p_i of the gapped pattern appear in the text. Clearly *socc* $\leq nk$.

Rahman et al. [24] suggest two algorithms for this problem. In the first, they build an Aho-Corasick [1] pattern matching machine from all the subpatterns and use it to go over the text. Validation of subpatterns appearances with respect to the limits of the gaps is performed using binary search over previous subpatterns locations. Their second algorithm uses a suffix array built over the text to locate occurrences of all subpatterns. For the validation of occurrences of subpattern, they use Van Emde Boas data structure [26] containing ending positions of previous occurrences. In order to report occurrences of the gapped pattern in both algorithms, they build a graph representing legal appearances of consecutive subpatterns. Traversing the graph yields all possible appearances of the pattern.

Their first algorithm works in time $O(n + m + socc \log(\max_j gap_j))$ where m is the length of the pattern (not including the gaps), *socc* is the total number of occurrences of the subpatterns in the text and $gap_j = \beta_j - \alpha_j$. The time requirements of their second algorithm is $O(n + m + socc \log \log n)$ where n is the length of the text, m is the length of the pattern (not including the gaps) and *socc* is the total number of occurrences of the subpatterns in the text. The DFS traversal on the subpatterns occurrences graph, reporting all the occurrences is done in $O(k \cdot occ)$ where *occ* is the number of occurrences of the complete pattern P in the text.

Bille et al. [6] also consider string matching with variable length gaps. They present an algorithm using sorted lists of disjoint intervals, after traversing the text with Aho-Corasick automaton. Their time complexity is $O(n \log k + m + socc)$ and space $O(m + A)$, where A is the sum of the lower bounds of the lengths of the gaps in the pattern P and $socc$ is the total number of occurrences of the substrings in P within T.

Kucherov and Rusinowitch [16] and Zhang et al. [29] solved the problem of matching a set of patterns with variable length of don't cares. They considered the question of determining whether one of the patterns of the set matches the text and report a leftmost occurrence of a pattern if there exists one. The algorithm of [16] has run time of $O((|t| + |D|) \log |P|)$, where $|D|$ is the total length of keywords in every pattern of the dictionary D. The algorithm of [29] takes $O((|t| + dk) \log dis / \log \log dis)$ time, where dk is the total number of keywords in every pattern of P, and dis is the number of distinct keywords in D.

There is a renewed current interest in the gapped matching problem stemming from a crucial modern concern - cyber security. Network intrusion detection systems perform protocol analysis, content searching and content matching, in order to detect harmful software. Such malware may appear on several packets, and thus the need for gapped matching [15]. However, the problem becomes more complex since there is a large list of such gapped patterns that all need to be detected. This list is called a *dictionary*. Dictionary matching has been amply researched in computer science (see e.g. [1,4,3,9,5,2,10]). We are concerned with a new dictionary matching paradigm - *dictionary matching with gaps*. Formally:

Definition 3. The Dictionary Matching with gaps (DMG) Problem:
*Input: A text T of length n over alphabet Σ and a dictionary D over
        alphabet Σ consisting of d gapped patterns $P_1, \ldots, P_d$.*
Output: All locations in T, where a pattern P_i, for $1 \leq i \leq d$ ends.

The DMG problem has not been sufficiently studied yet. Haapasalo et al. [13] give an on-line algorithm for the general problem. Their algorithm is based on locating "keywords" of the patterns in the input text, that is, maximal substrings of the patterns that contain only input characters. Matches of prefixes of patterns are collected from the keyword matches, and when a prefix constituting a complete pattern is found, a match is reported. In collecting these partial matches they avoid locating those keyword occurrences that cannot participate in any prefix of a pattern found thus far. Their experiments show that this algorithm scales up well, when the number of patterns increases. They report at most one occurrence for each pattern at each text position. The time required for their algorithm is $O(n \cdot SUF + occ \cdot PREF)$, where n is the size of the text, SUF is the maximal number of suffixes of a keyword that are also keywords, and $PREF$ denotes the number of occurrences in the text of pattern prefixes ending with a keyword.

Nevertheless, more research on this problem is needed. First, in many applications it is necessary to report all patterns appearances. Moreover, as far as we know [27], these Aho-Corasick automaton based methods fail when applied to real data security, which contain many overlaps in the dictionary patterns, due to overhead in the computation when run on several ports in parallel. Therefore, other methods should be developed and combined with the existing ones in order to·design efficient practical solutions.

Results. In this paper, we indeed suggest other directions for solving the problem. We focus on the DMG problem where the gapped patterns in the dictionary D have only a single gap, i.e., we consider the case of $k = 1$ implying each pattern P_i consists of two subpatterns $P_{i,1}, P_{i,2}$. In addition, we consider the same gaps limits, α and β, apply to all patterns in $P_i \in D$, $1 \le i \le d$. We prove:

Theorem 1. *The dictionary matching with a single gap problem can be solved in:*

1. *Preprocessing time: $O(d \log d + |D|)$.*
   Space: $O(d \log^\varepsilon d + |D|)$, for arbitrary small ε.
   Query time: $O(n(\beta - \alpha) \log \log d \log^2 \min\{d, \log |D|\} + occ)$, where occ is the number of patterns found.
2. *Preprocessing time: $O(d^2 \cdot ovr + |D|)$, where ovr is the maximal number of subpatterns including each other as a prefix or as a suffix.*
   Space: $O(d^2 + |D|)$.
   Query time: $O(n(\beta - \alpha) + occ)$, where occ is the number of patterns found.

Note that, $|D|$ is the sum of lengths of all patterns in the dictionary, *not including* the gaps sizes.

The paper is organized as follows. In Sect. 2 we describe our basic method based on suffix trees and prove the first part of Theorem 1. In Sect. 3 we describe how this algorithm query time can be improved while doing more work in the preprocessing time and prove the second part of Theorem 1. We also discuss efficient implementations of both algorithms using text splitting in Subsect. 3.1. Section 4 concludes the paper and poses some open problems.

2   Bidirectional Suffix Trees Algorithm

The basic observation used by our algorithm is that if a gapped pattern P_i appears in T, then searching to the left from the start position of the gap we should find the reverse of the prefix $P_{i,1}$, and searching to the right from the end position of the gap we should find the suffix $P_{i,2}$. A similar observation was used by Amir et al. [5] to solve the dictionary matching with one mismatch problem. Their problem is different from the DMG problem, since they consider a single mismatch while in our problem a gap may consist of several symbols. Moreover, a mismatch symbol appears both in the dictionary pattern and in the text, while in the DMG problem the gap implies skipping symbols only in the text. Nevertheless, we show that their idea can also be adopted to the solve the DMG problem.

Amir et al. [5] use two suffix trees: one for the concatenation of the dictionary and the other for the reverse of the concatenation of the dictionary. Combining this with set intersection on tree paths, they solved the dictionary matching with one mismatch problem in time $O(n \log^{2.5} |D| + occ)$ where n is the length of the text, $|D|$ is the sum of the lengths of the dictionary patterns, and occ is the number of occurrences of patterns in the text. Their preprocessing requires $O(|D| log|D|)$. We use the idea to design an algorithm to the DMG problem.

A naive method is to consider matching the prefixes $P_{i,1}$ for all $1 \leq i \leq d$, and then look for the suffixes subpatterns $P_{i,2}$, $1 \leq i \leq d$, after the appropriate gap and intersect the occurrences to report the dictionary patterns matchings. However, as some of the patterns may share subpatterns and some subpatterns may include other subpatterns, there may be several distinct subpatterns occurring at the same text location, each of different length. Therefore, we need to search for the suffixes $P_{i,2}$, $1 \leq i \leq d$, after several gaps, each beginning at the end of a matched prefix $P_{i,1}$, $1 \leq i \leq d$. To avoid multiple searches we search all subpatterns $P_{i,1}$, $1 \leq i \leq d$ that *end* at a certain location. Note that in order to find all subpatterns ending at a certain location of the text we need to look for them backwards and find their reverse $P_{i,1}^R$.

In the preprocessing stage we concatenate the subpatterns $P_{i,2}$, $1 \leq i \leq d$ of the dictionary separated by the symbol $\$ \notin \Sigma$ to form a single string S. We repeat the procedure for the subpatterns $P_{i,1}$, $1 \leq i \leq d$, to form a single string F. We construct a suffix tree T_S of the string S, and another suffix tree T_{F^R} of the string F^R, which is the reverse of the string F.

We then traverse the text by inserting suffixes of the text to the T_S suffix tree. When we pass the node of T_S for which the path from the root is labeled $P_{i,2}$, it implies that this subpattern occurs in the text starting from the beginning of the text suffix. We then should find whether $P_{i,1}$ also appears in the text within the appropriate gap from the occurrences of the $P_{i,2}$ subpatterns. To this end, we go backward in the text skipping as many locations as the gap requires and inserting the reversed prefix of the text to T_{F^R}. If a node representing $P_{i,1}$ is encountered, we can output that P_i appears in the text.

Note that several dictionary subpatterns representative nodes may be encountered while traversing the trees. Therefore, we should report the intersection between the subpatterns found from each traversal. We do that by efficient intersection of labels on tree paths, as done in Amir et al. [5], using range queries on a grid. However, since some patterns may share subpatterns, we do not label the tree nodes by the original subpatterns they represent, as done in [5]. Instead, we mark the nodes representing subpatterns numerically in a certain order, hereafter discussed, regardless to the origin of the subpatterns ending at those nodes.

In order to be able to trace the identity of the patterns from the nodes marking, we keep two arrays A_F and A_S, both of size d. The arrays contain linked lists that identify when subpatterns are shared among several dictionary patterns. These arrays are filled as follows: $A_F[g] = i$ if $P_{i,1}^R$ is represented by node labelled g in T_{F^R} and $A_S[h] = i$ if $P_{i,2}$ is represented by the node labelled by h in T_S. In addition, $A_F[g]$ is linked to h and vice versa, if g, h represent two subpatterns of the same pattern of the dictionary.

Every pattern P_i is represented as a point i on a grid of size $d \times d$, denoted by $< g, A_F[g].link >$, that is, the x-coordinate is the mark of the node representing $P_{i,1}^R$ in T_{FR}, and the y-coordinate is the mark of the node representing $P_{i,2}$ in T_S. Now, if we mark the nodes representing the end of subpatterns so that the marks on a path are consecutive numbers, then the problem of intersection of labels on tree paths can be reduced to range queries on a grid in the following way. Let the first and last mark on the relevant path in T_{FR} be g, g' and, similarly, on the path in T_S let the first and last mark be h, h'. Thus, points $< x, y >$ on the grid where $g \leq x \leq g'$ and $h \leq y \leq h'$, represent patterns in the dictionary for which both subpatterns appear at the current check. A range query can be solved using the algorithm of [11].

We use the decomposition of a tree into vertical path for the nodes marking, suggested by [5], though we use it differently. A vertical path is defined as follows.

Definition 4. *[5] A vertical path of a tree is a tree path (possibly consisting of a single node) where no two nodes on the path are of the same height.*

After performing the decomposition, we can traverse the vertical paths and mark by consecutive order all tree nodes representing the end of a certain subpattern appearing in T_S or its reverse appearing in T_{FR}. Since some subpatterns may be shared by several dictionary patterns, there are *at most d* marked nodes at each of the suffixes trees.

Note that due to the definition of vertical path there may be several vertical paths that have a non empty intersection with the unique path from the root to a specific node. Hence, when considering the intersection of marked nodes on the path from the root till a certain node marked by g in T_{FR} and the marked nodes on the path from the root till a node marked by h in T_S, we actually need to check the intersection of all vertical paths that are included in the path from the root to g with all vertical paths that are included in the path from the root to h.

The algorithm appears in Figure 1.

Lemma 1. *The intersection between the subpatterns appearing at location t_ℓ and the reversed subpatterns ending at $t_{\ell-gap-1}$ can be computed in time $O(occ + \log\log d \log^2 \min\{d, \log|D|\})$, where occ is the number of patterns found. The preprocessing requires $O(|D| + d\log d)$ time and $O(|D| + d\log^\varepsilon d)$ space, for arbitrary small ε.*

At each of the $O(n)$ relevant locations of the text, the algorithm inserts the current suffix of the text to T_S using Weiner's algorithm [28]. For each of the prefixes defined by all $\beta - \alpha + 1$ possible specific gaps we insert its reverse to T_{FR}. As explained in [5], the navigation on the suffix tree and reverse suffix tree can be done in amortized $O(1)$ time per character insertion. Note that each character is inserted to T_S once and to T_{FR} $O(\beta - \alpha)$ times. This concludes the proof of the first part of Theorem 1.

SINGLE_ GAP_ DICTIONARY $(T, D$)	
Preprocessing:	
1	$F = P_{1,1}\$P_{2,1}\$ \cdots P_{d,1}$.
2	$S = P_{1,2}\$P_{2,2}\$ \cdots P_{d,2}$.
3	$T_S \leftarrow$ a suffix tree of S.
4	$T_{FR} \leftarrow$ a suffix tree of the F^R.
5	**For** every edge $(u,v) \in \{T_S, T_{FR}\}$ with label $y\$z$, where $y, z \in \Sigma^*$
6	Break (u,v) into (u,w) and (w,v) labelling (u,w) with y and (w,v) with $\$z$.
7	Decompose T_{FR} into vertical paths.
8	Mark the nodes representing $P_{i,1}$ on the vertical paths of T_{FR}.
9	Decompose T_S into vertical paths.
10	Mark the nodes representing $P_{i,2}$ on the vertical paths of T_S.
11	Preprocess the points according to the patterns for range queries.
Query:	
12	**For** $\ell = \min_i\{|P_{i,1}|\} + \alpha$ to n
13	Insert $t_\ell t_{\ell+1} \ldots t_n$ to T_S.
14	$h \leftarrow$ node in T_S representing suffix $t_\ell t_{\ell+1} \ldots t_n$.
15	**For** $f = \ell - \alpha - 1$ to $\ell - \beta - 1$
16	Insert $t_f t_{f-1} \ldots t_1$ to T_{FR}.
17	$g \leftarrow$ node in T_{FR} representing $t_f t_{f-1} \ldots t_1$.
18	**For** every vertical path on the the path p from the root to h
19	**For** every vertical path p' on the path from the root to g
20	Perform a range query on a grid with the first and last marks of p and p'
21	Report appearance for every P_i where point i appears in the specified range.

Fig. 1. Dictionary matching with a single gap algorithm

3   Algorithm with Lookup Table

If a very fast query time is crucial and we are willing to pay in preprocessing time, we can solve the problem of intersection between the appearances of subpatterns on the paths of T_S, T_{FR} using an intersection lookup table.

We build a table of size $d \times d$ which we call *inter*, where $inter[g, h]$ includes all indices i of patterns P_i such that $P_{i,1}^R$ appears on the path from the root of T_{FR} till the node marked by g and $P_{i,2}$ appears on the path from the root of T_S till the node marked by h. We can use the marking of the nodes representing subpatterns, given by the trees decompositions, or any other markings guaranteeing that nodes closer to the root are marked by smaller numbers than nodes farther from the root, such as the BFS order.

The table construction is done in the preprocessing using dynamic programming. Such a procedure is possible since the marking method guarantees that a node representing a subpattern $P_{i,j}$ is marked by a lower number than that representing $P_{i',j'}$ which includes $P_{i,j}$.

Saving at every entry all the relevant pattern indices causes redundancy in case a certain pattern includes another in both its prefix and suffix. In order to save every possible occurrence only once in the table, during the preprocessing we add *include* links at $A_F[g], A_S[h]$ arrays, where $A_F[g]$ has an *include* link to $A_F[g']$ in case the subpattern represented by g', $P_{i',1}$ is a suffix of the subpattern represented by node g, $P_{i,1}$, and $P_{i',1}$, $P_{i,1}$ share the same second part, i.e., $P_{i',2} = P_{i,2}$. Similarly, $A_S[h]$ has an *include* link to $A_S[h']$ in case the subpattern represented by h', $P_{i',2}$ is a prefix of the subpattern represented by node h, $P_{i,2}$ and $P_{i',1} = P_{i,1}$. These links can be easily computed while constructing T_S, T_{FR}.

Having these *include* links we can save at $inter[g, h]$ merely 4 elements:

1. A link to the cell $inter[prev(g), prev(h)]$, where $prev(g)$ is the closest marked predecessor of the node marked by g and similarly for $prev(h)$.
2. The index i of pattern P_i such that $P_{i,1}$ is represented by node g and $P_{i,2}$ is represented by node h.
3. A link to $A_F[g]$ in case the subpattern represented by node g is $P_{i,1}$ and a predecessor of node h represents $P_{i,2}$.
4. A link to $A_S[h]$ in case the the subpattern represented by node h is $P_{i,2}$ and a predecessor of node g represents $P_{i,1}$.

Given g, h as above, we denote by $prev(g)$ the closest marked predecessor of the node marked by g, and by $prev^*(g)$ the closest marked predecessor of the node marked by g for which $A_F[prev^*(g)] = A_S[h'_y]$ for h'_y a predecessor of the node marked by h. $prev^*(h)$ is similarly defined. h' is a predecessor of node h and g' is a predecessor of node g. The recursive rule for constructing the table is described in the following lemma.

Lemma 2. *The Recursive Rule*

$$inter[g, h] = \bigcup \begin{cases} a \ link \ to \ inter[prev^*(g), prev^*(h)], \\ i, \textbf{\textit{if}} \ A_F[g] = A_S[h] = i \\ a \ link \ to \ A_F[g], \ \textbf{\textit{if}} \ A_F[g] = A_S[h'] = i' \ and \ h' \ is \ maximal \\ a \ link \ to \ A_S[h], \ \textbf{\textit{if}} \ A_F[g'] = A_S[h] = i'' \ and \ g' \ is \ maximal \end{cases}$$

Lemma 3 gives the preprocessing time and space guarantee. Lemma 4 gives the query time guarantee.

Lemma 3. *Preprocessing to build the inter table requires $O(|D| + d^2 ovr)$ time, where ovr is the maximal number of subpatterns including each other as a prefix or as a suffix.*

Lemma 4. *Having the inter table, the intersection between the subpatterns appearing at location t_ℓ and the reversed subpatterns ending at $t_{\ell - gap - 1}$ can be computed in time $O(occ)$, where occ is the number of patterns found.*

This proves the second part of Theorem 1.

3.1   Splitting the Text

Usually, the input text is very long and arrives on-line. This makes the query algorithm requirement to insert all suffixes of the text unreasonable. Transforming this algorithm into an online algorithm seems a difficult problem. The main difficulty is working with on-line suffix trees construction in a sliding window.

While useful constructions based on Ukkonen's [25] and McCgright's [17] suffix trees constructions exist (see [21]), no such results are known for Weiner's suffix tree construction, which our reversed prefixes tree construction depends on.

Nevertheless, we do not need to know the whole text in advance and we can process only separate chunks of it each time. To do this we take $m = \beta - \alpha + \max_i \sum_j |P_{i,j}|$ and split the text twice to pieces of size $2m$: first starting form the beginning of the text and the second starting after m symbols. We then apply the algorithms for a single gap or k-gaps for each of the pieces separately for both text splits. Note that any appearance of a dictionary pattern can still be found by the algorithms on the splitted text.

4   Conclusions and Open Problems

We showed that combinatorial string methods other than Aho-Corasick automaton can be applied to the DMG problem to yield efficient algorithms. In this paper we focused on solving DMG, where a single gap exists in all patterns in the dictionary. We also relaxed the problem so that all patterns in the dictionary have the same gap bounds. It is an interesting open problem to study the general problem without these relaxations.

References

1. Aho, A.V., Corasick, M.J.: Efficient string matching: an aid to bibliographic search. Comm. ACM 18(6), 333–340 (1975)
2. Amir, A., Calinescu, G.: Alphabet independent and dictionary scaled matching. J. of Algorithms 36, 34–62 (2000)
3. Amir, A., Farach, M., Giancarlo, R., Galil, Z., Park, K.: Dynamic dictionary matching. Journal of Computer and System Sciences 49(2), 208–222 (1994)
4. Amir, A., Farach, M., Idury, R.M., La Poutré, J.A., Schäffer, A.A.: Improved dynamic dictionary matching. Information and Computation 119(2), 258–282 (1995)
5. Amir, A., Keselman, D., Landau, G.M., Lewenstein, M., Lewenstein, N., Rodeh, M.: Indexing and dictionary matching with one error. In: Dehne, F., Gupta, A., Sack, J.-R., Tamassia, R. (eds.) WADS 1999. LNCS, vol. 1663, pp. 181–192. Springer, Heidelberg (1999)
6. Bille, P., Gørtz, I.L., Vildhøj, H.W., Wind, D.K.: String matching with variable length gaps. Theoretical Computer Science (443), 25–34 (2012)
7. Bille, P., Thorup, M.: Faster regular expression matching. In: Albers, S., Marchetti-Spaccamela, A., Matias, Y., Nikoletseas, S., Thomas, W. (eds.) ICALP 2009, Part I. LNCS, vol. 5555, pp. 171–182. Springer, Heidelberg (2009)
8. Bille, P., Thorup, M.: Regular expression matching with multi-strings and intervals. In: Proc. 21st Annual ACM-SIAM Symposium on Discrete Algorithms (SODA), pp. 1297–1308 (2010)
9. Brodal, G.S., Gasieniec, L.: Approximate dictionary queries. In: Hirschberg, D.S., Meyers, G. (eds.) CPM 1996. LNCS, vol. 1075, pp. 65–74. Springer, Heidelberg (1996)
10. Cole, R., Gottlieb, L., Lewenstein, M.: Dictionary matching and indexing with errors and don't cares. In: Proc. 36th Annual ACM Symposium on the Theory of Computing (STOC), pp. 91–100. ACM Press (2004)

11. Chan, T.M., Larsen, K.G., Pătraşcu, M.: Orthogonal range searching on the ram, revisited. In: Proc. 27th ACM Symposium on Computational Geometry (SoCG), pp. 1–10 (2011)
12. Fredriksson, K., Grabowski, S.: Efficient algorithms for pattern matching with general gaps, character classes and transposition invariance. Inf. Retr. 11(4), 338–349 (2008)
13. Haapasalo, T., Silvasti, P., Sippu, S., Soisalon-Soininen, E.: Online Dictionary Matching with Variable-Length Gaps. In: Pardalos, P.M., Rebennack, S. (eds.) SEA 2011. LNCS, vol. 6630, pp. 76–87. Springer, Heidelberg (2011)
14. Hofmann, K., Bucher, P., Falquet, L., Bairoch, A.: The PROSITE database. Nucleic Acids Res. (27), 215–219 (1999)
15. Krishnamurthy, M., Seagren, E.S., Alder, R., Bayles, A.W., Burke, J., Carter, S., Faskha, E.: How to Cheat at Securing Linux. Syngress Publishing, Inc., Elsevier, Inc., 30 Corporate Dr., Burlington, MA 01803 (2008), e-edition: http://www.sciencedirect.com/science/book/9781597492072
16. Kucherov, G., Rusinowitch, M.: Matching a set of strings with variable length don't cares. Theoret. Comput. Sci. 178(12), 129–154 (1997)
17. McCreight, E.M.: A Space-Economical Suffix Tree Construction Algorithm. Journal of the ACM 23(2), 262–272 (1976)
18. Morgante, M., Policriti, A., Vitacolonna, N., Zuccolo, A.: Structured motifs search. J. Comput. Bio. 12(8), 1065–1082 (2005)
19. Myers, G.: A four-russian algorithm for regular expression pattern matching. J. ACM 39(2), 430–448 (1992)
20. Myers, G., Mehldau, G.: A system for pattern matching applications on biosequences. CABIOS 9(3), 299–314 (1993)
21. Naa, J.C., Apostolico, A., Iliopoulos, C.S., Park, K.: Truncated suffix trees and their application to data compression. Theoretical Computer Science 304(3), 87–101 (2003)
22. Navarro, G., Raffinot, M.: Fast and simple character classes and bounded gaps pattern matching, with applications to protein searching. J. Comput. Bio. 10(6), 903–923 (2003)
23. Navarro, G., Raffinot, M.: New techniques for regular expression searching. Algorithmica 41(2), 89–116 (2004)
24. Rahman, M.S., Iliopoulos, C.S., Lee, I., Mohamed, M., Smyth, W.F.: Finding patterns with variable length gaps or don't cares. In: Chen, D.Z., Lee, D.T. (eds.) COCOON 2006. LNCS, vol. 4112, pp. 146–155. Springer, Heidelberg (2006)
25. Ukkonen, E.: On-line construction of suffix trees. Algorithmica 14(3), 249–260 (1995)
26. van Emde Boas, P.: Preserving order in a forest in less than logarithmic time. In: Proceedings of the 16th Annual Symposium on Foundations of Computer Science, pp. 75–84 (1975)
27. Verint. Packet intrusion detection. Personal communication (2013)
28. Weiner, P.: Linear pattern matching algorithm. In: Proc. 14 IEEE Symposium on Switching and Automata Theory, pp. 1–11 (1973)
29. Zhang, M., Zhang, Y., Hu, L.: A faster algorithm for matching a set of patterns with variable length don't cares. Inform. Process. Letters 110, 216–220 (2010)

Approximate On-line Palindrome Recognition, and Applications

Amihood Amir[1,2,*] and Benny Porat[1,**]

[1] Department of Computer Science, Bar-Ilan University, Ramat-Gan 52900, Israel
[2] Department of Computer Science, Johns Hopkins University, Baltimore, MD 21218
amir@cs.biu.ac.il, bennyporat@gmail.com

Abstract. Palindrome recognition is a classic problem in computer science. It is an example of a language that can not be recognized by a deterministic finite automaton and is often brought as an example of a problem whose decision by a single-tape Turing machine requires quadratic time.

In this paper we re-visit the palindrome recognition problem. We define a novel fingerprint that allows recognizing palindromes on-line in linear time with high probability. We then use group testing techniques to show that the fingerprint can be adapted to recognizing approximate palindromes on-line, i.e. it can recognize that a string is a palindrome with no more than k mismatches, where k is given.

Finally, we show that this fingerprint can be used as a tool for solving other problems on-line. In particular we consider approximate pattern matching by non-overlapping reversals. This is the problem where two strings S and T are given and the question is whether applying a sequence of non-overlapping reversals to S results in string T.

1 Introduction

Palindrome recognition is one of the fundamental problems in computer science. One of the features of palindrome recognition is that it is almost trivially solved in linear time on a RAM (or a 2-tape Turing machine), yet is more challenging as an online problem. Manacher [19] and Galil [16] showed how to use DPDAs for recognizing palindrome prefixes of a string that is input online. Yet no effort has been invested historically in seeking *approximate* palindrome prefixes of a string that is being input online. The approximation we are interested in is the *Hamming distance*. We say that a prefix of the string is a palindrome with k mismatches if changing k locations in the prefix will make it a palindrome.

---

* Partly supported by NSF grant CCR-09-04581, ISF grant 347/09, and BSF grant 2008217.
** Partly supported by a Bar Ilan University President Fellowship. This work is part of Benny Porat's Ph.D. thesis.

A.S. Kulikov, S.O. Kuznetsov, and P. Pevzner (Eds.): CPM 2014, LNCS 8486, pp. 21–29, 2014.

The Contributions of this Paper:

1. We define a fingerprint that recognizes a palindrome with high probability. This fingerprint is extended in constant time per symbol. The palindrome recognition algorithm proceeds in an analogous manner to the Karp-Rabin pattern matching algorithm [18].
2. We adapt the palindrome fingerprint, via group testing techniques, to provide an online solution to finding every palindrome prefix with k-mismatches, of an input stream.

Applications: Palindromes have been shown recently [6] to have a connection with *pattern matching by reversals*.

Consider a text $T = t_0 \cdots t_{n-1}$ and pattern $P = p_0 \cdots p_{m-1}$, both over an alphabet Σ. A recent new paradigm – *pattern matching with rearrangements*, assumes that there are no changes in the text *content*, but the *order* of the symbols may have been changed. We seek text locations where the pattern matches with a small number of rearrangements. For a survey on the rearrangement paradigm see [5].

Some of these rearrangement problems are fueled by biological challenges. During the course of evolution areas of the genome may be shifted from one location to another. Considering the genome as a string over the alphabet of genes, these problems represent a situation where the difference between the original string and the resulting one is in the locations of the different elements. Several works have considered specific versions of this biological setting, primarily focusing on the sorting problem (*sorting by reversals* [8, 10], *sorting by transpositions* [7], and *sorting by block interchanges* [11]).

The *Min-SBR* problem, gets a permutation $\pi = \pi[1], ..., \pi[n]$ of $\{1, ..., n\}$ as its input, and its task is to sort the permutation using the minimum number of possible *reversal* operations. A reversal operation reverses the order of a substring of the permutation, i.e.

$$rev_{(i,j)}(\pi[1], ...\pi[i-1], \pi[\mathbf{i}], \pi[i+1], ..., \pi[j-1], \pi[\mathbf{j}], \pi[j+1], ..., \pi[n]) =$$

$$\pi[1], ...\pi[i-1], \pi[\mathbf{j}], \pi[j-1], ..., \pi[i+1], \pi[\mathbf{i}], \pi[j+1], ..., \pi[n].$$

The Min-SBR problem is $\mathcal{NP}$-hard and has a long history. If the reversals are signed, there exist polynomial algorithms [8, 17]. For unsigned reversals there are efficient approximation algorithms [12, 9].

In [6] the Min-SBR problem was studied in a general alphabet setting, rather than as a sorting problem, i.e. symbols may occur in the strings multiple times. Consider a set A and let x and y be two m-tuples over A. One wishes to convert x to y through a sequence of reversal operations. Since the problem is $\mathcal{NP}$-hard, an additional constraint was added – *disjointness* – overlapping reversals are forbidden throughout the algorithm's operation. This type of constraint has been considered in a number of problems in the context of the rearrangement model [20, 2, 4, 13, 3, 1]. Formally:

Definition 1. *The* Reversal Matching Problem *is the General Alphabet Matching with Reversals problem with the disjointness constraint, i.e., each symbol can participate in at most one reversal.*

We say that string S reversal matches string T if string T results from performing a sequence of disjoint reversal operations on string S.

In [6], the reversal matching problem was studied. A linear-time algorithm was shown for the exact matching version, and a quadratic time and space algorithm was introduced for the *approximate pattern matching with non-overlapping reversals* problem, i.e. where mismatch errors are introduced. If only k mismatch errors are allowed, the approximate pattern matching with non-overlapping reversals problem was solved in time $O(n^2 \log k)$ and space $O(nk)$.

The approximate palindrome fingerprint scheme discussed in this paper can be used for solving the reversals matching problem with k mismatches **online** in space $O(n + k^2)$ and a slight degradation in time, $O(n^2 \log k + nk)$ w.h.p.

2   Palindrome Fingerprint

We need the following notation in order to define the key property enabling our algorithm's on-line efficiency.

Notation 1. *Let $s = s[0], s[1], s[2], \cdots, s[m-1]$ be a string over $\mathbb{N}$.*
We denote its reverse $s[m-1], \cdots, s[1], s[0]$ by s^R.
Let $0 \le i \le j < m$. Denote by $s[i:j]$ the substring $s[i], s[i+1], ..., s[j-1], s[j]$.
Let $0 \le j < m$. We denote the length-$j+1$ prefix of s, $s[0:j]$ by $Pre_j(s)$*. We denote the* suffix *of s starting at position j, $s[j:m-1]$, by $Suf_j(s)$.*

Definition 2. palindrome fingerprint*: Given a string $s = s[0], s[1], ...s[m-1]$ over $\mathbb{N}$, and some random number $r \in F_p$, we define the* palindrome fingerprint, *or for brevity the* fingerprint *of s to be:*

$$\phi(s) = (r^1 s[0] + r^2 s[1] + ...r^m s[m-1]) \bmod p$$

We define the reversal fingerprint *of s, to be: $\phi^R(s) = (r^{-1} s[0] + r^{-2} s[1] + ...r^{-m} s[m-1]) \bmod p$*

Lemma 1. *Given a string s of length m, $\phi(s) = (r^{m+1}\phi^R(s)) \bmod p$ iff s is a palindrome ($s = s^R$) w.h.p.*

Corollary 1. *Let s and t be two strings of length m over $\mathbb{N}$.*
Then $s = t^R$ iff $\phi(s) = (r^{m+1}\phi^R(t)) \bmod p$ w.h.p.

Proof: $\phi(s) = (r^1 s[0] + r^2 s[1] + ...r^m s[m-1]) \bmod p$ and $\phi^R(t) = (r^{-1}t[0] + r^{-2}t[1] + ...r^{-m}t[m-1]) \bmod p$

Multiply $\phi^R(t)$ by r^{m+1} we got: $(r^1 t[m-1] + ...r^{m-1}t[1] + r^m t[0]) \bmod p$
Which is equal to $\phi(s)$ iff $s = t^R$ w.h.p   □

The online palindrome recognition algorithm is now clear. Simply start reading the string s and at every input symbol i compute both $\phi(s[0:i])$ and $\phi^R(s[0:i])$. This computation merely involves adding $r^{i+1}s[i]$ to $\phi(s[0:i-1])$ and adding $r^{-(i+1)}s[i]$ to $\phi^R(s[0:i-1])$. To decide whether the prefix is a palindrome, the comparison $\phi(s) = (r^{m+1}\phi^R(s)) \bmod p$ is checked. All these operations take constant time per input symbol.

In the next section we show how to find the palindrome distance with 1 mismatch. We will later use group testing techniques (as in e.g. [21]) to calculate the palindrome distance with up to k mismatches. The paper culminates with using the palindrome fingerprint with errors for approximate online pattern matching with reversals.

3   Palindrome up to 1−mismatch

3.1   The Idea

Let $s = s[0], s[1], ...s[m-1]$ be the string and let $q_1, q_2, ...q_\ell$ be ℓ prime numbers such that $\prod_{i=1}^{\ell} q_i > m$. We create ℓ lists of partition strings of s in the following way. For the i-th list, partition the string to q_i partition strings. Denote the j-th partition string of the i-th list of the string by $s_{q_i,j}$. It consists of all characters $s[\tau]$ such that $\tau \bmod q_i = j$. Then $|s_{q_i,j}| = \lfloor \frac{m}{q_i} \rfloor + 1$, for $j \leq m \bmod q_i$, and $|s_{q_i,j}| = \lfloor \frac{m}{q_i} \rfloor$, for $j > m \bmod q_i$. **Example:** For $s = s[0]s[1]s[2]s[3]s[4]s[5]s[6]$ ($m = 7$) and $q_i = 3$, the partition strings are: $s_{3,0} = s[0]s[3]s[6]$, $s_{3,1} = s[1]s[4]$, $s_{3,2} = s[2]s[5]$.

Now, calculate the fingerprint and the reversal fingerprint for each of the partition strings. i.e, for each $1 \leq i \leq \ell$, and for each $0 \leq j < q_i$, we calculate the fingerprint $\phi(s_{q_i,j})$ and the reversal fingerprint $\phi^R(s_{q_i,j})$.

Note that s is a palindrome iff $s = s^R$. This means that we need to check equalities of $s[0]$ with $s[m-1]$, of $s[1]$ with $s[m-2]$, and, in general, of $s[i]$ with $s[m-i-1]$. We have partitioned the string to the partition strings. The following lemma, which is immediate from the definition, identifies the partition strings that need to be compared so that the above alignments are maintained.

Lemma 2. *Let s be a string. Then $s = s^R$ iff $\forall i,j \ \ 1 \leq i \leq \ell, 0 \leq j < q_i$ we have that $s_{q_i,j} = s^R_{q_i,(m-1-j) \bmod q_i}$.*

Example: In the above example, $s = s^R$ iff $s_{3,0} = s[0]s[3]s[6] = s^R_{3,0} = s[6]s[3]s[0]$ and $s_{3,1} = s[1]s[4] = s^R_{3,2} = s[5]s[2]$. The following lemma is the key property of the algorithm.

Lemma 3. *A string s is a palindrome with $1-$ mismatch iff for each list q_i, there is exactly one $0 \leq j < q_i$ such that $\phi(s_{q_i,j}) \neq q_i^{|s_{q_i,j}|+1} \phi^R(s_{q_i,(m-1-j) \bmod q_i})$.*

The proof of this lemma will appear in the full paper.

We are now ready to present an algorithm for deciding whether a given string s is a palindrome with one mismatch.

Algorithm – Palindrome with One Mismatch

1. **Initialization:**
   (a) Let $s = s[0], s[1], ...s[m-1]$ be the string and let $q_1, q_2, ...q_\ell$ be ℓ prime numbers such that $\prod_{i=1}^{\ell} q_i > m$. Create the ℓ lists of partition strings of s.
   (b) Compute the fingerprint and the reversal fingerprint of every partition string.
2. For $i = 1$ to ℓ
   (a) Set $counter \leftarrow 0$.
   (b) For $j = 1$ to mod q_i
      i. If $\phi(s_{q_i,j}) \neq q_i^{|s_{q_i,j}|+1} \phi^R(s_{q_i,(m-1-j) \bmod q_i})$ then $counter \leftarrow counter + 1$
   (c) If $counter > 1$ then **no 1-mismatch palindrome** and HALT.
3. **1-mismatch palindrome**
4. Compute mismatch location by Chinese Remainder Theorem and HALT.

end Algorithm

Lemma 4. *Let s be a string over $\mathbb{N}$. If s is a palindrome with exactly 1 mismatch then our algorithm finds the exact position of that mismatch.*

Proof: Immediate from the above algorithm and discussion.   □

Theorem 1.   *The space complexity of the algorithm for discovering a palindrome up to $1-$mismatch is $O(\frac{\log^2 m}{\log \log m})$.*
The running time of the algorithm for discovering a palindrome up to $1-$mismatch is bounded by $O(\frac{\log m}{\log \log m})$ per character.

A complete analysis of the running time and the space complexity will appear in the full paper.

3.2   The Online Version of Palindrome with 1-mismatch

Assume that we are given in advance some upper bound that is a polynomial of m. We can precompute the $\ell = O(\frac{\log m}{\log \log m})$ prime numbers $\{q_i\}_{i=1}^{\ell}$. The string s arrives online one symbol at a time. We would like to decide online for every prefix $s[0], ..., s[i]$ whether it is a palindrome with 1-mismatch.

The fingerprint and reversal fingerprint calculations can all be done online with a constant time calculation for each partition string. At every i, the equality between the appropriate fingerprint and reversal fingerprint overall partitions can be checked in $O(\frac{\log m}{\log \log m})$ time.

4   Palindrome Distance with up to $k-$mismatches

We now use the algorithm from the last subsection, finding if a string is a palindrome with $1-$mismatch, in order to determine if a string is a palindrome up to

$k-$mismatches. Our algorithm uses group testing combined with the 1-mismatch algorithm from section 3.

The group testing problem can be described as follows. Consider a set of n items, each of which can be defective or non-defective. The task is to identify the defective items using a minimum number of tests. Each test works on a group of items simultaneously and returns whether that group contains at least one defective item or not. If a defective item is present in a group then the test result is said to be positive, otherwise it is negative.

Group testing has a long history dating back to at least 1943 [14]. In this early work the problem of detecting syphilitic men for induction into the United State military using the minimum number of laboratory tests was considered. Subsequently a large literature has built up around the subject and we refer the interested reader to [15] for a comprehensive survey of the topic. We use a slightly different version of the group testing problem. In our version the tests return positive if there is exactly one defective item in the group, and negative otherwise. We obtain the test result by the 1-mismatch algorithm presented in section 3.

In our terminology, we are given string $s = s[0], s[1], ...s[m-1]$, and our interest is to know if $s[0] = s[m-1]$, if $s[1] = s[m-2]$,..., if $s[i] = s[m-1-i]$,..., and to count the mismatches.

Each of our groups will be a partition of S (for example $s[3], s[5], s[8], s[18], ...$), and each of our tests will take two groups (partitions) and distinguish between three cases: The first: the two partitions match. The second: there is exactly one mismatch between those two partitions. The third: there is more than one mismatch between those two partitions. For these tests we will use the algorithm for Palindrome with 1-mismatch from section 3.

In the remainder of this section we show how the groups are chosen, and how the palindrome distance up to $k-$mismatches at every location can be determined as a result of our group testing.

Let $q_1, q_2, ...q_\ell$ be ℓ prime numbers such that $\prod_{i=1}^{\ell} q_i > m^k$. In a similar manner to the method described in the previous section, for each $q_i \in \{q_1, q_2, ...q_\ell\}$ we create q_i disjoint partitions for $S = s[0], s[1], ..., s[m-1]$. Each of these partition is denoted by $s_{q_i,j}$ for $0 \le j < q_i$. The partition $s_{q_i,j}$ consists of all the characters $s[\tau]$ such that $\tau \bmod q_i = j$. Overall we build $\sum_{i=1}^{\ell} q_i$ partitions of the string S.

Definition 3. Let $s_1 = s[i_1], s[i_2], ..s[i_k]$ be a partition of $s = s[0], s[1], ..., s[m-1]$, we define the reversal pair of s_1 to be $s_2 = s[m-1-i_k], ...s[m-1-i_2], s[m-1-i_1]$. Notice that if s_1 is the reversal pair of s_2 then s_2 is also the reversal pair of s_1.

Lemma 5. Let $t_0, t_1, ...t_k$ be k disjoint partitions of $s = s[0], s[1], ...s[m-1]$ that cover s (i.e. for each $0 \le i < m$ $s[i] \in t_j$ for some $0 \le j \le k$) $s = s^R$ iff t_i match by reverse to his reversal pair for all $0 \le i \le k$

Lemma 6. The reversal pair of $s_{q_i,j}$ is $s^R_{q_i,(m-1-j) \bmod q_i}$

Returning to the group testing terminology we will "test" each partition with its reversal pair. For our tests we use the palindrome with $1-$mismatch algorithm from section 3.

Attention 1. *The algorithm for detecting a palindrome with $1-$mismatch has a single input string, which it uses as s and s^R. We use a slightly different version, where we give the algorithm two strings, one used as s and the other used as s^R.*

Theorem 2. *The palindrome distance up to k mismatches can be calculated using the group testing results.*

We present the complete algorithm before proving the theorem.

Algorithm – Palindrome Distance up to $k-$Mismatches

1. **Initialization:**
   (a) Let $s = s[0], s[1], ...s[m-1]$ be the string and let $q_1, q_2, ...q_\ell$ be ℓ prime numbers such that $\prod_{i=1}^{\ell} q_i > m^k$. Create the partition $s_{q_i,j}$ for each $1 \le i \le \ell$ and $0 \le j < q_i$.
   (b) Initialize an empty set D.
2. For each $1 \le i \le \ell$ and $0 \le j < q_i$.
   (a) **Test($s_{q_i,j}$):** Run the palindrome distance with one mismatch algorithm with $s = s_{q_i,j}$ and $s^R = s_{q_i,(m-1-j) \bmod q_i}$
      i. If the test returns exactly one mismatch, then insert the mismatch location into D.
3. If the size of D is greater than $k \rightarrow$ return **"More than k mismatches"**.
4. For all Tests($s_{q_i,j}$) that return "More than one mismatch":
   (a) If for any $\tau \in D$, $\tau \bmod q_i \ne j \rightarrow$ return **"More than k mismatches"**
5. S is a palindrome with $|D|$ mismatches, and D is all the mismatches location.

end Algorithm

The following two lemmas together will provide the proof for Theorem 2.

Lemma 7. *Let $s[j]$ be some character from the string. For each group of k indices, $i_1, i_2, ..., i_k$, $s[j]$ maps to at least one partition in which none of $i_1, i_2, ..., i_k$ occurs.*

Conclusion 1. *If there are at most k mismatches, each mismatch will be mapped to at least one partition where all the other mismatches do not belong. So, using the 1-mismatch algorithm on these partitions, all the mismatches will be discovered.*

Lemma 8. *It is possible to detect if the string is not a palindrome with up to k mismatches.*

The proofs of both of these lemmas will appear in the full paper.

This ends the proof of the correctness of our algorithm. We now proceed to present the space and the running time of the algorithm.

Theorem 3

> *The space complexity of the palindrome distance up to $k-$mismatch algorithm is $O(\frac{k \log^2 m}{\log^2 \log m})$.*
> *The running time of the palindrome distance up to $k-$mismatch algorithm is bounded by $O(\frac{k^2 \log^4 m}{\log^2 \log m})$ per character.*

A complete analysis of the running time and the space complexity will appear in the full paper.

5   Application: Reversal Matching

Theorem 4. *The exact reversal matching of strings s_1 and s_2 of length n can be computed on-line in time $O(n)$ w.h.p.*

Theorem 5. *Using the palindrome fingerprint idea, it is possible to solve the k-reversal matching problem online in time $O(n^2 \log k + nk)$ and space $O(n + k^2)$.*

The proofs for both theorems will appear in the full paper.

6   Conclusion and Further Work

We have shown a novel signature – the palindrome fingerprint – that allows to design efficient algorithms for detecting approximate palindromes online. The palindrome fingerprint can be applied for computing the approximate reversal distance of two strings when a disjointness condition is imposed. It would be interesting to explore the palindrome fingerprint and see whether it can aid other applications.

References

[1] Amir, A., Aumann, Y., Benson, G., Levy, A., Lipsky, O., Porat, E., Skiena, S., Vishne, U.: Pattern matching with address errors: rearrangement distances. J. Comp. Syst. Sci. 75(6), 359–370 (2009)

[2] Amir, A., Aumann, Y., Landau, G., Lewenstein, M., Lewenstein, N.: Pattern matching with swaps. Journal of Algorithms 37, 247–266 (2000) (Preliminary version appeared at FOCS 1997)

[3] Amir, A., Cole, R., Hariharan, R., Lewenstein, M., Porat, E.: Overlap matching. Information and Computation 181, 57–74 (2003)

[4] Amir, A., Landau, G.M., Lewenstein, M., Lewenstein, N.: Efficient special cases of pattern matching with swaps. Information Processing Letters 68(3), 125–132 (1998)

[5] Amir, A., Levy, A.: String rearrangement metrics: A survey. In: Elomaa, T., Mannila, H., Orponen, P. (eds.) Ukkonen Festschrift. LNCS, vol. 6060, pp. 1–33. Springer, Heidelberg (2010)

[6] Amir, A., Porat, B.: Pattern matching with non overlapping reversals - approximation and on-line algorithms. In: Cai, L., Cheng, S.-W., Lam, T.-W. (eds.) ISAAC 2013. LNCS, vol. 8283, pp. 55–65. Springer, Heidelberg (2013)

[7] Bafna, V., Pevzner, P.A.: Sorting by transpositions. SIAM J. on Discrete Mathematics 11, 221–240 (1998)

[8] Berman, P., Hannenhalli, S.: Fast sorting by reversal. In: Hirschberg, D.S., Meyers, G. (eds.) CPM 1996. LNCS, vol. 1075, pp. 168–185. Springer, Heidelberg (1996)

[9] Berman, P., Hannenhalli, S., Karpinski, M.: 1.375-approximation algorithm for sorting by reversals. In: Möhring, R.H., Raman, R. (eds.) ESA 2002. LNCS, vol. 2461, pp. 200–210. Springer, Heidelberg (2002)

[10] Carpara, A.: Sorting by reversals is difficult. In: Proc. 1st Annual International Conference on Research in Computational Biology (RECOMB), pp. 75–83. ACM Press (1997)

[11] Christie, D.A.: Sorting by block-interchanges. Information Processing Letters 60, 165–169 (1996)

[12] Christie, D.A.: A 3/2-approximation algorithm for sorting by reversals. In: Proc. 9th Annual ACM-SIAM Symposium on Discrete Algorithms (SODA), pp. 244–252 (1998)

[13] Cole, R., Hariharan, R.: Randomized swap matching in o(m logm log|σ|) time, Tech. Report TR1999-789, New York University, Courant Institute (September 1999)

[14] Dorfman, R.: The detection of defective members of large populations. The Annals of Mathematical Statistics 14(4), 436–440 (1943)

[15] Du, D.-Z., Hwang, F.K.: Combinatorial group testing and its applications, 2nd edn. Series on Applied Mathematics, vol. 12. World Scientific (2000)

[16] Galil, Z.: On converting on-line algorithms into real-time and on real-time algorithms for string matching and palindrome recognition. SIGACT News, 26–30 (1975)

[17] Kaplan, H., Shamir, R., Tarjan, R.E.: A faster and simpler algorithm for sorting signed permutations by reversals. SIAM J. Comp. 29(3), 880–892 (1999)

[18] Karp, R.M., Rabin, M.O.: Efficient randomized pattern-matching algorithms. IBM Journal of Res. and Dev., 249–260 (1987)

[19] Manacher, G.: A new linear-time "on-line" algorithm for finding the smallest initial palindrome of a string. Journal of the ACM 22(3), 346–351 (1975)

[20] Muthukrishnan, S.: New results and open problems related to non-standard stringology. In: Galil, Z., Ukkonen, E. (eds.) CPM 1995. LNCS, vol. 937, pp. 298–317. Springer, Heidelberg (1995)

[21] Porat, B., Porat, E.: Exact and approximate pattern matching in the streaming model. In: Proc. 50th IEEE Symposium on the Foundation of Computer Science (FOCS), pp. 315–323 (2009)

Computing Minimal and Maximal
Suffixes of a Substring Revisited

Maxim Babenko[1], Paweł Gawrychowski[2],
Tomasz Kociumaka[3], and Tatiana Starikovskaya[1,*]

[1] National Research University Higher School of Economics (HSE)
[2] Max-Planck-Institut für Informatik
[3] Institute of Informatics, University of Warsaw

Abstract. We revisit the problems of computing the maximal and the minimal non-empty suffixes of a substring of a longer text of length n, introduced by Babenko, Kolesnichenko and Starikovskaya [CPM'13]. For the minimal suffix problem we show that for any $1 \leq \tau \leq \log n$ there exists a linear-space data structure with $\mathcal{O}(\tau)$ query time and $\mathcal{O}(n \log n / \tau)$ preprocessing time. As a sample application, we show that this data structure can be used to compute the Lyndon decomposition of any substring of the text in $\mathcal{O}(k\tau)$ time, where k is the number of distinct factors in the decomposition. For the maximal suffix problem we give a linear-space structure with $\mathcal{O}(1)$ query time and $\mathcal{O}(n)$ preprocessing time, i.e., we manage to achieve both the optimal query and the optimal construction time simultaneously.

1 Introduction

Computing the lexicographically maximal and minimal suffixes of a string is both an interesting problem on its own and a crucial ingredient in solutions to many other problems. As an example of the former, a well-known result by Duval [5] is that the maximal and the minimal suffixes of a string can be found in linear time and constant additional space. As an example of the latter, the famous constant space pattern matching algorithm of Crochemore-Perrin is based on the so-called critical factorizations, which can be derived from the maximal suffixes [4].

We consider a natural generalization of the problems. We assume that the string we are asked to compute the maximal or the minimal suffixes for is actually a substring of a text T of length n given in advance. Then one can preprocess T and subsequently use this information to significantly speed up the computation of the desired suffixes of a query string. This seems to be a very natural setting whenever we are thinking about storing large collections of text data.

The problems of computing the minimal non-empty and the maximal suffixes of a substring of T were introduced in [1]. The authors proposed two linear-space data structures for T. Using the first data structure, one can compute the minimal suffix of any substring of T in $\mathcal{O}(\log^{1+\varepsilon} n)$ time. The second data structure

---

* Tatiana Starikovskaya was partly supported by Dynasty Foundation.

A.S. Kulikov, S.O. Kuznetsov, and P. Pevzner (Eds.): CPM 2014, LNCS 8486, pp. 30–39, 2014.
© Springer International Publishing Switzerland 2014

allows to compute the maximal suffix of a substring of T in $\mathcal{O}(\log n)$ time. Here we improve upon both of these results. We first show that for any $1 \leq \tau \leq \log n$ there exists a linear-space data structure with $\mathcal{O}(\tau)$ query time and $\mathcal{O}(n \log n / \tau)$ preprocessing time solving the minimal suffix problem. Secondly, we describe a linear-space data structure for the maximal suffix problem with $\mathcal{O}(1)$ query time. The data structure can be constructed in linear time. Computing the minimal or the maximal suffix is a fundamental ingredient of complex algorithms, so our results can hopefully be used to efficiently solve other problems in such setting, i.e., when we are working with substrings of some long text T. As a particular application, we show how to compute the Lyndon decomposition [3] of a substring of T in $\mathcal{O}(k\tau)$ time, where k is the number of distinct factors in the decomposition.

2 Preliminaries

We start by introducing some standard notation and definitions. A *border* of a string T is a string that is both a prefix and a suffix of T but differs from T. A string T is called *periodic with period* ρ if $T = \rho^s \rho'$ for an integer $s \geq 1$ and a (possibly empty) proper prefix ρ' of ρ. When this leads to no confusion, the length of ρ will also be called a period of T. Borders and periods are dual notions; namely, if T has period ρ then it has a border of length $|T| - |\rho|$, and vice versa (see, e.g., [4]). A string T_1 is lexicographically smaller than T_2 ($T_1 \prec T_2$) if either (i) T_1 is a proper prefix of T_2; or (ii) there exists $0 \leq i < \min(|T_1|, |T_2|)$ such that $T_1[1..i] = T_2[1..i]$, and $T_1[i+1] < T_2[i+1]$.

Consider a fixed string T. For $i < j$ let $Suf[i,j]$ denote $\{T[i..], \ldots, T[j..]\}$. The set $Suf[1, |T|]$ of all non-empty suffixes of T is also denoted as Suf. The suffix array and the inverse suffix array of a string T are denoted by SA and ISA respectively. Both SA and ISA occupy linear space and can be constructed in linear time (see [6] for a survey). For strings x, y we denote the length of their longest common prefix by $\mathrm{lcp}(x, y)$. SA and ISA can be enhanced in linear time [4,2] to answer the following queries in $\mathcal{O}(1)$ time:

(a) Given substrings x, y of T compute $\mathrm{lcp}(x, y)$ and determine if $x \prec y$.
(b) Given indices i, j compute the *maximal* and *minimal* suffix in $Suf[i, j]$.

Moreover, the enhanced suffix array can also be used to answer the following queries in constant time. Given substrings x, y of T compute the largest integer α such that x^α is a prefix of y. Indeed, it suffices to note that if x is a prefix of $y = T[i..j]$, then $(\alpha - 1)|x| \leq \mathrm{lcp}(T[i..j], T[i+|x|..j]) < \alpha|x|$. Queries on the enhanced suffix array of T^R, the reverse of T, are also meaningful in terms of T. In particular for a pair of substrings x, y of T we can compute their longest common suffix $\mathrm{lcs}(x, y)$ and the largest integer α such that x^α is a suffix of y.

3 Minimal Suffix

Consider a string T of length n. For each position j we select $\mathcal{O}(\log n)$ substrings $T[k..j]$, which we call *canonical*. By S_j^ℓ we denote the ℓ-th shortest canonical substring ending at the position j. For a pair of integers $1 \leq i < j \leq n$, we define $\alpha(i, j)$ to be the largest integer ℓ such that S_j^ℓ is a proper suffix of $T[i..j]$. We require the following properties of canonical substrings:

(a) $S_j^1 = T[j..j]$ and for some $\ell = \mathcal{O}(\log n)$ we have $S_j^\ell = T[1..j]$,
(b) $|S_j^{\ell+1}| \leq 2|S_j^\ell|$ for any ℓ,
(c) $\alpha(i, j)$ and $|S_j^\ell|$ are computable in $\mathcal{O}(1)$ time given i, j and ℓ, j respectively.

Our data structure works for any choice of canonical substrings satisfying these properties, including the simplest one with $|S_j^\ell| = \min(2^{\ell-1}, j)$. Our solution is based on the following lemma:

Lemma 1. *The minimal suffix of $T[i..j]$ is either equal to*

(a) $T[p..j]$, where p is the starting position of the minimal suffix in $Suf[i, j]$,
(b) or the minimal suffix of $S_j^{\alpha(i,j)}$.

Moreover, p can be found in $\mathcal{O}(1)$ time using the enhanced suffix array of T.

Proof. By Lemma 1 in [1] the minimal suffix is either equal to $T[p..j]$ or to its shortest non-empty border. Moreover, in the latter case the length of the minimal suffix is at most $\frac{1}{2}|T[p..j]| \leq \frac{1}{2}|T[i..j]|$. On the other hand, from the property (b) of canonical substrings we have that the length of $S_j^{\alpha(i,j)}$ is at least $\frac{1}{2}|T[i..j]|$. Thus, in the second case the minimal suffix of $T[i..j]$ is the minimal suffix of $S_j^{\alpha(i,j)}$. Note that for $i = j$ the values $\alpha(i, j)$ are not well-defined, but then case (a) holds. To prove the final statement, recall that finding the minimal suffix in $Suf[i, j]$ is one of the basic queries supported by the enhanced suffix array. □

The data structure, apart from the enhanced suffix array, contains, for each $j = 1, \ldots, n$, a bit vector B_j of length $\alpha(1, j)$. We set $B_j[\ell] = 1$ if and only if the minimal suffix of S_j^ℓ is longer than $|S_j^{\ell-1}|$. For $\ell = 1$ we always set $B_j[1] = 1$, as S_j^1 is the minimal suffix of itself. Recall that the number of canonical substrings for each j is $\mathcal{O}(\log n)$, so each B_j fits into a constant number of machine words, and the data structure takes $\mathcal{O}(n)$ space.

3.1 Queries

Assume we are looking for the minimal suffix of $T[i..j]$ with $\alpha(i, j) = \ell$. Our approach is based on Lemma 1. If case (a) holds, the lemma lets us compute the answer in $\mathcal{O}(1)$ time. In general we find the minimal suffix of S_j^ℓ, compare it with $T[p..j]$, and return the smaller of them.

We use both Lemma 1 and the bit vector B_j to compute the minimal suffix of S_j^ℓ. Let ℓ' be the largest index not exceeding ℓ such that $B_j[\ell'] = 1$. Note that such an index always exists (as $B_j[1] = 1$) and can be found in constant time using standard bit-wise operations. For any index $\ell'' \in \{\ell' + 1, \ldots, \ell\}$ we have $B_j[\ell''] = 0$, i.e., case (b) of Lemma 1 holds for $S_j^{\ell''}$. This inductively implies that the minimal suffix of S_j^ℓ is actually the minimal suffix of $S_j^{\ell'}$. On the other hand $B_j[\ell'] = 1$, so for the latter we have a guarantee that case (a) holds, which lets us find the minimal suffix of S_j^ℓ in constant time. This completes the description of an $\mathcal{O}(1)$-time query algorithm.

3.2   Construction

A simple $\mathcal{O}(n \log n)$-time construction algorithm also relies on Lemma 1. It suffices to show that, once the enhanced suffix array is built, we can determine B_j in $\mathcal{O}(\log n)$ time. We find the minimal suffix of S_j^ℓ for consecutive values of ℓ. Once we know the answer for $\ell - 1$, case (a) of Lemma 1 gives us the second candidate for the minimal suffix of S_j^ℓ, and the enhanced suffix array lets us choose the smaller of these two candidates. We set $B_j[\ell] = 1$ if the smaller candidate is not contained in $S_j^{\ell-1}$. Therefore we obtain the following result.

Theorem 2. *A string T of length n can be stored in an $\mathcal{O}(n)$-space structure that enables to compute the minimal suffix of any substring of T in $\mathcal{O}(1)$ time. This data structure can be constructed in $\mathcal{O}(n \log n)$ time.*

The construction described above is simple and works for any choice of canonical substrings, but, unfortunately, it cannot be used to achieve a trade-off between the query and the construction time. Below we consider a specific choice of canonical substrings and give an alternative construction method. The intuition behind such a choice is that given a string of length k we can compute the minimal suffix for each of its prefixes in $\mathcal{O}(k)$ total time. Hence it would be convenient to have many S_j^ℓ which are prefixes of each other. Then a natural choice is $|S_j^\ell| = 2^{\ell-1} + (j \bmod 2^{\ell-1})$, as then $S_{\alpha 2^{\ell-1}}^\ell, S_{\alpha 2^{\ell-1}+1}^\ell, \ldots, S_{(\alpha+1)2^{\ell-1}-1}^\ell$ are all prefixes of $S_{(\alpha+1)2^{\ell-1}-1}^\ell$. Unfortunately, such choice does not fulfill the condition $|S_j^\ell| \le 2|S_j^{\ell-1}|$, and we need to tweak it a little bit.

For $\ell = 1$ we define $S_j^1 = T[j..j]$. For $\ell > 1$ we set $m = \lfloor \ell/2 \rfloor - 1$ and define S_j^ℓ so that

$$|S_j^\ell| = \begin{cases} 2 \cdot 2^m + (j \bmod 2^m) & \text{if } \ell \text{ is even,} \\ 3 \cdot 2^m + (j \bmod 2^m) & \text{otherwise.} \end{cases}$$

Note that if $2 \cdot 2^m \le j < 3 \cdot 2^m$, then $T[1..j] = S_j^{2m+2}$ while if $3 \cdot 2^m \le j < 4 \cdot 2^m$, then $T[1..j] = S_j^{2m+3}$. Clearly the number of such substrings ending at j is therefore $\mathcal{O}(\log n)$. The following facts show that the above choice of canonical substrings satisfies the remaining required properties.

Fact 3. *For any S_j^ℓ and $S_j^{\ell+1}$ with $\ell \ge 1$ we have $|S_j^{\ell+1}| < 2|S_j^\ell|$.*

Proof. For $\ell = 1$ the statement holds trivially. Consider $\ell \ge 2$. Let m, as before, denote $\lfloor \ell/2 \rfloor - 1$. If ℓ is even, then $\ell + 1$ is odd and we have

$$|S_j^{\ell+1}| = 3 \cdot 2^m + (j \bmod 2^m) < 4 \cdot 2^m \le 2 \cdot (2 \cdot 2^m + (j \bmod 2^m)) = 2|S_j^\ell|$$

while for odd ℓ

$$|S_j^{\ell+1}| = 2 \cdot 2^{m+1} + (j \bmod 2^{m+1}) < 3 \cdot 2^{m+1} \le 2 \cdot (3 \cdot 2^m + (j \bmod 2^m)) = 2|S_j^\ell|.$$

Fact 4. *For $1 \leq i < j \leq n$, the value $\alpha(i, j)$ can be computed in constant time.*

Proof. Let $m = \lfloor \log |T[i..j]| \rfloor$. Observe that

$$\left|S_j^{2m-1}\right| = 3 \cdot 2^{m-2} + (j \bmod 2^{m-2}) < 2^m \leq |T[i..j]|$$
$$\left|S_j^{2m+2}\right| = 2 \cdot 2^m + (j \bmod 2^m) \geq 2^{m+1} > |T[i..j]|.$$

Thus $\alpha(i, j) \in \{2m - 1, 2m, 2m + 1\}$, and we can verify in constant time which of these values is the correct one. $\qquad\square$

After building the enhanced suffix array, we set all bits $B_j[1]$ to 1. Then for each $\ell > 1$ we compute the minimal suffixes of the substrings S_j^ℓ as follows. Fix $\ell > 1$ and split T into *chunks* of size 2^m each (with $m = \lfloor \ell/2 \rfloor - 1$). Now each S_j^ℓ is a prefix of a concatenation of at most four such chunks. Recall that given a string, a variant of Duval's algorithm (Algorithm 3.1 in [5]) takes linear time to compute the lengths of minimal suffixes of all its prefixes. We divide T into chunks of length 2^m (with $m = \lfloor \ell/2 \rfloor - 1$) and run this algorithm for each four (or less at the end) consecutive chunks. This gives the minimal suffixes of S_j^ℓ for all $1 \leq j \leq n$, in $\mathcal{O}(n)$ time. The value $B_j[\ell]$ is determined by comparing the length of the computed minimal suffix of S_j^ℓ with $|S_j^{\ell-1}|$. We have $\mathcal{O}(\log n)$ phases, which gives $\mathcal{O}(n \log n)$ total time complexity and $\mathcal{O}(n)$ total space consumption.

3.3 Trade-Off

To obtain a data structure with $\mathcal{O}(n \log n/\tau)$ construction and $\mathcal{O}(\tau)$ query time, we define the bit vectors in a slightly different way. We set B_j to be of size $\lfloor \alpha(1, j)/\tau \rfloor$ with $B_j[k] = 1$ if and only if $k = 1$ or the minimal suffix of $S_j^{\tau k}$ is longer than $|S_j^{\tau(k-1)}|$. This way we need only $\mathcal{O}(\log n/\tau)$ phases in the construction algorithm, so it takes $\mathcal{O}(n \log n/\tau)$ time.

Again, assume we are looking for the minimal suffix of $T[i..j]$ with $\alpha(i, j) = \ell$. As before, the difficult part is to find the minimal suffix of S_j^ℓ, and our goal is to find $\ell' \leq \ell$ such that the minimal suffix of S_j^ℓ is actually the minimal suffix of $S_j^{\ell'}$, but is longer than $|S_j^{\ell'-1}|$. If $\ell = \tau k$ for an integer k, we could find the largest $k' \leq k$ such that $B[k'] = 1$ and we would know that $\ell' \in (\tau(k'-1), \tau k']$. In the general case, we choose the largest k such that $\tau k \leq \ell$, and then we know that we should consider $\ell' \in (\tau k, \ell]$ and $\ell' \in (\tau(k'-1), \tau k']$, with k' defined as in the previous special case. In total we have $\mathcal{O}(\tau)$ possible values of ℓ', and we are guaranteed that the suffix we seek can be obtained using case (a) of Lemma 1 for $S_j^{\ell'}$ for one of these values. We simply generate all these candidates and use the enhanced suffix array to find the smallest suffix among them. In total, the query algorithm works in $\mathcal{O}(\tau)$ time, which gives the following result.

Theorem 5. *For any $1 \leq \tau \leq \log n$, a string T of length n can be stored in an $\mathcal{O}(n)$-space data structure that allows to compute the minimal suffix of any substring of T in $\mathcal{O}(\tau)$ time. This data structure can be constructed in $\mathcal{O}(n \log n/\tau)$ time.*

3.4   Applications

As a corollary we obtain an efficient data structure for computing Lyndon decompositions of substrings of T. A string w is a *Lyndon word* if it is strictly smaller than its proper cyclic rotations. For a nonempty string x a decomposition $x = w_1^{\alpha_1} w_2^{\alpha_2} \ldots w_k^{\alpha_k}$ is called a *Lyndon decomposition* if and only if $w_1 > w_2 > \ldots > w_k$ are Lyndon words [3]. The last factor w_k is the minimal suffix of x [5] and from the definition we easily obtain that $w_k^{\alpha_k}$ is the largest power of w_k which is a suffix of x. Also, $w_1^{\alpha_1} w_2^{\alpha_2} \ldots w_{k-1}^{\alpha_{k-1}}$ is the Lyndon decomposition of the remaining prefix of x, which gives us the following corollary.

Corollary 6. *For any $1 \le \tau \le \log n$ a string T of length n can be stored in an $\mathcal{O}(n)$-space data structure that enables to compute the Lyndon decomposition of any substring of T in $\mathcal{O}(k\tau)$ time, where k is the number of distinct factors in the decomposition. This data structure can be constructed in $\mathcal{O}(n \log n / \tau)$ time.*

4   Maximal Suffix

Our data structure for the maximal suffix problem is very similar to the one we have developed for the minimal suffix. However, in contrast to that problem, the properties specific to maximal suffixes will let us design a linear time construction algorithm.

Observe that the only component of Section 3 which cannot be immediately adapted to the maximal suffix problem is Lemma 1. While its exact counterpart is not true, in Section 4.1 we prove the following lemma which is equivalent in terms of the algorithmic applications. Canonical substrings S_j^ℓ are defined exactly as before.

Lemma 7. *Consider a substring $T[i..j]$. Using the enhanced suffix array of T, one can compute in $\mathcal{O}(1)$ time an index p ($i \le p \le j$) such that the maximal suffix $T[\mu..j]$ of $T[i..j]$ is either equal to*

(a) $T[p..j]$, or
(b) the maximal suffix of $S_j^{\alpha(i,j)}$.

Just as the data structure described in Section 3, our data structure, apart from the enhanced suffix array, contains bit vectors B_j, $j \in [1, n]$, with $B_j[\ell] = 1$ if $\ell = 1$ or the maximal suffix of S_j^ℓ is longer than $|S_j^{\ell-1}|$. The query algorithm described in Section 3.1 can be adapted in an obvious way, i.e., so that it uses Lemma 7 instead of Lemma 1 and chooses the larger of the two candidates as the answer. This shows the following theorem:

Theorem 8. *A string T of length n can be stored in an $\mathcal{O}(n)$-space structure that enables to compute the maximal suffix of any substring of T in $\mathcal{O}(1)$ time.*

The $\mathcal{O}(n \log n)$-time construction algorithms and the trade-off between query and construction time, described in Sections 3.2 and 3.3, also easily adapt to the maximal suffix problem. In this case, however, we can actually achieve $\mathcal{O}(n)$ construction time, as presented in Section 4.2.

4.1   Proof of Lemma 7

Below we describe a constant-time algorithm, which returns a position $p \in [i,j]$. If the maximal suffix $T[\mu..j]$ of $T[i..j]$ is shorter than $S_j^{\alpha(i,j)}$ (case (b) of Lemma 7), the algorithm can return any $p \in [i,j]$. Below we assume that $T[\mu..j]$ is longer than $S_j^{\alpha(i,j)}$ and show that under this assumption the algorithm will return $p = \mu$. From the assumption and the properties of canonical substrings it follows that $\mu \in [i,r]$, where $r = j - |S_j^{\alpha(i,j)}|$, and that the lengths of the suffixes of $T[i..j]$ starting at positions in $[i,r]$ differ by up to a factor of two.

We start with locating the maximal suffix $T[p_1..]$ in $Suf[i,j]$. Then the maximal suffix $T[\mu..j]$ of $T[i..j]$ must start with $T[p_1..j]$, so $\mu \leq p_1$. To check if $\mu < p_1$, we locate the maximal suffix $T[p_2..]$ in $Suf[i, p_1 - 1]$. If $T[p_1..j]$ is not a prefix of $T[p_2..j]$, one can see that $\mu = p_1$. More formally, we have the following lemma, which appears as Lemma 2 in [1].

Lemma 9.   *Let $P_1 = T[p_1..j]$ be a prefix of $T[\mu..j]$ and let $P_2 = T[p_2..j]$, where $T[p_2..]$ is the maximal suffix in $Suf[i, p_1 - 1]$. If P_1 is not a prefix of P_2, then $\mu = p_1$. Otherwise, P_2 is also a prefix of $T[\mu..j]$.*

We check if $P_1 = T[p_1..j]$ is a prefix of $P_2 = T[p_2..j]$. If not, we return p_1. If P_1 is a prefix of P_2, we know that $\mu \leq p_2$. To check if $\mu < p_2$ we could repeat the above step, i.e., locate the maximal suffix of $T[p_3..]$ in $Suf[i, p_2 - 1]$ and check if P_2 is a prefix of $P_3 = T[p_3..j]$. If not, we return $\mu = p_2$. If P_2 is a prefix of P_3, we again repeat the whole step. Unfortunately the number of repetitions could then be very large. Therefore we use the property that $2|P_1| \geq |P_2|$ to quickly jump to the last repetition. Informally, we apply the periodicity lemma to show that the situation must look like the one in Fig. 1. We prove this in two lemmas, which are essentially Lemmas 4 and 5 of [1]. We give their proofs here because we use different notation.

Lemma 10.   *The substring $\rho = T[p_2..p_1 - 1]$ is the shortest period of P_2, i.e., ρ is the shortest string such that for some $s \geq 1$ one has $P_2 = \rho^s \rho'$.*

Proof.   Since P_1 is a border of P_2, $\rho = T[p_2..p_1 - 1]$ is a period of P_2. It remains to prove that no shorter period is possible. So, consider the shortest period γ, and assume that $|\gamma| < |\rho|$. Then $|\gamma| + |\rho| \leq 2|\rho| \leq |T[p_2..j]|$, and by the periodicity lemma substring P_2 has another period $\gcd(|\gamma|, |\rho|)$. Since γ is the shortest period, $|\rho|$ must be a multiple of $|\gamma|$, i.e., $\rho = \gamma^k$ for some $k \geq 2$.

Suppose that $T[p_1..] \prec \gamma T[p_1..]$. Then prepending both parts of the latter inequality by copies of γ gives $\gamma^{\ell-1} T[p_1..] \prec \gamma^{\ell} T[p_1..]$ for any $1 \leq \ell \leq k$, so from the transitivity of $\prec$ we get that $T[p_1..] \prec \gamma^k T[p_1..] = T[p_2..]$, which contradicts the maximality of $T[p_1..]$ in $Suf[i, r]$. Therefore $T[p_1..] \succ \gamma T[p_1..]$, and consequently $\gamma^{k-1} T[p_1..] \succ \gamma^k T[p_1..]$. But $\gamma^{k-1} T[p_1..] = T[p_2 + |\gamma|..]$ and $\gamma^k T[p_1..] = T[p_2..]$, so $T[p_2 + |\gamma|..]$ is larger than $T[p_2..]$ and belongs to $Suf[i, p_1 - 1]$, a contradiction. $\square$

Lemma 11.   *Suppose that $P_2 = \rho P_1 = \rho^s \rho'$. The maximal suffix $T[\mu..j]$ is the longest suffix of $T[i..j]$ equal to $\rho^t \rho'$ for some integer t. (See also Fig. 1.)*

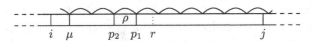

Fig. 1. A schematic illustration of Lemma 11

Proof. Clearly P_2 is a border of $T[\mu..j]$. From $P_2 = \rho P_1$ and $|T[\mu..j]| \le 2|P_1|$ we have $|T[\mu..j]| + |\rho| \le 2|P_1| + |\rho| \le 2|P_2|$. Consequently the occurrences of P_2 as a prefix and as a suffix of $T[\mu..j]$ have an overlap with at least $|\rho|$ positions. As $|\rho|$ is a period of P_2, this implies that $|\rho|$ is also a period of $T[\mu..j]$. Thus $T[\mu..j] = \rho'' \rho^r \rho'$, where r is an integer and ρ'' is a proper suffix of ρ. Moreover ρ^2 is a prefix of $T[\mu..j]$, since it is a prefix of P_2, which is a prefix of $T[\mu..j]$. Now $\rho'' \ne \varepsilon$ would imply a non-trivial occurrence of ρ in ρ^2, which contradicts ρ being primitive, see [4]. Thus $T[\mu..j] = \rho^r \rho'$. If $t > r$, then $\rho^t \rho' \succ \rho^r \rho'$, so $T[\mu..j]$ is the longest suffix of $T[i..j]$ equal to $\rho^t \rho'$ for some integer t.   □

Proof (of Lemma 7). Let $T[p_1..]$ be the maximal suffix in $Suf[i, r]$ and $T[p_2..]$ be the maximal suffix in $Suf[i, p_1 - 1]$. We first compute p_1 and p_2 in constant time using the enhanced suffix array. Then we check if $T[p_1..j]$ is a prefix of $T[p_2..j]$. If it is not, we return $p = p_1$. Otherwise we compute the largest integer r such that ρ^r (for $\rho = T[p_2..p_1 - 1]$), is a suffix of $T[i..p_1 - 1]$ using the method described in Section 2, and return $p = p_1 - r|\rho|$. From the lemmas above it follows that if $T[\mu..j]$ is longer than $S_j^{\alpha(i,j)}$, then $p = \mu$.   □

4.2   Construction

Our algorithm is based on the following notion. For $1 \le p \le j \le n$ we say that a position p is j-*active* if there is no position $p' \in [p + 1, j]$ such that $T[p..j] \prec T[p'..j]$. Equivalently, p is j-active exactly when $T[p..j]$ is its own maximal suffix. The maximal suffix of any string is its own maximal suffix, so from the definition it follows that the starting position of the maximal suffix of $T[i..j]$ is the minimum j-active position in $[i, j]$. Therefore, for $\ell > 1$ we have $B_j[\ell] = 1$ if and only if there is at least one j-active position within the range $R_j^\ell = [j - |S_j^\ell| + 1, j - |S_j^{\ell-1}|]$. We set $R_j^1 = [j, j]$ so that this equivalence also holds for $\ell = 1$ (since j is always j-active).

Example 12. If $T[1..8] = \texttt{dcccabab}$, the 8-active positions are $1, 2, 3, 4, 6, 8$.

The construction algorithm iterates over j ranging from 1 to n, maintaining the list of active positions and computing the bit vectors B_j. We also maintain the ranges R_j^ℓ for the choice of canonical substrings defined in Section 3.2, which form a partition of $[1, j]$. The following two results describe the changes of the list of j-active positions and the ranges R_j^ℓ when we increment j.

Lemma 13. *If the list of all $(j - 1)$-active positions consists of $p_1 < p_2 < \ldots < p_z$, the list of j-active positions can be created by adding j, and repeating the following procedure: if p_k and p_{k+1} are two neighbours on the current list, and $T[j + p_k - p_{k+1}] < T[j]$, remove p_k from the list.*

Proof. First we prove that if a position $1 \leq p \leq j-1$ is not $(j-1)$-active, then it is not j-active either. Indeed, if p is not $(j-1)$-active, then by the definition there is a position $p < p' \leq j-1$ such that $T[p..j-1] \prec T[p'..j-1]$. Consequently, $T[p..j] = T[p..j-1]T[j] \prec T[p'..j-1]T[j] = T[p'..j]$ and p is not j-active. Hence, the only candidates for j-active positions are $(j-1)$-active positions and j.

Secondly, note that if $1 \leq p \leq j-1$ is a $(j-1)$-active position and $T[p'..j-1]$ is a prefix of $T[p..j-1]$, then p' is $(j-1)$-active too. If not, then there exists a position p'', $p' < p'' < j-1$, such that $T[p'..j-1] \prec T[p''..j-1]$, and it follows that $T[p..j-1] = T[p'..j-1]T[j+p-p'..j-1] \prec T[p''..j-1]$, a contradiction.

A $(j-1)$-active position p is not j-active only if (1) $T[p] < T[j]$ or (2) there exists $p < p' \leq j-1$ such that $T[p'..j-1]$ is a prefix of $T[p..j-1]$, i.e., p' is $(j-1)$-active, and $T[p'..j] \succ T[p..j]$, or, equivalently, $T[j+p-p'] < T[j]$. Both of these cases are detected by the deletion procedure.   □

Example 14. If $T[1..9] = $ dcccababb, the 8-active positions are $1, 2, 3, 4, 6, 8$, and the 9-active positions are $1, 2, 3, 4, 8, 9$, i.e., we add 9 and delete 6.

Fact 15. *Let* $j \in [1, n]$ *and assume* 2^k *is the largest power of two dividing* j.

(a) *If* $\ell = 1$, *then* $R_j^\ell = [j, j]$.
(b) *If* $2 \leq \ell < 2k+4$, *then* $R_j^\ell = R_{j-1}^{\ell-1}$.
(c) *If* $\ell = 2k+4$, *then* $R_j^\ell = R_{j-1}^\ell \cup R_{j-1}^{\ell-1}$.
(d) *If* $\ell > 2k+4$, *then* $R_j^\ell = R_{j-1}^\ell$.

Proof. Observe that we have $R_j^1 = [j, j]$ and $R_j^2 = [j-1, j-1]$, while for $\ell > 2$

$$R_j^\ell = \begin{cases} [2^m(\lfloor \frac{j}{2^m} \rfloor - 2) + 1, 2^{m-1}(\lfloor \frac{j}{2^{m-1}} \rfloor - 3)] & \text{if } \ell \text{ is even,} \\ [2^m(\lfloor \frac{j}{2^m} \rfloor - 3) + 1, 2^m(\lfloor \frac{j}{2^m} \rfloor - 2)] & \text{otherwise,} \end{cases}$$

where $m = \lfloor \ell/2 \rfloor - 1$. Also note that

$$2^m(\lfloor \tfrac{j}{2^m} \rfloor - 3) = \begin{cases} 2^m(\lfloor \frac{j-1}{2^m} \rfloor - 2) & \text{if } 2^m \mid j \\ 2^m(\lfloor \frac{j-1}{2^m} \rfloor - 3) & \text{otherwise,} \end{cases}$$

$$2^m(\lfloor \tfrac{j}{2^m} \rfloor - 2) = \begin{cases} 2^{m-1}(\lfloor \frac{j-1}{2^{m-1}} \rfloor - 3) & \text{if } 2^m \mid j \\ 2^m(\lfloor \frac{j-1}{2^m} \rfloor - 2) & \text{otherwise.} \end{cases}$$

Moreover, $2^m \mid j \Longleftrightarrow \ell \leq 2k+3$ and $2^{m-1} \mid j \Longleftrightarrow \ell \leq 2k+5$, which makes it easy to check the claimed formulas. Note that it is possible that R_j^ℓ is defined only for values ℓ smaller than $2k+4$. This is exactly when the number of ranges grows by one, otherwise it remains unchanged.   □

Fig. 2. The partitions of $[1, j]$ into R_j^ℓ for $j = 27$ and $j = 28$. As for $j = 28$ we have $k = 2$ and $2k+4 = 8$, R_{27}^7 and R_{27}^8 are merged into R_{28}^8.

We scan the positions of T from left to right computing the bit vectors. We maintain the list of active positions and the partition of $[1, j]$ into ranges R_j^ℓ. Additionally, for every such range we have a counter storing the number of active positions inside. Recall that $B_j[\ell] = 1$ exactly when the ℓ-th counter is nonzero.

To efficiently update the list of $(j-1)$-active position and turn it into the list of j-active positions, we also maintain for every j' a list of pointers to pairs of neighbouring positions. Whenever a new pair of neighbouring positions p_k, p_{k+1} appears, we compute $L = \text{lcp}(T[p_k..], T[p_{k+1}..])$, and insert a pointer to the pair p_k, p_{k+1} into the list of a position $j' = p_{k+1} + L$. When we actually reach $j = j'$, we follow the pointer and check if p_k and p_{k+1} are still neighbours. If they are and $T[j + p_k - p_{k+1}] < T[j]$, we remove p_k from the list of active positions. Otherwise we do nothing. From Lemma 13 it follows that the two possible updates of the list under transition from $j-1$ to j are either adding j or deleting some position from the list. This guarantees that the process of deletion from Lemma 13 and the process we have just described are equivalent.

Suppose that we already know the list of $(j-1)$-active positions, the bit vector B_{j-1}, and the number of $(j-1)$-active positions in each range R_{j-1}^ℓ. First we update the list of $(j-1)$-active positions. When a position is deleted from the list, we find the range it belongs to, and decrement the counter of active positions there. If a counter drops down to zero, we clear the corresponding bit of the bit vector. Then we start updating the partition: first we append a new range $[j, j]$ to the partition of $[1..j-1]$ and initialize the counter of active positions there to one. Then, k being the largest power of 2 dividing j, we update the first $2k + 4$ ranges using Fact 15, including the counters and the bit vector. This takes $\mathcal{O}(k)$ time which amortizes to $\mathcal{O}(\sum_{k=1}^{\infty} \frac{k}{2^k}) = \mathcal{O}(1)$ over all values of j.

Theorem 16. *A string T of length n can be stored in an $\mathcal{O}(n)$-space structure that allows computing the maximal suffix of any substring of T in $\mathcal{O}(1)$ time. The data structure can be constructed in $\mathcal{O}(n)$ time.*

References

1. Babenko, M., Kolesnichenko, I., Starikovskaya, T.: On minimal and maximal suffixes of a substring. In: Fischer, J., Sanders, P. (eds.) CPM 2013. LNCS, vol. 7922, pp. 28–37. Springer, Heidelberg (2013)
2. Bender, M.A., Farach-Colton, M.: The LCA problem revisited. In: Gonnet, G.H., Panario, D., Viola, A. (eds.) LATIN 2000. LNCS, vol. 1776, pp. 88–94. Springer, Heidelberg (2000)
3. Chen, K.T., Fox, R.H., Lyndon, R.C.: Free differential calculus, IV. The quotient groups of the lower central series. The Annals of Mathematics 68(1), 81–95 (1958)
4. Crochemore, M., Hancart, C., Lecroq, T.: Algorithms on Strings. Cambridge University Press (2007)
5. Duval, J.-P.: Factorizing words over an ordered alphabet. J. Algorithms 4(4), 363–381 (1983)
6. Puglisi, S.J., Smyth, W.F., Turpin, A.: A taxonomy of suffix array construction algorithms. ACM Comput. Surv. 39(2) (2007)

Compressed Subsequence Matching
and Packed Tree Coloring

Philip Bille*, Patrick Hagge Cording, and Inge Li Gørtz*

Technical University of Denmark, DTU Compute
{phbi,phaco,inge}@dtu.dk

Abstract. We present a new algorithm for subsequence matching in grammar compressed strings. Given a grammar of size n compressing a string of size N and a pattern string of size m over an alphabet of size σ, our algorithm uses $O(n + \frac{n\sigma}{w})$ space and $O(n + \frac{n\sigma}{w} + m \log N \log w \cdot occ)$ or $O(n + \frac{n\sigma}{w} \log w + m \log N \cdot occ)$ time. Here w is the word size and occ is the number of occurrences of the pattern. Our algorithm uses less space than previous algorithms and is also faster for $occ = o(\frac{n}{\log N})$ occurrences. The algorithm uses a new data structure that allows us to efficiently find the next occurrence of a given character after a given position in a compressed string. This data structure in turn is based on a new data structure for the tree color problem, where the node colors are packed in bit strings.

1 Introduction

In the *compressed subsequence matching problem* we are given a grammar S of size n compressing a string S of size N and a pattern string P of size m over an alphabet of size σ, and the goal is to find and report the index of all minimal substrings of S that contain P as a subsequence. A substring is said to be minimal if shortening it implies that P is no longer a subsequence of that substring. In this paper we present a new algorithm for compressed subsequence matching which is space efficient and is faster than the previously fastest algorithm for a bounded number of occurrences. Our algorithm relies on a method that is different from the ones used by previous algorithms.

Subsequence matching is useful when searching sequential log data for a sequence of events that may be separated by other events. Say for instance that we are running a webserver and we want to know how often a visitor has found her way to subpage C through page A and then B. We then set $P = ABC$ and apply a subsequence matching algorithm to the contents of the log file. Many applications will automatically compress log data to save space, and so the bottleneck of the procedure becomes decompression of the data. In this case, processing the data without fully decompressing it, is crucial. Subsequence matching was also considered in relation to knowledge discovery and data mining [20].

* Supported in part by the The Danish Council for Independent Research | Natural Sciences grant DFF 1323–00178.

A.S. Kulikov, S.O. Kuznetsov, and P. Pevzner (Eds.): CPM 2014, LNCS 8486, pp. 40–49, 2014.
© Springer International Publishing Switzerland 2014

Several algorithms have been presented for uncompressed strings [6,10,12,14, 15,20,27]. The fastest of these is due to Das et al. [15]. Since it is an online algorithm we may apply it to the compressed version without having to store the entire decompressed string, and we get an algorithm with running time $O(\frac{Nm}{\log m})$ that uses $O(n + m)$ space. The first algorithm with time complexity independent of the size of the string was presented by Cegielski et al. [11] in 2006. Its runnning time is $O(nm^2 \log m + occ)$ time and it uses $O(nm^2)$ space. Using a different approach, Tiskin improved the running time to $O(nm^{1.5} + occ)$ [25] and later even further to $O(nm \log m + occ)$ [26]. The space usage of his algorithms is $O(nm)$. The most recent improvement is due to Yamamoto et al. [28] who present an algorithm based on the ideas of Cegielski et al. that runs in $O(nm + occ)$ time and $O(nm)$ space.

Assume without loss of generality that the compressed string is given as a Straight Line Program (SLP). An SLP is an acyclic grammar in Chomsky normal form, i.e., a grammar where each nonterminal production rule expands to two other rules and generates one string only. SLPs are widely studied because they model many well-known compression schemes, such as LZ77 [29], LZ78 [30], and Re-Pair [19] with little overhead [13,22]. The following theorem is the main result of this work.

Theorem 1. *Given an SLP S of size n compressing a string S of size N and a pattern P of size m over an alphabet of size σ, compressed subsequence matching can be solved in $O(n + \frac{n\sigma}{w})$ words of space and time*

(i) $O(n + \frac{n\sigma}{w} + m \log N \log w \cdot occ)$, *or*
(ii) $O(n + \frac{n\sigma}{w} \log w + m \log N \cdot occ)$

in the word RAM model with word size $w \geq \log N$, and where occ is the number of minimal occurrences of P in S.

Our new algorithm uses less space (linear in n if $\sigma \leq w$) and is also faster than the previously fastest algorithm for $o(\frac{n}{\log N})$ occurrences when $\sigma \leq m$. Note that we can guarantee that the latter requirement always holds by bounding $\sigma = O(m)$ using hashing in return for using $O(m)$ additional extra space.

The algorithm is based on the idea of a simple algorithm for subsequence matching in uncompressed strings which basically scans the string for occurrences of the pattern. We speed up the scanning on compressed strings by introducing the first data structure for SLPs that supports labelled successor queries. The answer to a labelled successor query $\text{LS}(i, c)$ on a string is the index of the first character c occurring after position i in the string. An essential part of this data structure is a new data structure for the tree color problem. This problem is to preprocess a tree where each node is colored by zero or more colors, such that given a node v and a color c, we may efficiently answer a first colored ancestor query, i.e., compute the lowest ancestor of v with color c. Additionally, this data structure also supports a new type of query we call the last colored ancestor. Here the query is two nodes u and v and a color c, and the answer is the highest node on the path from u to v with color c. These results may be of independent interest.

2   Preliminaries

Bit Strings. We will use bit strings to represent sets. In a bit string $B = b_1 b_2 \ldots b_u$ representing a set $\mathcal{B}$ of elements from a universe of size u, $b_i = 1$ iff element i is in $\mathcal{B}$. $B = [0]^u$ denotes the empty set. The operators $\wedge$, $\vee$, and $\oplus$ denote the bitwise AND, OR, and exclusive OR (XOR) of two bit strings. The negation of a bit string B is $\overline{B}$. A *summary* B_s of k bit strings $B_1, B_2, \ldots, B_k$ of equal length is $B_s = B_1 \vee B_2 \vee \ldots \vee B_k$. For a bit string of length w we assume that the mask of any constant can be computed in $O(1)$ time. Given a bit string B of length w, the index of the least significant set bit can be found in $O(1)$ time from $\log_2((B-1) \oplus \overline{B} \wedge B)$. Finding the most significant set bit is more elaborate, but can also be done $O(1)$ time [18]. An $n \times m$ bit matrix may be transposed in $O(w \log w)$ time if $n \leq w$ and $m \leq w$ [24].

Trees. In this paper all trees are rooted, ordered, and have labels on the nodes. The number of nodes in a tree T is t. We denote by $T(v)$ the subtree rooted in v containing all descendants of v. The size $|T(v)|$ is the number of nodes in the subtree $T(v)$ including v. If u is a node in the subtree $T(v)$ we write $u \in T(v)$. If T is a binary tree we denote the left and right child of a node v by $left(v)$ and $right(v)$.

A heavy path decomposition [23] decomposes T into disjoint paths. Nodes are classified as either heavy or light and the decomposition is defined as follows. The root is light. For each internal node v, its heavy child w is the node for which $T(w)$ is of maximum size among the subtrees rooted in children of v. The other children of v are light. Edges are also classified as heavy and light. An edge going into a heavy node is heavy and likewise for light nodes. The heavy path decomposition ensures the property that $\frac{1}{2}|T(v)| > |T(u)|$ for any light child u of v. This means that there are $O(\log t)$ light edges on any path from the root to a leaf. The heavy path decomposition can be computed in $O(t)$ time and space.

Given a binary tree T rooted in a node r, $t > 1$, and a parameter $1 \leq x \leq t$, we may partition T into at most t/x clusters such that for a fixed constant c, the size of any cluster is at most cx [3,5] (see also [1] for a full proof). Two clusters overlap in at most one node, and a node is called a boundary node if it is part of more than one cluster. Any cluster has at most two boundary nodes, and a boundary node is either a leaf or the root in the subtree that is the cluster. The tree obtained by repeatedly contracting edges between two nodes if one of them is not a boundary node is called the macro tree. In other words, the macro tree is the tree consisting only of boundary nodes. A cluster partition can be found in $O(t)$ time.

The answer to a level ancestor query $\text{LA}(v, d)$ on T is the ancestor of v with depth d. A linear space data structure that answers an LA query in $O(1)$ time can be computed for T in $O(t)$ time [16] (see also [2,7,8]).

Straight Line Programs. A Straight Line Program $\mathcal{S}$ is a context-free grammar in Chomsky normal form with n production rules that unambiguously derives a string S of length N. We represent the SLP as a rooted, ordered, and node-labelled directed acyclic graph (DAG) with outdegree 2 and we will refer to production rules as nodes in the DAG. A depth-first left-to-right traversal start-ing from a node v in the DAG produces the string $S(v)$ of length $|S(v)|$. The tree that emerges from the traversal we call the derivation tree. We denote the left and right children of v for $left(v)$ and $right(v)$, respectively. Furthermore, the height of the SLP is the length of the longest path going from the root to a terminal node and is denoted by h.

We may access a character $S[i]$ in $O(h)$ time by storing $|S(v)|$ for each node v in the SLP, and simulate a top-down search of the derivation tree. Doing so yields a unique path from the root of S to the terminal node labelled $S[i]$. There is also a linear space data structure that supports random access in SLPs in $O(\log N)$ time [9]. A key technique used in this data structure is the extension of the heavy path decomposition of trees to SLPs which we will also use in our data structure. For each node $v \in S$, we select the child of v that derives the longest string to be a heavy node. The other child is light. Heavy and light edges are defined as in the decomposition of trees. Whereas applying this technique to a tree results in a decomposition into disjoint paths, it will result in a decomposition into disjoint trees when applied to an SLP. We denote this set of trees by the heavy forest $\mathcal{H}$ of the SLP. This decomposition ensures that the number of light edges on any path from the root to a terminal node is $O(\log N)$. Hence, on any path from the root of the SLP to a terminal node, we visit at most $\log N$ trees from $\mathcal{H}$. When accessing a character using the data structure of [9] we may also report the entry and exit nodes for each tree visited on the unique root-to-terminal path that emerges from the query.

3   Packed Tree Color Problems

In a colored tree, each node is colored by zero or more colors from the set $\{1, \ldots, \sigma\}$. A packed colored tree is a colored tree where the colors of each node v is given as a bit string $C(v)$ where $C(v)[c] = 1$ iff v is colored c. In this section we consider the *packed tree color problem* which is to preprocess a packed colored tree T to support first and last colored ancestor queries. The answer to a first colored ancestor query FIRSTCOLOR(v, c) is the lowest ancestor of v with color c, and the answer to a last colored ancestor query LASTCOLOR(u, v, c) is the highest node with color c on the path from u to v, where we always assume that u is an ancestor of v. Throughout this section we will use the following notation to distinguish results. If a data structure requires $p(t)$ time to build, uses $s(t)$ space, and supports FIRSTCOLOR and LASTCOLOR queries in $q(t)$ time, then the the triple $\langle p(t), s(t), q(t) \rangle$ refers to the solution.

Solutions to the tree color problem for trees that are not packed may be applied to packed trees. All known solutions focus entirely on supporting FIRSTCOLOR queries [4,16,17,21]. A simple solution that supports FIRSTCOLOR queries in $O(1)$ time is to store the answer for every color in every node. This yields a $\langle O(t\sigma), O(t\sigma), O(1) \rangle$ solution. The currently best known trade-off for the tree color problem is $\langle O(t + D), O(t + D), O(\log w) \rangle$ [21], where $D = \sum_{v \in T} \sum_{i=1}^{\sigma} C(v)[i]$ is the accumulated number of colors used.

Our motivation for revisiting this problem is twofold. First we have that $D = O(t\sigma)$ in our application and we are striving for a space bound that is in $o(t\sigma)$. Second we want to support LASTCOLOR queries.

Due to lack of space, the analyses of the first two data structures are omitted.

3.1   A $\langle O(t\sigma), O(t\sigma), O(1)\rangle$ Solution

We store the result of a FIRSTCOLOR(v, c) query for every node and color. For each color, let the induced c-colored subtree be the tree obtained by deleting all nodes that are not colored by color c except the root. Build a levelled ancestor data structure for each induced colored subtree.

The result of a FIRSTCOLOR query is precomputed. A LASTCOLOR(u, v, c) query is answered as follows. If FIRSTCOLOR$(v, c) =$ FIRSTCOLOR(u, c) then there is not a node with color c on the path from u to v. If FIRSTCOLOR$(v, c) \neq$ FIRSTCOLOR(u, c) then let u' and v' be the nodes corresponding to u and v in the induced c-colored subtree. The answer to LASTCOLOR(u, v, c) is then the answer to LA$(v', depth(u') - 1)$ in the induced c-colored subtree.

Lemma 1. *The packed tree color problem can be solved using $O(t\sigma)$ preprocessing time and space, and $O(1)$ query time.*

3.2   A $\langle O(t + \frac{t\sigma}{w}), O(t + \frac{t\sigma}{w}), O(\log t)\rangle$ Solution

We fix a heavy path decomposition of T. For each path p in the heavy path decomposition of T we build a balanced binary tree T_p having the nodes of p as leaves. For each node v in T_p we store a summary $B(v)$ of the colors of its children. For each heavy path $p = v_1, v_2, \ldots, v_k$, where v_1 is the highest node on the path, we store a summary $P(v_i)$ of colors on the path prefix $v_1 \ldots v_i$ for every v_i on p.

For answering a FIRSTCOLOR(v, c) query, let $p = v_1, v_2, \ldots, v_k$ be the heavy path containing v and let $v_i = v$ for some $1 \leq i \leq k$. If $P(v_i)[c] = 1$ we find the lowest ancestor v_a of v_i in T_p for which $B(left(v_a))[c] = 1$ and $v_i \notin T_p(left(v_a))$. The answer to the query is then the rightmost leaf in $T_p(left(v_a))$ with color c. If $P(v_i)[c] = 0$ we repeat the procedure with $v_i = parent(v_1)$, i.e., we jump to the previous heavy path, until we find the first colored ancestor or we reach the root of T.

A LASTCOLOR(u, v, c) query is handled in a similar way. We first find the highest light node w on the path from u to v for which $P(parent(w))[c] = 1$. Let p be the heavy path containing $parent(w)$. Now there are three cases. If u is not on p, the answer to the query is the leftmost leaf in T_p that has color c. If p contains u, the answer is the leftmost leaf with color c to the right of u in T_p, if such a node exists. If it does not exist, we repeat the first step for the second highest light node w' between u and v for which $P(parent(w'))[c] = 1$.

Lemma 2. *The packed tree color problem can be solved using $O(t + \frac{t\sigma}{w})$ preprocessing time and space, and $O(\log t)$ query time.*

3.3   A $\langle O(t + \frac{t\sigma \log w}{w}), O(t + \frac{t\sigma}{w}), O(\frac{t}{w})\rangle$ Solution

Let $v_1, \ldots, v_t$ be the nodes of T in pre-order. We will represent T as a $\sigma \times t$ bit matrix M. Let c be a color from the set of colors $\{1, \ldots, \sigma\}$. In row c of M we store a bit string where bit i is 1 iff v_i has color c. For each node v_i we also store a bit string $A(i)$ where bit j is 1 iff v_j is an ancestor of v_i.

We construct this data structure from a packed colored tree as follows. Assume that the bit strings representing the node colorings form a $t \times \sigma$ matrix where row i is the colorings of node v_i. We transpose this matrix to get M. To do this we partition the matrix into a $\frac{t}{w} \times \frac{\sigma}{w}$ matrix (assume w.l.o.g. that w divides t and σ), transpose each $w \times w$ submatrix as described in [24], and transpose the $\frac{t}{w} \times \frac{\sigma}{w}$ matrix to get M.

To compute the ancestor bit strings first set $A(root(T)) = [0]^t$. For all other nodes v_i, where v_j is the parent of v_i, set $A(v_i) = A(v_j) \vee 2^j$.

We answer a FIRSTCOLOR(v, c) as follows. Let $R = M[c] \wedge A(v)$. Now R is a bit string representing the set of ancestors of v with color c. Since the nodes have pre-order indices, the answer to the query is v_i, where i is the index of the least significant set bit in R.

To answer a LASTCOLOR(v, u, c) query we start by computing R the same way as above. We then set the first $i - 1$ bits of R to 0, where i is the index of u. The answer to the query is the most significant set bit of R.

The $\sigma \times t$ bit matrix M can be packed in words and therefore uses $O(\frac{t\sigma}{w})$ space. The same is evident for the ancestor bit strings. Transposing a $w \times w$ matrix takes $O(w \log w)$ time, and since there are $\frac{t\sigma}{w^2}$ submatrices of this size in the color bit matrix, the total time spent for all submatrices is $O(\frac{t\sigma \log w}{w})$. Transposing the $\frac{t}{w} \times \frac{\sigma}{w}$ matrix takes $O(\frac{t\sigma}{w})$ time. Computing the ancestor bit strings clearly takes $O(\frac{t\sigma}{w})$ time.

The size of R is $O(\frac{t}{w})$, so finding the first non-zero word takes $O(\frac{t}{w})$ time. Determining the least or most significant set bit of a word is done in $O(1)$ time. Thus, the query time for both a FIRSTCOLOR and a LASTCOLOR query is $O(\frac{t}{w})$.

Lemma 3. *The packed tree color problem can be solved using $O(t + \frac{t\sigma \log w}{w})$ preprocessing time, $O(t + \frac{t\sigma}{w})$ space, and $O(\frac{t}{w})$ query time.*

3.4   Combining the Solutions

We now show how to combine the previously described solutions to get $\langle O(t + \frac{n\sigma}{w}), O(t + \frac{n\sigma}{w}), O(\log w)\rangle$ and $\langle O(t + \frac{t\sigma \log w}{w}), O(t + \frac{t\sigma}{w}), O(1)\rangle$ trade-offs. This is achieved by doing a cluster partitioning of the tree.

First we convert T to a binary tree T'. Then we partition T' into $O(\frac{t}{w})$ clusters, i.e., each cluster has size $O(w)$. For each cluster C, where one boundary node is a leaf in the cluster and the other is the root of the cluster, we make a summary of the colors of the nodes on the path from the root to the leaf. The summary is stored in the macro tree node that corresponds to the leaf boundary node of C. Apply the $\langle O(t\sigma), O(t\sigma), O(1)\rangle$ solution to the macro tree, and apply either the $\langle O(\frac{t\sigma}{w}), O(\frac{t\sigma}{w}), O(\log t)\rangle$ solution or the $\langle O(\frac{t\sigma \log w}{w}), O(\frac{t\sigma}{w}), O(\frac{t}{w})\rangle$ solution to each cluster using the original colors.

Here is how we answer a FIRSTCOLOR(v, c) query. Let C_v be the cluster containing v. First we ask for FIRSTCOLOR(v, c) in C_v. If the answer is a node in C_v, we are done. If it is undefined, we find the node r in the macro tree corresponding to the root of C_v. We check if r has color c in the macro tree and otherwise ask for $w =$ FIRSTCOLOR(r, c) in the macro tree. In the cluster C_w having w as a leaf boundary node we then check if w has color c and otherwise ask for FIRSTCOLOR(w, c) in C_w.

We answer a LASTCOLOR(u, v, c) query as follows. Assume that $u \neq v$ and let C_u and C_v be the clusters containing u and v. If $C_u = C_v$ then the answer is LASTCOLOR(u, v, c) in the cluster containing u and v. If $C_u \neq C_v$, let w be the leaf boundary node of C_u where $v \in T(w)$. We now proceed in three steps. First, we ask for LASTCOLOR(u, w, c) in C_u. If the query returns a node, this is also the answer to the LASTCOLOR(u, v, c) query. If the answer in the first step is undefined we ask for $z =$ LASTCOLOR($w, root(C_v), c$) in the macro tree to locate the highest cluster with a node with color c between u and v. The answer to the query is then LASTCOLOR($root(C_z), z, c$) on C_z. If the first two steps fail, the answer to a query is LASTCOLOR($root(C_v), v, c$).

The cluster partition can be computed in linear time, and the cluster path summaries are computed in $O(\frac{t\sigma}{w})$ time. Since the macro tree has $O(\frac{t}{w})$ nodes the preprocessing time and space to apply the $\langle O(t\sigma), O(t\sigma), O(1) \rangle$ solution becomes $O(\frac{t\sigma}{w})$. To answer a query we perform a constant number of FIRSTCOLOR and LASTCOLOR queries on the macro tree and clusters. Therefore the total time to perform queries on the macro tree is $O(1)$ time. To get (i) we apply the $\langle O(t + \frac{t\sigma}{w}), O(t + \frac{t\sigma}{w}), O(\log t) \rangle$ solution to clusters. Since a cluster has size $O(w)$ we use a total of $O(\log w)$ time performing queries on clusters. To get (ii) we apply the $\langle O(\frac{t\sigma \log w}{w}), O(\frac{t\sigma}{w}), O(\frac{t}{w}) \rangle$ solution to clusters. Again, since clusters have size $O(w)$ we use a total of $O(1)$ time performing queries on clusters. Preprocessing time and space for the cluster data structures follow because $\sum_{C \in CS} |C| = O(t)$.

Theorem 2. *The packed tree color problem can be solved using $O(t + \frac{t\sigma}{w})$ space,*

(i) $O(t + \frac{t\sigma}{w})$ preprocessing time, and $O(\log w)$ query time, or
(ii) $O(t + \frac{t\sigma}{w} \log w)$ preprocessing time, and $O(1)$ query time.

4   Labelled Successor Data Structure for SLPs

The answer to a labelled successor LS(i, c) query on a string S is the index of the first occurrence of the character c after position i in S. More formally, the answer to LS(i, c) is an index j such that $S[j] = c$, $j > i$, and $S[k] \neq c$ for $k = i + 1, \ldots, j - 1$.

In this section we present a data structure that supports LS(i, c) queries on an SLP. This is the first data structure dedicated to solving this problem on SLPs. Alternatively, we may build the random access data structure of [9] and then answer an LS(i, c) query by doing a random access query for position i followed by a linear scan to find the first occurrence of c. This yields a query time of $O(\log N + j - i)$ while using $O(n)$ space for the data structure.

Theorem 3. *There is a data structure supporting labelled successor (and predecessor) queries on a string of size N over an alphabet of size σ compressed by an SLP of size n in the word RAM model with word size $w \geq \log N$ using $O(n + \frac{n\sigma}{w})$ space and*

(i) $O(n + \frac{n\sigma}{w})$ preprocessing time, and $O(\log N \log w)$ query time, or

(ii) $O(n + \frac{n\sigma}{w} \log w)$ preprocessing time, and $O(\log N)$ query time.

We first apply the construction of [9], and let $\mathcal{H}$ be the heavy forest obtained from the heavy path decomposition of $\mathcal{S}$. For each node v in $\mathcal{S}$ with children $left(v)$ and $right(v)$ we store two bit strings $L(v)$ and $R(v)$ summarizing the characters in $S(left(v))$ and $S(right(v))$. If v and $left(v)$ are in the same tree in $\mathcal{H}$ then $L(v) = [0]^\sigma$ and similarly for $right(v)$ and $R(v)$. For each tree in $\mathcal{H}$ we build two data structures for the packed tree color problem. One where the L bit strings serve as colors and one where the R bit strings serve as colors.

We answer an LS(i, c) query as follows. First we access the character $S[i]$ using the random access data structure. We now have the entry and exit points of the heavy trees in $\mathcal{H}$ on the unique path p describing $S[i]$. Let $T_1, \ldots, T_k \in \mathcal{H}$ be a sequence of trees on p in the order they are visited when starting from the root and ending in the terminal generating $S[i]$, and let $(v_1, u_1), \ldots, (v_k, u_k)$ be the entry and exit nodes for each tree in the sequence. Using the packed tree color data structure for the R colors, we repeat LASTCOLOR(u_i, v_i, c) for $i = k$ down to some j until LASTCOLOR(u_j, v_j, c) is not undefined. Let $w = right(\text{LASTCOLOR}(u_j, v_j, c))$. We now search for the first occurrence of c in $S(w)$. Let T_i be the tree in $\mathcal{H}$ that contains the node w, then the search proceeds in three steps. First, we ask for $v = \text{FIRSTCOLOR}(w, c)$ in T_i in the data structure for L colors and restart the search from $left(v)$. If the query FIRSTCOLOR(w, c) is undefined we continue to the next step. In the second step we check if $root(T_i)$ generates c. If it does, we now have a unique set of entry and exit nodes in the trees of $\mathcal{H}$ that constitutes a path to a terminal that generates the first c after position i. The answer to the LS(i, c) query is the index of this c which we retrieve using the random access data structure. Finally, if $root(T_i)$ does not generate c we ask for $v = \text{LASTCOLOR}(w, root(T_i), c)$ in T_i in the data structure for R colors, and restart the search from $right(v)$.

The data structure uses $O(n + \frac{n\sigma}{w})$ space because the random access data structure uses linear space and the bit strings L and R use $O(\frac{n\sigma}{w})$ space. The random access data structure, including the heavy path decomposition, takes $O(n)$ time to compute and the L and R values are computed using $O(\frac{n\sigma}{w})$ OR operations in a bottom up fashion. Therefore, this part of the data structure is computed in $O(n + \frac{n\sigma}{w})$ time.

To get Theorem 3 (i) we use the packed tree color data structure of Theorem 2 (i) for the trees in $\mathcal{H}$ and likewise for (ii). Since the trees are disjoint, the preprocessing time and space becomes as in the Theorem 3.

For the query, we first do one random access query that takes $O(\log N)$ time, then we perform at most $\log N$ LASTCOLOR queries walking up the SLP and at most $2 \log N$ FIRSTCOLOR and LASTCOLOR queries locating the labelled successor. Finally, retrieving the index also takes $O(\log N)$ time using the random access data structure.

5   Subsequence Matching

We will now use the labelled successor data structure to obtain a subsequence matching algorithm for SLPs. Our algorithm is based on the folklore algorithm for subsequence matching which works as follows (see also [15,20]). First we find the minimal prefix $S[1..j]$ that contains P as a subsequence. This is done by reading S left to right while searching for the characters of P one at a time. We then find the minimal suffix $S[i..j]$ of the prefix $S[1..j]$ that contains P. Similarly, this is done by scanning the prefix right to left. Now $S[i..j]$ is the first minimal occurrence of P. To find the next minimal occurrence we repeat this process for the suffix $S[i + 1..N]$. It can be shown that this algorithm finds all minimal occurrences of P in $O(Nm)$ time.

By using our labelled successor data structure described in the previous section we speed up the procedure of finding some specific character of P. Assume we have matched $P[1..k]$ to $S[1..j]$ such that $P[k] = S[j]$. Instead of doing a linear scan of $S[j + 1..N]$ to find $P[k + 1]$ we ask for the next occurrence of $P[k + 1]$ using $\text{LS}(j, P[k + 1])$.

For each occurrence of P we perform $O(m)$ labelled successor (and labelled predecessor) queries, and we also have to construct the data structures to support these. By applying the results of Theorem 3 we get Theorem 1.

References

1. Abiteboul, S., Alstrup, S., Kaplan, H., Milo, T., Rauhe, T.: Compact labeling scheme for ancestor queries. SIAM J. Comput. 35(6), 1295–1309 (2006)
2. Alstrup, S., Holm, J.: Improved algorithms for finding level ancestors in dynamic trees. In: Welzl, E., Montanari, U., Rolim, J.D.P. (eds.) ICALP 2000. LNCS, vol. 1853, pp. 73–84. Springer, Heidelberg (2000)
3. Alstrup, S., Holm, J., de Lichtenberg, K., Thorup, M.: Minimizing diameters of dynamic trees. In: Degano, P., Gorrieri, R., Marchetti-Spaccamela, A. (eds.) ICALP 1997. LNCS, vol. 1256, pp. 270–280. Springer, Heidelberg (1997)
4. Alstrup, S., Husfeldt, T., Rauhe, T.: Marked ancestor problems. In: Proc. 39th FOCS, pp. 534–543 (1998)
5. Alstrup, S., Secher, J.P., Spork, M.: Optimal on-line decremental connectivity in trees. Inform. Process. Lett. 64(4), 161–164 (1997)
6. Baeza-Yates, R.A.: Searching subsequences. Theoret. Comput. Sci. 78(2), 363–376 (1991)
7. Bender, M.A., Farach-Colton, M.: The level ancestor problem simplified. Theoret. Comput. Sci. 321(1), 5–12 (2004)
8. Berkman, O., Vishkin, U.: Finding level-ancestors in trees. J. Comput. System Sci. 48(2), 214–230 (1994)
9. Bille, P., Landau, G.M., Raman, R., Sadakane, K., Satti, S.R., Weimann, O.: Random access to grammar-compressed strings. In: Proc. 22nd SODA, pp. 373–389 (2011)

10. Boasson, L., Cegielski, P., Guessarian, I., Matiyasevich, Y.: Window-accumulated subsequence matching problem is linear. In: Proc. 18th PODS, pp. 327–336 (1999)
11. Cégielski, P., Guessarian, I., Lifshits, Y., Matiyasevich, Y.V.: Window subsequence problems for compressed texts. In: Grigoriev, D., Harrison, J., Hirsch, E.A. (eds.) CSR 2006. LNCS, vol. 3967, pp. 127–136. Springer, Heidelberg (2006)
12. Cégielski, P., Guessarian, I., Matiyasevich, Y.: Multiple serial episodes matching. Inform. Process. Lett. 98(6), 211–218 (2006)
13. Charikar, M., Lehman, E., Liu, D., Panigrahy, R., Prabhakaran, M., Sahai, A., Shelat, A.: The smallest grammar problem. IEEE Trans. Inf. Theory 51(7), 2554–2576 (2005)
14. Crochemore, M., Melichar, B., Troníček, Z.: Directed acyclic subsequence graph-overview. J. Discrete Algorithms 1(3), 255–280 (2003)
15. Das, G., Fleischer, R., Gasieniec, L., Gunopulos, D., Kärkkäinen, J.: Episode matching. In: Hein, J., Apostolico, A. (eds.) CPM 1997. LNCS, vol. 1264, pp. 12–27. Springer, Heidelberg (1997)
16. Dietz, P.F.: Finding level-ancestors in dynamic trees. In: Dehne, F., Sack, J.-R., Santoro, N. (eds.) WADS 1991. LNCS, vol. 519, pp. 32–40. Springer, Heidelberg (1991)
17. Ferragina, P., Muthukrishnan, S.: Efficient dynamic method-lookup for object oriented languages. In: Díaz, J. (ed.) ESA 1996. LNCS, vol. 1136, pp. 107–120. Springer, Heidelberg (1996)
18. Fredman, M.L., Willard, D.E.: Surpassing the information theoretic bound with fusion trees. J. Comput. System Sci. 47(3), 424–436 (1993)
19. Larsson, N.J., Moffat, A.: Off-line dictionary-based compression. Proc. IEEE 88(11), 1722–1732 (2000)
20. Mannila, H., Toivonen, H., Verkamo, A.I.: Discovery of frequent episodes in event sequences. Data Min. Knowl. Discov. 1(3), 259–289 (1997)
21. Muthukrishnan, S., Müller, M.: Time and space efficient method-lookup for object-oriented programs. In: Proc. 7th SODA, pp. 42–51 (1996)
22. Rytter, W.: Application of Lempel–Ziv factorization to the approximation of grammar-based compression. Theoret. Comput. Sci. 302(1), 211–222 (2003)
23. Sleator, D.D., Endre Tarjan, R.: A data structure for dynamic trees. J. Comput. System Sci. 26(3), 362–391 (1983)
24. Thorup, M.: Randomized sorting in $O(n \log \log n)$ time and linear space using addition, shift, and bit-wise boolean operations. J. Algorithms 42(2), 205–230 (2002)
25. Tiskin, A.: Faster subsequence recognition in compressed strings. J. Math. Sci. 158(5), 759–769 (2009)
26. Tiskin, A.: Towards approximate matching in compressed strings: Local subsequence recognition. In: Kulikov, A., Vereshchagin, N. (eds.) CSR 2011. LNCS, vol. 6651, pp. 401–414. Springer, Heidelberg (2011)
27. Troníček, Z.: Episode matching. In: Amir, A., Landau, G.M. (eds.) CPM 2001. LNCS, vol. 2089, pp. 143–146. Springer, Heidelberg (2001)
28. Yamamoto, T., Bannai, H., Inenaga, S., Takeda, M.: Faster subsequence and don't-care pattern matching on compressed texts. In: Giancarlo, R., Manzini, G. (eds.) CPM 2011. LNCS, vol. 6661, pp. 309–322. Springer, Heidelberg (2011)
29. Ziv, J., Lempel, A.: A universal algorithm for sequential data compression. IEEE Trans. Inf. Theory 23(3), 337–343 (1977)
30. Ziv, J., Lempel, A.: Compression of individual sequences via variable-rate coding. IEEE Trans. Inf. Theory 24(5), 530–536 (1978)

Reversal Distances for Strings with Few Blocks or Small Alphabets

Laurent Bulteau[1,*], Guillaume Fertin[2], and Christian Komusiewicz[2,**]

[1] Institut für Softwaretechnik und Theoretische Informatik, TU Berlin
l.bulteau@campus.tu-berlin.de
[2] Université de Nantes, LINA - UMR CNRS 6241, France
{guillaume.fertin,christian.komusiewicz}@univ-nantes.fr

Abstract. We study the STRING REVERSAL DISTANCE problem, an extension of the well-known SORTING BY REVERSALS problem. STRING REVERSAL DISTANCE takes two strings S and T as input, and asks for a minimum number of reversals to obtain T from S. We consider four variants: STRING REVERSAL DISTANCE, STRING PREFIX REVERSAL DISTANCE (in which any reversal must include the first letter of the string), and the signed variants of these problems, namely SIGNED STRING REVERSAL DISTANCE and SIGNED STRING PREFIX REVERSAL DISTANCE. We study algorithmic properties of these four problems, in connection with two parameters of the input strings: the number of blocks they contain (a block being maximal substring such that all letters in the substring are equal), and the alphabet size Σ. For instance, we show that SIGNED STRING REVERSAL DISTANCE and SIGNED STRING PREFIX REVERSAL DISTANCE are NP-hard even if the input strings have only *one* letter.

1 Introduction

Many problems studied in the realm of comparative genomics concern genome rearrangements, in which, given two genomes G_1 and G_2 and a set $\mathcal{S}$ of operations (called rearrangements) on genomes, the question is to compute the smallest number of rearrangements that allows to obtain G_2, starting from G_1, and using only operations from $\mathcal{S}$ (see for instance [8] for an extensive survey). One of the most studied, and historically one of the firstly described [14] such rearrangement is the *reversal*, which consists in taking a contiguous sequence of a genome, reverse its order, and reincorporate it at the same location. This gave rise to the SORTING BY REVERSALS (SBR) (resp. SORTING BY SIGNED REVERSALS (SBSR)) problem, in which a genome is represented by a permutation (resp. signed permutation) whose elements are genes. In the signed version of the problem each position of the permutation is additionally labeled with a sign $+$ or $-$ and a reversal ρ not only reverses the order, but also inverts the

---

* Supported by the Alexander von Humboldt Foundation.
** Supported by a post-doctorial grant funded by the Région Pays de la Loire.

A.S. Kulikov, S.O. Kuznetsov, and P. Pevzner (Eds.): CPM 2014, LNCS 8486, pp. 50–59, 2014.

signs of all the elements involved in ρ. The main complexity results concerning these two problems are the following: SBR is NP-hard [4] and the best current approximation ratio is 1.375 [2], while SBSR is polynomial [1]. Another variant consists in using *prefix reversals* only, that is, each reversal must contain the first letter of the string it is applied to. The unsigned (resp. signed) corresponding problem is called SBPR (resp. SBSPR). The SBPR problem has been recently shown to be NP-hard [3], and the best current approximation ratio is 2 [9]. On the other hand, the complexity of SBSPR is still open.

In biological applications, however, genomes of related species often contain many homologous genes. In this case, the genomes cannot be modeled by permutations, but rather by (signed) strings [5]. Hence, a natural (and more biologically relevant) extension of SBR is the STRING REVERSAL DISTANCE problem, formally defined (in its decision version) as follows:

STRING REVERSAL DISTANCE
Input: Two strings S and T over alphabet Σ and an integer k.
Question: Can S be transformed into T by applying at most k reversals?

Besides, SIGNED STRING REVERSAL DISTANCE will denote the signed version of the above problem. If we allow prefix reversals only, the extension of SBPR to strings is defined as follows:

STRING PREFIX REVERSAL DISTANCE
Input: Two strings S and T over alphabet Σ and an integer k.
Question: Can S be transformed into T by applying at most k prefix reversals?

As above, SIGNED STRING PREFIX REVERSAL DISTANCE will denote the signed version of the problem. Any of these four problems is only nontrivially posed if S and T have the same letter content, that is, for each letter the number of its occurrences in S equals the number of its occurrences in T. We call such strings *balanced*. In the remainder of the paper, we assume that S and T are balanced. A *block* is a maximal substring such that all letters in the substring are equal (and have the same sign, if strings are signed). For any instance, we use b_{max} to denote the maximum of the number of blocks in S and T and b_{min} to denote its minimum. Unless stated otherwise, we assume that S has b_{max} blocks. Note that $n \geq b_{max} \geq b_{min} \geq |\Sigma|$.

In this paper, we study algorithmic and complexity issues for these four problems, in connection with two parameters of the input strings: the maximum number of blocks b_{max} they contain and the size of the alphabet Σ.

Known Results. Computing the reversal distance between binary strings is NP-hard via reduction from 3-PARTITION [6, 7]. A second proof uses a reduction from SORTING BY REVERSALS [13]. Computing the prefix reversal distance between binary strings is NP-hard [11]. Sorting a binary string S (to $0^p 1^{n-p}$) is solvable in polynomial time [6, 7]. This result was later generalized to the case of ternary strings [13]. Hurkens et al. [11] considered the problem of "grouping" a string by prefix reversals, that is, find a shortest sequence of reversals from a string S to any string that has $|\Sigma|$ blocks. Concerning the diameter, that is, the maximum distance between any length-n strings, it was first shown that the reversal diameter for binary strings of length n is $\lfloor n/2 \rfloor$ [6, 7]. Later, this result was generalized to fixed alphabets of arbitrary size. More precisely, the reversal diameter of strings of length n is $n - \max_{a \in \Sigma} \#(a)$ where $\#(a)$ denotes the number of occurrences of letter a in either input string. SIGNED STRING REVERSAL DISTANCE has applications in the identification of orthologous genes across two genomes [5, 10, 12]. SIGNED STRING REVERSAL DISTANCE is NP-hard if each letter occurs at most twice [5] and for binary signed strings [13].

Our Results. Our main algorithmic result is a fixed-parameter algorithm for the four problem variants and the parameter maximum number of blocks $b_{\max}$. This result relies mainly on diameter bounds that depend only on $b_{\max}$ and Σ and which we provide in Section 2. Then, in Section 3 we show the aforementioned algorithm for the parameter $b_{\max}$. In Section 4 we describe a reduction from SBR that yields several hardness results. First, it shows that SIGNED STRING REVERSAL DISTANCE and SIGNED STRING PREFIX REVERSAL DISTANCE remain NP-hard over unary alphabet. This strengthens the previous hardness result by Radcliffe et al. [13] who showed hardness for binary alphabets. Second, it shows that for STRING REVERSAL DISTANCE and STRING PREFIX REVERSAL DISTANCE we cannot make use of a bounded block length even if input strings are binary as both problems become NP-hard for the first nontrivial case, that is, if each 0-block has length one and each 1-block has length at most two. Finally, we show a simple algorithm that achieves a running time of $|\Sigma|^n \cdot \text{poly}(n)$ for many string distances including the ones under consideration in this work.

Preliminaries. We denote the i-th letter of a string S by $S[i]$. A reversal $\rho(i,j)$ in a string S of length n transforms $S[1] \cdots S[i-1]S[i]S[i+1] \cdots S[j-1]S[j]S[j+1] \cdots S[n]$ into $S[1] \cdots S[i-1]S[j]S[j-1] \cdots S[i+1]S[i]S[j+1] \cdots S[n]$. We denote the string that results from applying reversal ρ to S by $S \circ \rho$. We use $b(S)$ to denote the number of blocks of a string S. For a (signed) string S we denote by $\underline{S}$ the string that is obtained by the reversal $\rho(1, |S|)$. The following simple observation will be useful in a later part of this work.

Proposition 1. *There is a shortest sequence $\rho_1, \rho_2, \ldots, \rho_\ell$ of reversals from S to T such that the start and endpoint of each ρ_i have different letters.*

2   Upper Bounds on the Reversal Diameter

In this section, we upper-bound the reversal diameter of balanced strings based on the number of blocks in these strings. To this end, we first show an upper bound on the number of reversals needed to reach an arbitrary "grouped" string, that is, any string with $|\Sigma|$ blocks.

Lemma 1. *Let S be a string with b blocks over alphabet Σ. There exists a string S_g with $|\Sigma|$ blocks such that S can be transformed into S_g by at most $b - |\Sigma|$ reversals and by at most b prefix reversals.*

Proof. The claim is obviously true if $b = |\Sigma|$. Now, assume by induction that the claim holds for all strings S' with $b' < b$ blocks. Since $b > |\Sigma|$ there is one letter a that appears in two blocks. By applying the reversal to S that starts at the leftmost letter of the first block containing a and ends at the letter before the second block containing a, one obtains a string S' with $b - 1$ blocks. By induction, there is a grouped string S_g such that the reversal distance from S' to S_g is at most $b - 1$.

For prefix reversals, we use a similar greedy strategy. If the first letter a appears only in this block, then reverse the complete string (which is a prefix reversal) and remove the last block of the resulting string from the instance (or similarly, apply the following only on the substring that excludes this block). The removal of the last block reduces $|\Sigma|$ by one since a appears only in this block. Since a string with unary alphabet has one block, we perform this type of reversal at most $|\Sigma| - 1$ times. Note that this does not increase $b - |\Sigma|$ since b also decreases by one. If a appears in at least two blocks, then apply the prefix reversal whose endpoint is the rightmost letter before the second block that contains a. This prefix reversal reduces the number b of blocks by one. The overall number of prefix reversals that are applied until a grouped string is reached is thus at most $b - |\Sigma| + |\Sigma| - 1$. □

We now use this upper bound to obtain an upper bound for the reversal distance between any strings. The approach is to transform each input string into some grouped string and then transform one grouped string into the other.

Theorem 1. *Two balanced strings S and T with b_{max} and b_{min} blocks, respectively, can be transformed into each other by at most $b_{max} + b_{min} - |\Sigma| - 1$ reversals and at most $b_{max} + b_{min} + 2|\Sigma| - 3$ prefix reversals.*

For the reversal case, we can also obtain a bound of the type $b_{max} + O(|\Sigma|^2)$.

Theorem 2. *Two balanced strings S and T each with at most b_{max} blocks can be transformed into each other by at most $b_{max} + |\Sigma|^2 - 2|\Sigma|$ reversals.*

Proof. If $b_{max} < |\Sigma|^2 - |\Sigma| + 2$, then the claim is true as $d(S, T) \leq b_{max} + |\Sigma|^2 - 2|\Sigma|$ by Theorem 1. Now, assume that $b_{max} \geq |\Sigma|^2 - |\Sigma| + 2$ and that the claim holds for all pairs of strings with $b'_{max} < b_{max}$. We show how to apply a constant number of reversals on S or on S and T that reduce the number of blocks sufficiently to obtain the bound.

Case 1: $b_{max} > b_{min}$. Assume that S has b_{max} blocks. Apply any reversal on S that reduces the number of blocks by one. Let S' be the resulting string. By the inductive hypothesis, $d(S', T) \leq b_{max} - 1 + |\Sigma|^2 - 2|\Sigma|$. Since $d(S, S') = 1$, the claim holds in this case.

Case 2: $b_{max} = b_{min}$. Any string U with at least $|\Sigma|^2 - |\Sigma| + 2$ blocks has the following property: there are two letters, say a and b, such that U contains the substring ab twice. This can be seen by considering a directed multigraph with the vertex set Σ in which we add an edge (u, v) for each substring uv, $u \neq v$ of U. Any block change corresponds to an edge in this graph. Now there are at least $|\Sigma|^2 - |\Sigma| + 1$ block changes in U. Hence, the multigraph has $|\Sigma|^2 - |\Sigma| + 1$ edges. Since a simple directed graph can have at most $|\Sigma|^2 - |\Sigma|$ edges, there is one pair of vertices u and v, for which the edge (u, v) is contained twice in this multigraph.

By the above, S has two letters, say a and b, such that there are i and $j > i+1$ with $S[i] = a$, $S[i + 1] = b$, $S[j] = a$ and $S[j + 1] = b$. The reversal $\rho(i + 1, j)$ produces a string S' with $b_{\max} - 2$ blocks. Similarly, there is some reversal that transforms T into a string T' with $b_{\max} - 2$ blocks. By the inductive hypothesis, $d(S', T') = b_{\max} - 2 + |\Sigma|^2 - 2|\Sigma|$. Together with the two additional reversals on S and T we obtain the bound on the number of reversals also in this case. □

3   An Algorithm for Strings with Few Blocks

We now show how to solve STRING REVERSAL DISTANCE in $(b_{\max})^{O(b_{\max})} \cdot$ poly(n) time. Our algorithm consists of two main steps: first, "guess" between which blocks each of the reversals takes place. These guesses fix the structure of the reversals and we will thus call a sequence of at most k of those guesses a *scaffold*. A scaffold would completely describe a sequence of reversals if one would additionally specify the precise positions at which the reversal starts or ends in each of the blocks. Since the blocks can be very long compared to the number of blocks, we can not branch into all possible cases for these positions. However, we guess whether the startpoint of the reversal is the first position of a block and whether the endpoint of the reversal is the last position of a block. This notion is defined as follows.

Definition 1. *A reversal $\rho(i, j)$ is called* left-breaking *if $S[i - 1] = S[i]$ and* right-breaking *if $S[j] = S[j + 1]$.*

We can now give a formal definition of a scaffold.

Definition 2. *A scaffold $((i_1, j_1, B_1), (i_2, j_2, B_2), \ldots, (i_k, j_k, B_k))$ is a tuple of triples, called* reversal-triples, *where $i_\ell, j_\ell \in \mathbb{N}$ and $B_\ell \subseteq \{L, R\}$, $1 \leq \ell \leq k$. A sequence of k reversals $\rho_1, \rho_2, \ldots, \rho_k$ from string S_1 to S_{k+1} with $S_i \rho_i = S_{i+1}$ respects a scaffold if for each ℓ, $1 \leq \ell \leq k$ we have that*

- *the startpoint of ρ_ℓ is in the i_ℓ-th block of S_ℓ,*
- *the endpoint of ρ_ℓ is in the j_ℓ-th block of S_ℓ,*
- *ρ_ℓ is left-breaking if and only if $L \in B_\ell$, and*
- *ρ_ℓ is right-breaking if and only if $R \in B_\ell$.*

For each possible scaffold, the algorithm now aims to compute whether one can assign two numbers to each reversal to obtain a sequence of reversals that respects the scaffold and transforms S into T. This is done by computing a maximum flow on an auxiliary graph.

First, we bound the number of different scaffolds that need to be considered. By Theorems 1 and 2, we can assume that $k < b_{\max} + b_{\min} - |\Sigma| \leq 2b_{\max} - 2$ and $k \leq b_{\max} + |\Sigma|^2 - 2|\Sigma| < b_{\max} + |\Sigma|^2$ since otherwise the instance is a yes-instance. Hence, every scaffold that is respected by an optimal solution has at most $2b_{\max} - 2$ reversal-triples. The algorithm branches for each such reversal-triple into the possible choices for i_ℓ and j_ℓ and whether the reversal shall be left- or right-breaking. By the above lemma, it needs to perform at most $2b_{\max} - 2$ branchings. Furthermore, the number of blocks in any "intermediate" string is bounded as shown below.

Lemma 2. *Let S' be a string such that there is an optimal sequence of reversals from S to T in which S' is one of the strings produced by this sequence. Then, S' has at most $\min\{2b_{max} + b_{min} - |\Sigma| - 1, 2b_{max} + |\Sigma|^2 - 2|\Sigma|\}$ blocks.*

Now, the algorithm creates for increasing $k' \le k$ all possible reversal scaffolds. By the above lemma, there are less than $3b_{\max}$ choices for each i_ℓ and j_ℓ. Hence, the overall number of reversal scaffolds that need to be considered is at most

$$(3b_{\max})^{2\cdot(2b_{\max}-2)} \cdot 4^{2b_{\max}-2} = O((6b_{\max})^{4b_{\max}})$$

in the case of arbitrary alphabets. For constant-size alphabets, we can use the bound on k given by Theorem 2 and thus the overall number of reversal scaffolds that need to be considered in this case is less than

$$(3b_{\max})^{2\cdot(b_{\max}+|\Sigma|^2)} \cdot 4^{b_{\max}+|\Sigma|^2} = (6b_{\max})^{2b_{\max}} \cdot \text{poly}(b_{\max}).$$

Consider one such scaffold, assume there is a sequence of reversals that respects the scaffold, and let $S_i := S_{i-1} \circ \rho_i$, $i \le k'+1$, denote the string obtained after the i-th reversal. We show that the number and order of blocks of each S_i is completely fixed by the reversal scaffold. First, consider $S_1 := S$ and let δ_ℓ denote the number of letters in the ℓ-th block of S_1, let σ_i denote the letter of the ℓ-th block and assume that S has $b_{\max}$ blocks. Furthermore, assume that i_1 is in the i-th block of S_1 and j_1 is in the j-th block of S_1. Then this reversal transforms the string

$$S_1 = (\sigma_1)^{\delta_1} \cdots (\sigma_i)^{\delta_i}(\sigma_{i+1})^{\delta_{i+1}} \cdots (\sigma_{j-1})^{\delta_{j-1}}(\sigma_j)^{\delta_j} \cdots (\sigma_{b_{\max}})^{\delta_{b_{\max}}}$$

into the following string (where x and y represent the number of elements to the left of the cut in the i-th and j-th blocks):

$$S_2 = (\sigma_1)^{\delta_1} \cdots (\sigma_i)^{x}(\sigma_j)^{y}(\sigma_{j-1})^{\delta_{j-1}} \cdots (\sigma_{i-1})^{\delta_{i-1}}(\sigma_i)^{\delta_i-x}(\sigma_j)^{\delta_j-y} \cdots (\sigma_{b_{\max}})^{\delta_{b_{\max}}}.$$

Recall that the scaffold fixes whether the reversal is left-breaking and whether it is right-breaking. In other words, it is known whether $x = 0$ or $x > 0$ and whether $y < \delta_i$ or $y = \delta_i$. Consequently, it is fixed whether the letter preceding the endpoint of the reversal in S_2 is σ_i or whether this letter is σ_{i-1}. Similarly, it is fixed whether the letter succeeding the startpoint of the reversal in S_2 is σ_j or whether it is σ_{j+1}. Therefore, we know whether the borders of the reversal are start or endpoints of new blocks in S_2 or whether they are "merged" with old blocks. Consequently, the number of blocks in S_2 and the letter for each block in S_2 is known. This is similarly true for $S_3 = S_2 \circ \rho_2$ up until $S_{k'+1} = S_k \circ \rho_k$. Hence, the number of blocks, their order, and the letter that each block contains is fixed in $S_{k'+1}$. Thus, if the number of blocks in T is different from the number of blocks in $S_{k'+1}$ or if the letter of the i-th block in T is different from the letter from the i-th block in $S_{k'+1}$, then we can discard the reversal scaffold. Thus, it now remains to check whether the reversal scaffold can produce blocks of the correct size.

One possible way of checking whether this is indeed true would be to introduce a variable for the length of each block in each S_i and then introduce equations that model the dependencies between the blocks. For instance if the reversal from S_i to S_{i+1} appears after the first block, then the lengths of the first blocks should be equal. Since the number of blocks and k' are bounded in functions of b_{max} this would yield an integer linear program whose number of variables depends only on b_{max} which implies fixed-parameter tractability with respect to b_{max}. In the following, we describe a more efficient approach that is based on computing maximum value flows. For each considered reversal scaffold we create one flow network with $O((b_{max})^2)$ vertices as follows.

Add two special vertices, the source s and the sink t. For each block i in each intermediate string S_ℓ add one vertex v_ℓ^i; we use $V_\ell := \{v_\ell^i \mid 1 \le i \le b(S_\ell)\}$ to denote the vertex set corresponding to S_ℓ. Now, add edges and capacities as follows. For each v_1^i add the edge (s, v_1^i). Set the capacity of $c(s, v_1^i)$ to be exactly the length of the i-th block in S. For each $\ell \le k'$ introduce directed edges between the vertices corresponding to blocks of S_ℓ to those representing blocks of $S_{\ell+1}$ as follows.

Assume that the reversal ρ is fixed to start within the i-th block of S_ℓ and end in the j-th block of S_ℓ. Furthermore, let β denote the difference between the number of blocks in $S_{\ell+1}$ and in S_ℓ. Then, add the following edges, with unbounded capacity, to G:

- for all $i' < i$ add the edge $(v_\ell^{i'}, v_{\ell+1}^{i'})$
- for all $i' > j$ add the edge $(v_\ell^{i'}, v_{\ell+1}^{\beta+i'})$
- if ρ is left-breaking:
  - add the edge $(v_\ell^i, v_{\ell+1}^i)$,
  - for each i' with $i \le i' \le j$ add the edges $(v_\ell^{i'}, v_{\ell+1}^{i+1+j-i'})$;
- if ρ is not left-breaking:
  - if the endpoint of ρ and the $(i-1)$-th block in S_ℓ have the same letter, then add for each i' with $i \le i' \le j$ the edges $(v_\ell^{i'}, v_{\ell+1}^{i-1+j-i'})$,
  - if they have different letters, then add for each i' with $i \le i' \le j$ the edges $(v_\ell^{i'}, v_{\ell+1}^{i+j-i'})$;
- if ρ is right-breaking, add the edge $(v_\ell^j, v_{\ell+1}^{j+\beta})$.

Note that for the case that ρ is left-breaking, we assume by Proposition 1 that the i-th and j-th block in S_ℓ have different letters and thus the endpoint of the reversal creates a new block in $S_{\ell+1}$. Note that for the right side of ρ we do not check explicitly whether the startpoint of ρ and the successor of its endpoint have the same letter, since this fact is completely determined when we know β (which can be directly deduced from the scaffold) and whether a block is "created" or "lost" at the left side of the reversal. The construction is completed by adding for each $v_{k'+1}^i$, the edge $(v_{k'+1}^i, t)$ and setting the capacity $c(v_{k'+1}^i, t)$ to be exactly the length of the i-th block in T.

Lemma 3. *Let $N = (V, E, c, s, t)$ be a flow network constructed from a reversal scaffold as described above. Then there is a sequence of reversals that transforms S into T and respects the reversal scaffold if and only if N admits a flow of value n.*

Theorem 3. STRING REVERSAL DISTANCE *can be solved in* $(6b_{max})^{4b_{max}} \operatorname{poly}(n)$ *time on arbitrary strings and in* $(6b_{max})^{2b_{max}} \operatorname{poly}(n)$ *time if* $|\Sigma|$ *is constant.*

One possible approach to improve the above result would be to show that there is always an optimal sequence such that the number of blocks of every intermediate string never exceeds b_{max}. However, the instance with $S := 011100100$ and $T := 110001001$ is a counterexample. An optimal solution contains exactly two reversals as shown by the example $011100100 \to 011\underline{100100} = 011001001 \to \underline{011}001001 = 110001001$. This solution creates an intermediate string with six blocks but the input strings have only five blocks. There is also no solution that has less than six blocks in an intermediate string which can be shown by a case distinction. The algorithm can be adapted to work for the other problem variants as well.

Theorem 4. STRING PREFIX REVERSAL DISTANCE *can be solved in* $(6b_{max})^{4b_{max}} \cdot \operatorname{poly}(n)$ *time;* SIGNED STRING REVERSAL DISTANCE *and* SIGNED STRING PREFIX REVERSAL DISTANCE *can be solved in* $(b_{max})^{O(b_{max})} \cdot \operatorname{poly}(n)$ *time.*

4 Reversals on Strings with Small Alphabet

Hardness Results for Restricted Cases. We describe two reductions, one for the signed case and one for the unsigned case, that show hardness of both the reversal and prefix reversal problems. For reversal problems the reduction is from SORTING BY REVERSALS, for prefix reversal problems the reduction is from SORTING BY PREFIX REVERSALS. Recall that both problems are NP-hard [3, 4].

Given an instance of SBR or SBPR, replace each permutation as follows.

Construction 1: Signed (Prefix) Reversals. For each integer $i > 0$, let $S_i := +a(-a)^{i+1}(+a)^{i+1} - a$. Note that each S_i has length $2i + 4$.

Construction 2: Unsigned (Prefix) Reversals. For each $i > 0$, let $S_i := 01001(01)^{i+1}0010$. Note that each S_i has length $2i + 11$.

In both constructions, let $S(\pi) := (S_{\pi(1)})^{2n}(S_{\pi(2)})^{2n} \cdots (S_{\pi(n)})^{2n}$ for any permutation π. Given an input to SBR or SBPR with permutation π, the reduction creates the two strings $S(I_n)$ and $S(\pi)$ as defined above (I_n denotes the identity permutation of length n).

The following observations apply to both constructions. First, the reversed string of S_i is S_i itself. Further, for any $i \neq j$, S_i is not a substring of S_j, and the longest suffix of S_i which is also prefix of S_j has length two in Construction

1 (it is $+a - a$) and length six in Construction 2 (it is 010010). In both cases, this is strictly smaller than half the length of S_i and S_j. Hence, if, for any integers $i, j_1, \ldots, j_\ell$, S_i is a substring of $S_{j_1} \cdot S_{j_2} \cdot \cdots \cdot S_{j_\ell}$, then $i \in \{j_1, \ldots, j_\ell\}$. Moreover, if $S_{j_1} \cdot S_{j_2} \cdot \cdots \cdot S_{j_\ell}$ contains substrings $S_{i_1}, S_{i_2}, \ldots, S_{i_h}$ in this order, then $(i_1, i_2, \ldots, i_h)$ is a subsequence of $(j_1, j_2, \ldots, j_\ell)$.

Theorem 5. SIGNED STRING [PREFIX] REVERSAL DISTANCE *is NP-hard even when* $|\Sigma| = 1$. STRING [PREFIX] REVERSAL DISTANCE *is NP-hard even when restricted to binary strings where all 0-blocks have length at most 2 and all 1-blocks have length 1.*

Proof. Let $S(I_n)$ and $S(\pi)$ be constructed from a permutation π as described above. Let k_S be the [prefix] reversal distance from $S(\pi)$ to $S(I_n)$, and k_π be the [prefix] reversal distance from π to I_n. We show that $k_\pi = k_S$.

First, we show $k_S \leq k_\pi$. Consider any sequence of k_π [prefix] reversals sorting π, we show that there exists a corresponding sequence of k_π [prefix] reversals sorting $S(\pi)$. For any [prefix] reversal ρ of the sub-permutation of π ranging from $\pi[i]$ to $\pi[j]$, we reverse the substring of $S(\pi)$ ranging from the first $S_{\pi[i]}$ to the last $S_{\pi[j]}$: the resulting string is $S(\rho \circ \pi)$. In the end, we obtain $S(I_n)$ after k_π [prefix] reversals.

We now show that $k_\pi \leq k_S$. Note that $k_S \leq k_\pi$ implies that $k_S < n$. Consider a sequence $\rho_1, \ldots, \rho_{k_S}$ of [prefix] reversals sorting $S(\pi)$. For each $1 \leq i \leq n$, among the $2n$ copies of $S_{\pi[i]}$ in $S(\pi)$, at least one does not contain an endpoint of any reversal: we thus assign to each i an *untouched* copy of $S_{\pi[i]}$. For each [prefix] reversal ρ_r, there exists i and j such that ρ_r reverses a string containing the ith to the jth untouched copies (ordered from left to right). Let $\rho'_r = (1, \ldots, i - 1, j, j - 1, \ldots, i + 1, i, j + 1, \ldots, n)$. Note that if ρ_r is a prefix reversal, then $i = 1$ and ρ'_r is also a prefix reversal.

Let $\tau = \rho'_{k_S} \circ \ldots \circ \rho'_1 \circ \pi$, then the final string $S(I_n)$ contains the untouched copies of $S(\tau[1])$, $S(\tau[2])$, $\ldots$, $S(\tau[n])$ as substrings in this order. By definition of $S(I_n)$ and using the property of strings S_i, sequence $(1, \ldots, 1, 2, \ldots, 2, \ldots, n, \ldots, n)$ contains $\tau[1], \tau[2], \ldots, \tau[n]$ as a subsequence. Since τ is a permutation, τ is the identity I_n. Thus, the sequence of k_S [prefix] reversals $\rho'_1, \ldots, \rho'_{k_S}$ transforms π into the identity, and $k_\pi \leq k_S$. □

An Algorithm for Small Alphabets. So far, none of the known exact algorithms for STRING REVERSAL DISTANCE or SIGNED STRING REVERSAL DISTANCE achieves a singly-exponential running time of $2^{O(n)}$. We show that such a running time can be achieved for constant size alphabets and a generic type of distance measures on strings. Call a string distance d *well-formed* if it has the following properties: 1) For each string S of length n, the set containing exactly the strings T with $d(S, T) = 1$ can be computed in poly(n) time. 2) All strings S and T with $d(S, T) = 1$ are balanced. 3) For any two strings S and T with $d(S, T) = k$ there exists a string S' with $d(S, S') = 1$ and $d(S', T) = k - 1$.

Theorem 6. *Let d be a well-formed string distance, and let S and T be two strings of length n over alphabet Σ. Then $d(S, T)$ can be computed in $|\Sigma|^n \cdot$ poly(n) time.*

5   Conclusion

Our work leads to several open questions. Is the reversal diameter for strings $b_{max} -$ 1? If yes, this would generalize the upper bound of $n - 1$ on the diameter for the reversal distance between permutations. If not, is an upper bound of $b_{max} + O(|\Sigma|)$ achievable? Further, does STRING REVERSAL DISTANCE admit a polynomial kernel for b_{max}, that is, can it be reduced in polynomial time to an equivalent instance of STRING REVERSAL DISTANCE with $n \leq \text{poly}(b_{max})$? Finally, can we solve STRING REVERSAL DISTANCE in time $O((|\Sigma| - \epsilon)^n)$? In particular, can we solve STRING REVERSAL DISTANCE on binary strings in $O(c^n)$ time for $c < 2$?

References

[1] Bafna, V., Pevzner, P.A.: Genome rearrangements and sorting by reversals. SIAM J. Comput. 25(2), 272–289 (1996)

[2] Berman, P., Hannenhalli, S., Karpinski, M.: 1.375-approximation algorithm for sorting by reversals. In: Möhring, R.H., Raman, R. (eds.) ESA 2002. LNCS, vol. 2461, pp. 200–210. Springer, Heidelberg (2002)

[3] Bulteau, L., Fertin, G., Rusu, I.: Pancake flipping is hard. In: Rovan, B., Sassone, V., Widmayer, P. (eds.) MFCS 2012. LNCS, vol. 7464, pp. 247–258. Springer, Heidelberg (2012)

[4] Caprara, A.: Sorting by reversals is difficult. In: Proc. 1st RECOMB, pp. 75–83 (1997)

[5] Chen, X., Zheng, J., Fu, Z., Nan, P., Zhong, Y., Lonardi, S., Jiang, T.: Assignment of orthologous genes via genome rearrangement. IEEE ACM T. Comput. Bi. 2(4), 302–315 (2005)

[6] Christie, D.A.: Genome Rearrangement Problems. PhD thesis, University of Glasgow (1998)

[7] Christie, D.A., Irving, R.W.: Sorting strings by reversals and by transpositions. SIAM J. Discrete Math. 14(2), 193–206 (2001)

[8] Fertin, G., Labarre, A., Rusu, I., Tannier, E., Vialette, S.: Combinatorics of Genome Rearrangements. Computational Molecular Biology. MIT Press (2009)

[9] Fischer, J., Ginzinger, S.W.: A 2-approximation algorithm for sorting by prefix reversals. In: Brodal, G.S., Leonardi, S. (eds.) ESA 2005. LNCS, vol. 3669, pp. 415–425. Springer, Heidelberg (2005)

[10] Fu, Z., Chen, X., Vacic, V., Nan, P., Zhong, Y., Jiang, T.: MSOAR: A high-throughput ortholog assignment system based on genome rearrangement. J. Comput. Biol. 14(9), 1160–1175 (2007)

[11] Hurkens, C.A.J., van Iersel, L., Keijsper, J., Kelk, S., Stougie, L., Tromp, J.: Prefix reversals on binary and ternary strings. SIAM J. Discrete Math. 21(3), 592–611 (2007)

[12] Jiang, T.: Some algorithmic challenges in genome-wide ortholog assignment. J. Comput. Sci. Technol. 25(1), 42–52 (2010)

[13] Radcliffe, A., Scott, A., Wilmer, E.: Reversals and transpositions over finite alphabets. SIAM J. Discrete Math. 19(1), 224 (2006)

[14] Watterson, G., Ewens, W., Hall, T., Morgan, A.: The chromosome inversion problem. J. Theor. Biol. 99(1), 1–7 (1982)

On Combinatorial Generation of Prefix Normal Words

Péter Burcsi[1], Gabriele Fici[2], Zsuzsanna Lipták[3], Frank Ruskey[4], and Joe Sawada[5]

[1] Department of Computer Algebra, Eötvös Loránd University, Budapest, Hungary
bupe@compalg.inf.elte.hu
[2] Dipartimento di Matematica e Informatica, University of Palermo, Italy
gabriele.fici@math.unipa.it
[3] Dipartimento di Informatica, University of Verona, Italy
zsuzsanna.liptak@univr.it
[4] Department of Computer Science, University of Victoria, Canada
ruskey@cs.uvic.ca
[5] School of Computer Science, University of Guelph, Canada
jsawada@uoguelph.ca

Abstract. A prefix normal word is a binary word with the property that no substring has more 1s than the prefix of the same length. This class of words is important in the context of binary jumbled pattern matching. In this paper we present an efficient algorithm for exhaustively listing the prefix normal words with a fixed length. The algorithm is based on the fact that the language of prefix normal words is a bubble language, a class of binary languages with the property that, for any word w in the language, exchanging the first occurrence of 01 by 10 in w results in another word in the language. We prove that each prefix normal word is produced in $O(n)$ amortized time, and conjecture, based on experimental evidence, that the true amortized running time is $O(\log(n))$.

1 Introduction

A binary word of length n is *prefix normal* if for all $1 \leq k \leq n$, no substring of length k has more 1s than the prefix of length k. For example, 1001010 is not prefix normal because the substring 101 has more 1s than the prefix 100. These words were introduced in [8], where it was shown that each binary word w has a canonical *prefix normal form* w' of the same length.

The study of prefix normal words and prefix normal forms is motivated by the string problem known as *binary jumbled pattern matching* (binary JPM). In that problem, we are given a text of length n over a binary alphabet, and two numbers x and y, and ask whether the text has a substring with exactly x 1s and y 0s. While the online version can be solved with a simple sliding window algorithm in $O(n)$ time, the offline version, where many queries are expected, has recently attracted much interest: here an index of size $O(n)$ can be generated which then allows answering queries in constant time [5]. However, the best construction algorithms up to date have running time $O(n^2 / \log n)$ [2, 13].

A.S. Kulikov, S.O. Kuznetsov, and P. Pevzner (Eds.): CPM 2014, LNCS 8486, pp. 60–69, 2014.

Several recent papers have yielded better results under specific assumptions, such as word level parallelism or highly compressible strings [1, 6, 10, 14], or for constructing an approximate index [7]; but the general case has not been improved. It was demonstrated in [1,8] that prefix normal forms of the text can be used to construct this index. JPM over an arbitrary alphabet has also been studied [4,5,11]. Moreover, several variants of the original problem have recently been introduced: approximate JPM [3], JPM in the streaming model [12], JPM on trees and graphs [6,9].

We note that the connection of the present paper to binary JPM is that of supplying a new approach: We do not present an improvement of the JPM problem, but we strongly believe that a better understanding of these words will eventually lead to better solutions for JPM.

Bubble languages are an interesting new class of binary languages defined by the following property: $\mathcal{L}$ is a bubble language if, for every word $w \in \mathcal{L}$, replacing the first occurrence of 01 (if any) by 10 results in another word in $\mathcal{L}$ [15, 16, 17]. A generic generation algorithm for bubble languages was given in [17], leading to Gray codes for each of these languages. The algorithm's efficiency depends only on a language-dependent subroutine, which in the best case leads to CAT (constant amortized time) generation algorithms. Many important languages are bubble languages, including binary necklaces and Lyndon words, and k-ary Dyck words.

In this paper, we show that prefix normal words form a bubble language and present an efficient generation algorithm which runs in $O(n)$ amortized time per word, and which yields a Gray code for prefix normal words. Generating these words naively takes $O(2^n \cdot n^2)$ time. Based on experimental evidence, we conjecture that the running time of our algorithm is in fact $\Theta(\log(n))$ amortized. We also give a new characterization of bubble languages in terms of a closure property in the computation tree of a certain generation algorithm for all binary words (Prop. 1). We prove new properties of prefix normal words and present a linear time testing algorithm for words which have been obtained from prefix normal words via a simple operation. We present several open problems in the last section.

Most proofs have been omitted for lack of space and will be contained in the full version of the paper.

2   Basics

A *binary word* (or *string*) $w = w_1 \cdots w_n$ over $\Sigma = \{0, 1\}$ is a finite sequence of elements from Σ. Its length n is denoted by $|w|$. We denote by Σ^n the words over Σ of length n, and by $\Sigma^* = \cup_{n \geq 0} \Sigma^n$ the set of finite words over Σ. The empty word is denoted by ε. Let $w \in \Sigma^*$. If $w = uv$ for some $u, v \in \Sigma^*$, we say that u is a *prefix* of w and v is a *suffix* of w. A *substring* of w is a prefix of a suffix of w. A *binary language* is any subset $\mathcal{L}$ of Σ^*.

In the following, we will often write binary words $w \neq 1^n$ in a canonical form $w = 1^s 0^t \gamma$, where $\gamma \in 1\{0, 1\}^* \cup \{\varepsilon\}$ and $s \geq 0, t \geq 1$. In other words, s is the length of the first, possibly empty, 1-run of w, t the length of the first 0-run, and γ the remaining, possibly empty, suffix. Note that this representation is unique.

We call 1^s0^t the *critical prefix* of w and $cr(w) = s + t$ the *critical prefix length* of w. We denote by $|w|_c$ the number of occurrences in w of character $c \in \{0, 1\}$, and by $\mathcal{B}_d^n$ the set of all binary strings w of length n such that $|w|_1 = d$ (the *density* of w is d). We denote by $swap(w, i, j)$ the string obtained from w by exchanging the characters in positions i and j.

2.1 Prefix Normal Words

Let $w \in \Sigma^*$. For $i = 0, \ldots, n$, we set

- $P(w, i) = |w_1 \cdots w_i|_1$, the number of 1s in the i-length prefix of w.
- $F(w, i) = \max\{|u|_1 : u \text{ is a substring of } w \text{ and } |u| = i\}$, the maximum number of 1s over all substrings of length i.

Definition 1. *A binary word w is* prefix normal *if, for all $1 \le i \le |w|$, $F(w, i) = P(w, i)$. In other words, a word is prefix normal if no substring contains more 1s than the prefix of the same length.*

We denote by $\mathcal{L}_{\mathrm{PN}}$ the language of prefix normal words. In [8] it was shown that for every word w there exists a unique word w', called its *prefix normal form*, or $\mathrm{PNF}(w)$, such that for all $1 \le i \le |w|$, $F(w, i) = F(w', i)$, and w' is prefix normal. Therefore, a prefix normal word is a word coinciding with its prefix normal form. In the following table we list all prefix normal words of length 5 followed by the set of binary words w such that $\mathrm{PNF}(w) = w'$ (i.e., its equivalence class):

$11111 \Rightarrow \{11111\}$	$11000 \Rightarrow \{11000, 011000, 00110, 00011\}$
$11110 \Rightarrow \{11110, 01111\}$	$10101 \Rightarrow \{10101\}$
$11101 \Rightarrow \{11101, 10111\}$	$10100 \Rightarrow \{10100, 01010, 00101\}$
$11100 \Rightarrow \{11100, 01110, 00111\}$	$10010 \Rightarrow \{10010, 01001\}$
$11011 \Rightarrow \{11011\}$	$10001 \Rightarrow \{10001\}$
$11010 \Rightarrow \{11010, 10110, 01101, 01011\}$	$10000 \Rightarrow \{10000, 01000, 00100, 00010, 00001\}$
$11001 \Rightarrow \{11001, 10011\}$	$00000 \Rightarrow \{00000\}.$

Several methods were presented in [8] for testing whether a word is prefix normal; however, all ran in quadratic time in the length of the word. One open problem given there was that of enumerating prefix normal words (counting). The number of prefix normal words of length n can be computed by checking for each binary word whether it is prefix normal, i.e. altogether in $O(2^n \cdot n^2)$ time. In this paper, we present an algorithm that is far superior in that it generates only prefix normal words, rather than testing every binary word; it runs in $O(n)$ time per word; and it generates prefix normal words in cool-lex order, constituting a Gray code (subsequent words differ by a constant number of swaps or flips).

2.2 Bubble Languages and Combinatorial Generation

Here we give a brief introduction to bubble languages, mostly summarising results from [15, 17]. We also give a new characterization of bubble languages in terms of the computation tree of a generation algorithm (Prop. 1).

Algorithm *Generate(s, t, γ)*
(∗ current string resides in array w ∗)
1.   **if** $s > 0$ and $t > 0$
2.   **then for** $i = 1, 2, \ldots, t$
3.   **do** $w \leftarrow swap(w, s, s + i)$
4. $Generate(s - 1, i, 10^{t-i}\gamma)$
5. $w \leftarrow swap(w, s, s + i)$
6.   *Visit()*

Fig. 1. The Recursive Swap Generation Algorithm

Definition 2. *A language* $\mathcal{L} \subseteq \{0,1\}^*$ *is called* a bubble language *if, for every word* $w \in \mathcal{L}$, *exchanging the first occurrence of* 01 *(if any) by* 10 *results in another word in* $\mathcal{L}$.

For example, the languages of binary Lyndon words and necklaces are bubble languages. A language $\mathcal{L} \subseteq \{0,1\}^n$ is a bubble language if and only if each of its fixed-density subsets $\mathcal{L} \cap \mathcal{B}_d^n$ is a bubble language [15]. This implies that for generating a bubble language, it suffices to generate its fixed-density subsets.

Next we consider combinatorial generation of binary strings. Let w be a binary string of length n. Let d be the number of 1s in w, and let $i_1 < i_2 < \ldots < i_d$ denote the positions of the 1s in w. Clearly, we can obtain w from the word $1^d 0^{n-d}$ with the following algorithm: first swap the last 1 with the 0 in position i_d, then swap the $(d-1)$st 1 with the 0 in position i_{d-1} etc. Note that every 1 is moved at most once, and in particular, once the k'th 1 is moved into the position i_k, the suffix $w_{i_k} \cdots w_n$ remains fixed for the rest of the algorithm.

These observations lead us to the following generation algorithm (Fig. 1), which we will refer to as *Recursive Swap Generation Algorithm* (like Alg. 1 from [17], but without the language-specific subroutine). It generates recursively all binary strings from $\mathcal{B}_d^n$ with fixed suffix γ, where $\gamma \in 1\{0,1\}^* \cup \{\varepsilon\}$, starting from the string $1^s 0^t \gamma$. The call $Generate(d, n - d, \varepsilon)$ generates all binary strings of length n with density d.

The algorithm swaps the last 1 of the first 1-run with each of the 0s of the first 0-run, thus generating a new string each, for which it makes a recursive call. During the execution of the algorithm, the current string resides in a global array w. In the subroutine *Visit()* we can print the contents of this array, or increment a counter, or check some property of the current string. The main point of *Visit()* is that it touches every object once.

Let T_d^n denote the recursive computation tree of $Generate(d, n - d, \varepsilon)$. As an example, Fig. 2 illustrates the computation tree T_4^7 (ignore for now the highlighted words, see Sec. 3). The depth of the tree equals d, the number of 1s, while the maximum degree is $n - d$, the number of 0s. In general, for the subtree rooted at $v = 1^s 0^t \gamma$, we have depth s and maximum degree t; the number of children of v is t, and v's ith child is $1^{s-1} 0^i 1 0^{t-i} \gamma$. Note that suffix γ remains unchanged in the entire subtree, that the computation tree is isomorphic to the computation tree of $1^s 0^t$, and that the critical prefix length strictly decreases along any downward path in the tree.

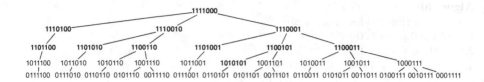

Fig. 2. The computation tree T_d^n for $n = 7, d = 4$. Prefix normal words in bold.

The algorithm performs a post-order traversal of the tree, yielding an enumeration of the strings of $\mathcal{B}_d^n$ in what is referred to as *cool-lex order* [15,17,19]. A pre-order traversal of the same tree, which implies moving line 4 of the algorithm before line 1, would yield an enumeration in *co-lex order*. A crucial property of cool-lex order is that any two subsequent strings differ by at most two swaps (transpositions), thus yielding a *Gray code* [15]. This can be seen in the computation tree T_d^n as follows. Note that in a post-order traversal of T_d^n, we have:

$$
next(u) = \begin{cases} parent(u) & \text{if } u \text{ is rightmost child} \\ \text{leftmost descendant of } u\text{'s right sibling} & \text{otherwise.} \end{cases}
$$

Let u, u' both be children of v. This means that for some $s, t, i, j \in \mathbb{N}$ and $\gamma \in 1\{0,1\}^* \cup \{\varepsilon\}$, we have $v = 1^s 0^t \gamma$, $u = 1^{s-1} 0^i 1 0^{t-i} \gamma$, and $u' = 1^{s-1} 0^j 1 0^{t-j} \gamma$. Let v' be a descendant of v along the leftmost path, i.e. $v' = 1^k 0 1^{s-k} 0^{t-1} \gamma$ for some k. Then $v = swap(u, s, s+i)$ (parent), $u' = swap(u, s+i, s+j)$ (sibling), and $v' = swap(v, k, s+1)$ (descendant along leftmost path).

The following proposition states a crucial property of bubble languages with respect to the Recursive Swap Generation Algorithm. The proof follows immediately from the definition of bubble languages:

Proposition 1. *A language $\mathcal{L}$ is a bubble language if and only if, for every $d = 0, \ldots, n$, its fixed-density subset $\mathcal{L} \cap \mathcal{B}_d^n$ is closed w.r.t. parents and left siblings in the computation tree T_d^n of the Recursive Swap Generation Algorithm. In particular, if $\mathcal{L} \cap \mathcal{B}_d^n \neq \emptyset$, then it forms a subtree rooted in $1^d 0^{n-d}$.*

Using this property, the Recursive Swap Generation Algorithm can be used to generate *any* fixed-density bubble language $\mathcal{L}$, as long as we have a way of deciding, for a node $w = 1^s 0^t \gamma$, already known to be in $\mathcal{L}$, which is its rightmost child (if any) that is still in $\mathcal{L}$. If such a child exists, and it is the kth child $u = 1^{s-1} 0^k 1 0^{t-k} \gamma$, then the bubble property ensures that all children to its left are also in $\mathcal{L}$. Thus, line 2. in the algorithm can simply be replaced by "for $i = 1, \ldots, k$". Moreover, this algorithm, which visits the words in the language in cool-lex order, will yield a Gray code, since because of this closure property, $next(u)$ will again either be the parent, or a node on the leftmost path of the right sibling, both of which are reachable within two swaps.

In [17], a generic generation algorithm was given which moves the job of finding this rightmost child k into a subroutine $Oracle(s, t, \gamma)$. If $Oracle(s, t, \gamma)$ runs in time $O(k)$, then we have a CAT algorithm. In general, this may not be possible, and a generic Oracle tests for each child from left to right whether it is in the language. Because of the bubble property, after the first negative test, we know that no more children will be in the language, and the running time of the algorithm is amortized that of the membership tester. The crucial trick is that we do not need a *general* membership tester, since all we want to know is which of the children of a node *already known to be in $\mathcal{L}$* are in $\mathcal{L}$; moreover, the membership tester is allowed to use other information, which it can build up iteratively while examining earlier nodes.

3   Combinatorial Generation of Prefix Normal Words

In this section we prove that the set of prefix normal words $\mathcal{L}_{PN}$ is a bubble language. Then, by providing some properties regarding membership testing, we can apply the cool-lex framework to generate all prefix normal words of a given length n and density d in $O(f(n))$-amortized time, where $f(n)$ is the average critical prefix length of a prefix normal word of length n. (Proofs omitted.)

Lemma 1. *The language $\mathcal{L}_{PN}$ is a bubble language.*

The computation tree in Fig. 2 highlights the subtree corresponding to $\mathcal{L}_{PN} \cap \mathcal{B}_4^7$. Since $\mathcal{L}_{PN}$ is a bubble language, by Prop. 1 it is closed w.r.t. left siblings and parents. However, we still have to find a way of identifying the rightmost child of a node that is still in $\mathcal{L}_{PN}$.

The following lemma states that, given a prefix normal word w, in order to decide whether one of its children in the computation tree is prefix normal, it suffices to check the PN-property for one particular length only: the critical prefix length of the child node. Moreover, this check can be done w.r.t. γ only.

Lemma 2. *Let $w \in \mathcal{L}_{PN}$, with $w = 1^s 0^t \gamma$, with $\gamma \in 1\{0,1\}^* \cup \{\varepsilon\}$. Let $\overline{\gamma} = \gamma 0^{s+t}$, i.e. γ padded with 0s to length n. Let $w' = swap(w, s, s+i)$. Then $w' \in \mathcal{L}_{PN}$, unless one of the following holds:*

1. *$\overline{\gamma}$ has a substring of length $s + i - 1$ with at least s 1s, or*
2. *the string $w'_{s+i} \cdots w'_{2(s+i-1)}$ has at least s 1s.*

Moreover, the latter is the case if and only if $P(\overline{\gamma}, s + 2(i-1) - t) \geq s - 1$ (where by convention, we regard a prefix of negative length as the empty word).

Corollary 1. *Given $w = 1^s 0^t \gamma \in \mathcal{L}_{PN}$. If we know $F(\gamma, j)$ and $P(\gamma, j)$ for all $j \leq s + t$, then it can be decided in constant time whether $w' = swap(w, s, s+i)$, for $i \leq t$, is prefix normal.*

Lemma 3. *Let $\gamma' = 1 0^r \gamma$, with $\gamma \in 1\{0,1\}^* \cup \{\varepsilon\}$. Then for all $i = 0, 1, \ldots, |\gamma'|$,*

$$F(\gamma', i) = \begin{cases} \max(P(\gamma', i), F(\gamma, i)) & \text{for } i \leq |\gamma| \\ \max(P(\gamma', i), F(\gamma, |\gamma|)) & \text{for } i > |\gamma|. \end{cases}$$

Corollary 2. *The F-function of γ for node $w = 1^s 0^t \gamma$, up to entry $s + t$, can be computed in time $O(s + t)$ based on the F-function of w's parent node.*

By applying these results, the algorithm *GeneratePN(d, n − d, ε)* will generate $\mathcal{L}_{\mathrm{PN}} \cap \mathcal{B}_d^n$ in cool-lex Gray code order, see Fig. 3. Starting from the left child and proceeding right (with respect to the computation tree T_n^d), the algorithm will make a recursive call finding a child which is not prefix normal. The membership test is done in the subroutine *isPN*, which uses the conditions of Lemma 2. The algorithm maintains an array F which contains the maximum number of 1s in i-length substrings of γ (the F-function of γ), and a variable z. Before testing the first child, in *update(F, s + t)*, it computes the current γ's F-function based on the parent's (Corollary 2). Note that it is not necessary to compute all of the F-function, since all nodes in the subtree have critical prefix length smaller than $s + t$, thus this update is done only up to length $s + t$. After the recursive calls to the children, the array is restored to its previous state in *restore(F, s + t)*. The variable z contains the number of 1s in the prefix of γ which is spanned by the substring of case 2. of Lemma 2, for the first child. It is updated in constant time after each successful call to *isPN*, to include the number of 1s in the two following positions in γ.

Algorithm *GeneratePN(s, t, γ)*
(* $w = 1^s 0^t \gamma$ must be prefix normal *)
1.   **if** $s > 0$ **and** $t > 0$
2.   **then** *update(F, s + t)*
3. $z \leftarrow P(\gamma, s - t)$
4. $i \leftarrow 1$
5.   **while** $i \leq t$ **and** *isPN(swap(w, s, s + i))*
6.   **do** $w \leftarrow swap(w, s, s + i)$
7.   *GeneratePN(s − 1, i, $10^{t-i}\gamma$)*
8.   *update(z)*
9. $i \leftarrow i + 1$
10. $w \leftarrow swap(w, s, s + i)$
11.   *restore(F, s + t)*
12. *Visit()*

Fig. 3. Algorithm generating all prefix normal words in the subtree rooted in $1^s 0^t \gamma$

By concatenating the lists of prefix normal words with densities $0, 1, \ldots, n$, we obtain an exhaustive listing of $\mathcal{L}_{\mathrm{PN}} \cap \Sigma^n$, see Fig. 4. As an example, *GeneratePN(5)* produces the following list of prefix normal words: 00000, 10000, 10100, 10010, 10001, 11000, 11010, 10101, 11001, 11100, 11011, 11101, 11110, 11111. Since the fixed-density listings are a cyclic Gray code (Theorem 3.1 from [15]), it follows that this complete listing is also a Gray code. In fact, if the fixed-density listings are listed by the odd densities (increasing), followed by the even densities (decreasing), the resulting listing would be a cyclic Gray code.

Theorem 1. *Algorithm* GeneratePN(n) *generates all prefix normal words of length n in amortized $O(f(n))$ time per word, where $f(n)$ is the average critical prefix length of prefix normal words of length n. In particular, $f(n) = O(n)$.*

Algorithm *GeneratePN(n)*
(∗ generates all prefix normal words of length n ∗)
1.   **for** $d = 0, 1, \ldots, n$
2.   **do** initialize F of length n with all 0s
3.   *GeneratePN(d, n − d, ε)*

Fig. 4. Algorithm generating all prefix normal words of length n

Proof. Since $1^d 0^{n-d}$ is prefix normal for every d, we only need to show that the correct subtrees of T_d^n are generated by the algorithm. By Lemma 2, only those children will be generated that are prefix normal; on the other hand, by the bubble property (Prop. 1), as soon as a child tests negative, no further children will be prefix normal. The running time of the recursive call on $w \in \mathcal{L}_{\mathrm{PN}}$ consists of (a) updating and restoring F (lines 2 and 9): the number of steps equals the critical prefix length $cr(w)$ of w; (b) computing z (line 3): again $cr(w)$ many steps; and (c) work within the while-loop (lines 5 to 8), which, for a word with k prefix normal children, consists of k positive and 1 negative membership tests, of k updates of z, and the recursive calls on the positive children. The membership tests take constant time by Corollary 1, so does the update of z. Since w has k prefix normal children, we charge the positive membership tests and the z-updates to the children, and the negative test to the current word. So for one word $w \in \mathcal{L}_{\mathrm{PN}}$, we get $3 \cdot cr(w) + O(1) + 2 \cdot O(1) = O(cr(w))$ work.   □

Next we present experimental evidence for the following conjecture:

Conjecture 1. $f(n) = \Theta(\log n)$.

4   Experimental Results

In this section we present some theoretical and numerical results about the number of prefix normal words and their structure. These have become available thanks to the algorithm presented, which allowed us to generate $\mathcal{L}_{\mathrm{PN}}$ up to length 50 on a home computer. Let $pnw(n) := |\mathcal{L}_{\mathrm{PN}} \cap \Sigma^n|$. The following lemma follows from the observation that $1^{\lceil n/2 \rceil} w$ is a prefix normal word of length n for all words w of length $\lfloor n/2 \rfloor$.

Lemma 4. *The number of prefix normal words grows exponentially in n. We have that $pnw(n) \geq 2^{\lfloor n/2 \rfloor}$.*

The first members of the sequence $pnw(n)$ are listed in [18], and these values suggest that the lower bound above is not sharp. We turn our attention to the growth rate of $pnw(n)$ as n increases. Note that $1 \leq pnw(n)/pnw(n-1) \leq 2$. The lower bound follows from the fact that all prefix normal words can be extended by adding a 0 to the end, and the upper bound is implied by the prefix-closed property of $\mathcal{L}_{\mathrm{PN}}$. Fig. 5 (left) shows the growth ratio for small values of n. The figure shows two interesting phenomena: the values seem to approach 2 slowly, i.e., the number of prefix normal words almost doubles as we increase the length

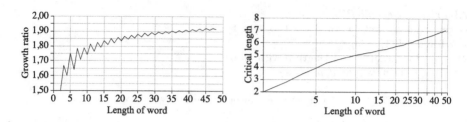

Fig. 5. The value of $pnw(n)/pnw(n-1)$ (left), and of $f(n) = E(cr(w))$ for prefix normal words w of length n, for $n \leq 50$ (right, loglinear scale)

by 1. Second, the values show on oscillation pattern between even and odd values. We have so far been unable to establish these observations theoretically.

The structure of prefix normal words is relevant for the generation algorithm, since the amortized running time of the algorithm is bounded from above by the average value $f(n)$ of the critical prefix length $cr(w)$, taken over all prefix normal words w. Note that this differs from the expected critical prefix length of the prefix normal form *of a uniformly random word*, for which we have the following:

Lemma 5. *Given a random word w of length n, let $w' = \mathrm{PNF}(w)$. Let Z' be the r.v. denoting the critical prefix length of w'. Then for the expected value of Z' we have $E(Z') = \Theta(\log n)$.*

In contrast, the average value $f(n)$ of critical prefix length for prefix normal words is shown in Fig. 5 (right) for $n \leq 50$. The linear alignment of the data points together with Lemma 5 supports the conjecture that also $f(n) = \Theta(\log n)$.

5   Conclusion and Open Problems

Based on the observation that prefix normal words form a bubble language, we presented a Gray code for all prefix normal words of length n in $O(n)$ amortized time per word. Moreover, we conjecture that our algorithm runs in time $\Theta(\log(n))$ per word. The number of words that are *not* prefix normal grows exponentially and greatly dominates prefix normal words (e.g., $pnw(30)/2^{30} < 0.05$), so the gain of any algorithm whose running time is proportional to the output size, over brute-force testing of all binary words, is considerable.

We gave a linear time testing algorithm for words which are derived from a word w already known to be prefix normal. We pose as an open problem to find a strongly subquadratic time testing algorithm *for arbitrary words*. Another open problem is the fast computation of prefix normal forms, which would lead immediately to an improvement for indexed binary jumbled pattern matching.

Acknowledgements. We thank two anonymous referees whose suggestions significantly improved the presentation of the paper.

References

1. Badkobeh, G., Fici, G., Kroon, S., Lipták, Zs.: Binary jumbled string matching for highly run-length compressible texts. Inf. Process. Lett. 113(17), 604–608 (2013)
2. Burcsi, P., Cicalese, F., Fici, G., Lipták, Zs.: On Table Arrangements, Scrabble Freaks, and Jumbled Pattern Matching. In: Boldi, P. (ed.) FUN 2010. LNCS, vol. 6099, pp. 89–101. Springer, Heidelberg (2010)
3. Burcsi, P., Cicalese, F., Fici, G., Lipták, Zs.: On approximate jumbled pattern matching in strings. Theory Comput. Syst. 50(1), 35–51 (2012)
4. Butman, A., Eres, R., Landau, G.M.: Scaled and permuted string matching. Inf. Process. Lett. 92(6), 293–297 (2004)
5. Cicalese, F., Fici, G., Lipták, Zs.: Searching for jumbled patterns in strings. In: Proc. of the Prague Stringology Conference 2009 (PSC 2009), pp. 105–117. Czech Technical University in Prague (2009)
6. Cicalese, F., Gagie, T., Giaquinta, E., Laber, E.S., Lipták, Zs., Rizzi, R., Tomescu, A.I.: Indexes for jumbled pattern matching in strings, trees and graphs. In: Kurland, O., Lewenstein, M., Porat, E. (eds.) SPIRE 2013. LNCS, vol. 8214, pp. 56–63. Springer, Heidelberg (2013)
7. Cicalese, F., Laber, E.S., Weimann, O., Yuster, R.: Near linear time construction of an approximate index for all maximum consecutive sub-sums of a sequence. In: Kärkkäinen, J., Stoye, J. (eds.) CPM 2012. LNCS, vol. 7354, pp. 149–158. Springer, Heidelberg (2012)
8. Fici, G., Lipták, Zs.: On prefix normal words. In: Mauri, G., Leporati, A. (eds.) DLT 2011. LNCS, vol. 6795, pp. 228–238. Springer, Heidelberg (2011)
9. Gagie, T., Hermelin, D., Landau, G.M., Weimann, O.: Binary jumbled pattern matching on trees and tree-like structures. In: Bodlaender, H.L., Italiano, G.F. (eds.) ESA 2013. LNCS, vol. 8125, pp. 517–528. Springer, Heidelberg (2013)
10. Giaquinta, E., Grabowski, S.: New algorithms for binary jumbled pattern matching. Inf. Process. Lett. 113(14-16), 538–542 (2013)
11. Kociumaka, T., Radoszewski, J., Rytter, W.: Efficient indexes for jumbled pattern matching with constant-sized alphabet. In: Bodlaender, H.L., Italiano, G.F. (eds.) ESA 2013. LNCS, vol. 8125, pp. 625–636. Springer, Heidelberg (2013)
12. Lee, L.-K., Lewenstein, M., Zhang, Q.: Parikh matching in the streaming model. In: Calderón-Benavides, L., González-Caro, C., Chávez, E., Ziviani, N. (eds.) SPIRE 2012. LNCS, vol. 7608, pp. 336–341. Springer, Heidelberg (2012)
13. Moosa, T.M., Rahman, M.S.: Indexing permutations for binary strings. Inf. Process. Lett. 110, 795–798 (2010)
14. Moosa, T.M., Rahman, M.S.: Sub-quadratic time and linear space data structures for permutation matching in binary strings. J. Discrete Alg. 10, 5–9 (2012)
15. Ruskey, F., Sawada, J., Williams, A.: Binary bubble languages and cool-lex order. J. Comb. Theory, Ser. A 119(1), 155–169 (2012)
16. Ruskey, F., Sawada, J., Williams, A.: De Bruijn sequences for fixed-weight binary strings. SIAM Journal of Discrete Mathematics 26(2), 605–517 (2012)
17. Sawada, J., Williams, A.: Efficient oracles for generating binary bubble languages. Electr. J. Comb. 19(1), P42 (2012)
18. Sloane, N.J.A.: The On-Line Encyclopedia of Integer Sequences. Sequence A194850, Available electronically at http://oeis.org
19. Williams, A.M.: Shift Gray Codes. PhD thesis, Univ. of Victoria, Canada (2009)

Permuted Scaled Matching

Ayelet Butman, Noa Lewenstein*, and J. Ian Munro**

[1] Department of Computer Science, Holon Institute of Technology, Israel
[2] Department of Computer Science, Netanya College, Israel
[3] Chertion School of Computer Science, University of Waterloo

Abstract. Scaled matching and permutation matching are two well known paradigms in the domain of pattern matching. Scaled matching refers to finding an occurrence of a pattern which is enlarged proportionally by some scale k within a larger text. Permutation matching is the problem of finding all substrings within a text where the character statistics of the substring and the pattern are the same. Permutation matching is easy, while scaled matching requires innovative solutions. One interesting setting of applications is the merge of the two. The problem of scaled permuted matching (i.e. first permuting and then scaling) has been addressed and solved optimally. However, it was left as an open problem whether there are efficient algorithms for permuted scaled matching. In this paper we solve the problem efficiently in a deterministic setting and optimally in a randomized setting.

1 Introduction

Classical pattern matching for exact matching has been very successfully approached lending a multitude of methods that work in linear time and limited space, e.g. [8, 14, 15, 18–20, 23]. However, the definition of match has been expanded in many directions over the years, taking into account the different types of problems that arise from different domains. Two domains will be of central interest in this paper. The first is that of images, where scaling is one of the central operations allowed. A scaling is an enlargement of a text in a proportional manner. Almost always in the setting of matching when we consider scaling we mean to enlarge, but it is also possible to consider scaling with the meaning to compress. However, one must note that when enlarging the image the result is lossless and when compressing the image the result is lossy. Therefore, one usually considers enlargements. Secondly, while one usually considers, in the context of images, two dimensional text it is also natural to consider scalings in one dimension and this has indeed been done. For example, if $S = abbca$ then the 3-scaling of S is $aaabbbbbbcccaaa$.

Note that this can be extended quite naturally to two dimensions. Scaled pattern matching, where one is given a pattern P and a text T and one seeks

---

* The second author did this research while on sabbatical in the U. of Waterloo.
** This research was supported by NSERC of Canada and the Canada Research Chairs Programme.

A.S. Kulikov, S.O. Kuznetsov, and P. Pevzner (Eds.): CPM 2014, LNCS 8486, pp. 70–78, 2014.
© Springer International Publishing Switzerland 2014

all substrings of T which match a k-scaling of P for some $k \geq 1$, was first considered in [7]. Efficient solutions were proposed in that paper also for the two dimensional version. More efficient algorithms for this problem were proposed in [6]. In [5] it was shown how to remove the alphabet dependence of the [7] solution. The scaling problem, of course, in real-life, works also with real scales. However, in the setting of matching a good definition is quite elusive and this has been an avenue of research that has been explored quite extensively. For the case of one dimensional real scaled matching there are two natural definitions. Both were explored in the setting of matchings [2, 4]. It is interesting to note that the different definitions lead to very different solutions. The two dimensional case for scaling with real numbers was examined in [3], where interesting solutions were presented.

The other domain is that of *permutation matching*, also known as *jumbled pattern matching* [9, 11] and in the case of indexing as *histogram indexing*. The problem of permutation matching is when one is given a pattern and a text and needs to find all substrings of a text for which the frequency count of the pattern character set is equal to the frequency count of the substring character set.

For example, if $S = abbca$ and the text is $acababbbcaacca$ there is a permutation match at location 2, because the substring $cabab$ has 2 a's, 2 b's and 1 c, exactly as it does in the pattern. The same is true for location 7 where the substring $bbcaa$ begins.

Permutations of data are used in various settings. For example, permutations are used when one wants to maintain security of files, see e.g. [22, 16]. In pattern matching this was initially used as a filter in approximate pattern matching algorithms [17]. The problem is actually quite simple to solve in linear time. However, there are several points of interest for this problem. In [12] it was shown how to obtain non-trivial efficient average-case behaviour. In [10] solutions for finding approximate permutation matching were considered. The indexing variant of the problem turns out to be very difficult. The interest in indexing was initiated by the work on text fingerprinting via Parikh mappings [1]. The hardness of the indexing variant is quite surprising because exact matching is harder than permutation matching in the pattern matching setting but seems to be much easier in the indexing setting. An initial result for indexing for permutation matching was presented in [21]. There the authors considered the binary alphabet case and only announced whether there was a match or not. They succeeded in shaving off a double log factor from the quadratic space necessary. In [13] a more sophisticated near-linear time construction was given for the problem with the same constraints using randomization.

The natural combination of scaled matching and permutation matching was originally considered in [11]. In that paper the combination was defined as first permuting the characters of the string and then allowing a k-scaling. However, first allowing a k-scaling and then permuting the characters is more natural. This makes the problem harder. This happens because the k-scaling gives a handle on the matching because of the repetition of the characters. So, if it is done as the second operation the shape of the k-scaling allows exploitation of this data.

But, if the k-scaling happens first then the permutation jumbles the k-scaling structure. This problem was left as an open question in [11]. This is the problem we consider in this paper. We formally define the problem in the preliminaries and show how to solve the problem efficiently in a deterministic setting and optimally in a randomized setting.

2   Preliminaries

Let S be a string. Denote with $\#_\sigma(S)$ the number of appearances of σ in S. Denote with $S[0, i]$ the $i+1$ length prefix of S, i.e. $s_0 s_1 \cdots s_i$. Denote with $S[i, j]$ the $j - i + 1$ length substring of S starting at i, i.e. $s_i s_{i+1} \cdots s_j$.

Let S be a string and let k be a natural number. A k-scaling of S is a string that is obtained by replacing every character σ of S with σ^k, where σ^k denotes σ repeated k times. For example, if $S = abca$ then the 4-scaling of S is $aaaabbbbccccaaaa$.

Let S be a string and k a natural number. We say that the string $\pi(S, k)$ is the permuted scale of order k of S if it can be generated by a permutation of the characters of a k-scaling of S.

Say we have a text T. A pattern P is said to have a *permuted k-scaled appearance at location i of* T if the substring $t_i \ldots t_{(i+|P|k-1)}$ is a permuted scale of order k of P.

Permuted Scaled Matching
INPUT: A text $T = t_0 t_1 \cdots t_{n-1}$ and a pattern $P = p_0 p_1 \cdots p_{m-1}$ over alphabet $\Sigma = \{\sigma_0, \sigma_1, \cdots, \sigma_{|\Sigma|-1}\}^1$.
OUTPUT: All permuted k-scaled appearances of P, for all possible k ($1 \leq k \leq \lfloor \frac{n}{m} \rfloor$).

A straightforward algorithm that solves the Permuted Scaled matching problem works as follows:

1. Construct a table R of size $(n+1) \times |\Sigma|$ such that $R(i, j) = \#_{\sigma_j}(T[0, i])$ for $i \geq 0$ and $R(-1, j) = 0$.
2. For every $0 \leq i < j \leq n - 1$ such that $j - i + 1 = km$ for some natural number $k \geq 1$ do:
   (a) Let $r(l) = \frac{R(j,l) - R(i-1,l)}{\#_{\sigma_l}(P)}$.
   (b) if $r(l) = k$ for each l, $0 \leq l \leq |\Sigma| - 1$, then announce that i is a k-scaled appearance.

The running time for this algorithm is $O(n^2 |\Sigma|)$ where $|T| = n$.

Correctness follows from the claim below that, in turn, follows directly from the definitions.

Claim. For every $0 \leq i \leq j \leq n - 1$, $\#_\sigma(T[i, j]) = \#_\sigma(T[0, j]) - \#_\sigma(T[0, i - 1])$.

---

[1] The alphabet is that of the pattern. Any superfluous characters in the text simply partition the text into smaller text pieces upon which any algorithm can be run separately.

3   Properties of Permuted Scaled Matches

Since we desire to find a more efficient algorithm we will begin with presenting some properties that occur for permuted scaled matching. We will later utilize these properties to obtain an efficient algorithm for the permuted scaled matching problem. The two properties that we obtain will be sufficient to characterize matches and efficiently verifying the substrings satisfying these properties will be the focal point of our algorithm.

Definition 1. *Let T be a text and P be a pattern. We say that locations i and j are* mod-equivalent *with respect to σ if $\#_\sigma(T[0, i-1]) \equiv \#_\sigma(T[0, j]) \bmod \#_\sigma(P)$.*

We note that if locations i and j are mod equivalent with respect to σ then $\#_\sigma(T[i, j])$ is a multiple of $\#_\sigma(P)$. Nevertheless, even if locations i and j are mod equivalent with respect to σ for *all* characters of the alphabet this is not sufficient to declare a k-scaled appearance (for some k). The reason is that even though each specific character σ satisfies that $\#_\sigma(T[i, j])$ is a multiple of $\#_\sigma(P)$ it may be with different multiples. Therefore, we still need to find a way to verify that all different characters have the same k as a multiple.

Definition 2. *Let T be a text and P a pattern. Let $a, b \in \Sigma$ be two different characters. We say that locations i and j satisfy the* equal-quotients *property for a and b at locations i and j if:*

$$\left\lfloor \frac{\#_a(T[0, i-1])}{\#_a(P)} \right\rfloor - \left\lfloor \frac{\#_b(T[0, i-1])}{\#_b(P)} \right\rfloor = \left\lfloor \frac{\#_a(T[0, j])}{\#_a(P)} \right\rfloor - \left\lfloor \frac{\#_b(T[0, j])}{\#_b(P)} \right\rfloor$$

We now show a central lemma which will give us a handle on obtaining efficient algorithms for the permuted scaled matching problem.

Lemma 1. *Let $T = t_0 t_1 \ldots t_{n-1}$ be a text and $P = p_0 p_1 \ldots p_{m-1}$ be a pattern. Let $\Sigma = \{\sigma_0, \sigma_1, \ldots, \sigma_{|\Sigma|-1}\}$ be the alphabet of the pattern. Let $0 \leq i \leq j \leq n-1$ then:*

$T[i, j]$ is a permuted k-scaling of P for some k $\Longleftrightarrow$

1. *Locations i and j of T are mod-equivalent with respect to σ_l for every $0 \leq l \leq |\Sigma| - 1$.*
2. *Locations i and j of T satisfy the equal-quotients property for each pair of characters σ_l and σ_{l+1} for $0 \leq l \leq |\Sigma| - 2$.*

Proof. ($\Rightarrow$) Let $T[i, j]$ be a permuted k-scaling of P. Obviously $j - i + 1 = km$. Let σ be a character of the alphabet. Therefore, the number of appearances of σ in $T[i, j]$ must be $k * \#_\sigma(P)$. By Claim 2 we have that $k * \#_\sigma(P) = \#_\sigma(T[0, j]) - \#_\sigma(T[0, i - 1])$. Therefore it follows from the mod function that locations i and j are mod-equivalent (with respect to σ). Hence, what needs to be shown is that for locations i and j the equal-quotients property is satisfied for consecutive characters of the alphabet. Let $0 \leq l \leq |\Sigma| - 2$.

$$\frac{\#_{\sigma_l}(T[i,j])}{\#_{\sigma_l}(P)} = k = \frac{\#_{\sigma_{l+1}}(T[i,j])}{\#_{\sigma_{l+1}}(P)}$$

It follows by Claim 2 that

$$\left\lfloor \frac{\#_{\sigma_l}(T[0,j])}{\#_{\sigma_l}(P)} \right\rfloor - \left\lfloor \frac{\#_{\sigma_l}(T[0,i-1])}{\#_{\sigma_l}(P)} \right\rfloor = \left\lfloor \frac{\#_{\sigma_{l+1}}(T[0,j])}{\#_{\sigma_{l+1}}(P)} \right\rfloor - \left\lfloor \frac{\#_{\sigma_{l+1}}(T[0,i-1])}{\#_{\sigma_{l+1}}(P)} \right\rfloor$$

By moving terms to the other side of the equation we can obtain the desired.
($\Leftarrow$)
Assume that conditions (1) and (2) are satisfied. We will show that $T[i,j] = \pi(P,k)$ (since, it is given that $j - i + 1 = km$).
By (2) we have that:

$$\left\lfloor \frac{\#_{\sigma_l}(T[0,i-1])}{\#_{\sigma_l}(P)} \right\rfloor - \left\lfloor \frac{\#_{\sigma_{l+1}}(T[0,i-1])}{\#_{\sigma_{l+1}}(P)} \right\rfloor = \left\lfloor \frac{\#_{\sigma_l}(T[0,j])}{\#_{\sigma_l}(P)} \right\rfloor - \left\lfloor \frac{\#_{\sigma_{l+1}}(T[0,j])}{\#_{\sigma_{l+1}}(P)} \right\rfloor$$

by moving terms to the other side of the equation we obtain:

$$\left\lfloor \frac{\#_{\sigma_l}(T[0,j])}{\#_{\sigma_l}(P)} \right\rfloor - \left\lfloor \frac{\#_{\sigma_l}(T[0,i-1])}{\#_{\sigma_l}(P)} \right\rfloor = \left\lfloor \frac{\#_{\sigma_{l+1}}(T[0,j])}{\#_{\sigma_{l+1}}(P)} \right\rfloor - \left\lfloor \frac{\#_{\sigma_{l+1}}(T[0,i-1])}{\#_{\sigma_{l+1}}(P)} \right\rfloor$$

From property (1) we know that for every l:
$\#_{\sigma_l}(T[0,i-1]) \equiv \#_{\sigma_l}(T[0,j]) \mod \#_{\sigma_l}(P)$. However, by rules of the mod function we know that:
if $a \equiv b \mod c$ then $\frac{a-b}{c}$ is a whole number and $\lfloor \frac{a}{c} \rfloor - \lfloor \frac{b}{c} \rfloor = \frac{a-b}{c}$. Therefore,

$$\frac{\#_{\sigma_l}(T[0,j]) - \#_{\sigma_l}(T[0,i-1])}{\#_{\sigma_l}(P)} = \frac{\#_{\sigma_{l+1}}(T[0,j]) - \#_{\sigma_{l+1}}(T[0,i-1])}{\#_{\sigma_{l+1}}(P)}$$

However, by Claim 2 it follows that:

$$\frac{\#_{\sigma_l}(T[i,j])}{\#_{\sigma_l}(P)} = \frac{\#_{\sigma_{l+1}}(T[i,j])}{\#_{\sigma_{l+1}}(P)}.$$

Since this equality is true for all $0 \le l \le |\Sigma| - 2$ and since $T[i,j]$ is of length km for some positive integer k it must be that for every σ_l: $\frac{\#_{\sigma_l}(T[i,j])}{\#_{\sigma_l}(P)} = k$.   □

4   Algorithm

In order to solve the Permuted Scaled Matching problem over alphabet $\Sigma = \{\sigma_0, \sigma_1, \ldots, \sigma_{r-1}\}$ we will use Lemma 1. Specifically, we will find the substrings $T[i,j]$ such that locations i and j are mod-equivalent with respect to every character and that satisfy the equal-quotients property for every pair of lexicographically consecutive characters. However, checking every substring directly for these properties will return us to the running time of $O(n^2|\Sigma|)$ of the naive algorithm. On the other hand, consider the mod-equivalent property. If we compute for each location r and σ the value: $\#_\sigma(T[0,r])$ mod $\#_\sigma(P)$ then by assigning locations (for each σ separately) to buckets according to their computed values we know that $i-1$ and j are in the same bucket for every σ iff locations i and j are mod-equivalent. Likewise, for the equal-quotients property we can compute for each location r and consecutive symbols σ_i and σ_{i+1} the values $\left\lfloor \frac{\#_{\sigma_i}(T[0,r])}{\#_{\sigma_i}(P)} \right\rfloor - \left\lfloor \frac{\#_{\sigma_{i+1}}(T[0,r])}{\#_{\sigma_{i+1}}(P)} \right\rfloor$. Once again by assigning locations (for each σ_i seperately) to buckets we partition those elements that satisfy the equal-quotients property.

However, this creates numerous tests (for each of the different σ values) which can revert to the naive running time. To rectify this situation we treat the values as vectors, i.e. $2|\Sigma|$ length vectors, where each scalar of the vector represents one of the values. Now, if we hash locations to buckets according to the vectors we have that locations i and j are mod-equivalent (for every σ) and are equal-quotient (for every consecutive pair of characters) iff $i-1$ and j are in the same bucket. To implement this "hash" we radix sort the vectors and partition them into equivalence classes for those that have the same vectors.

4.1 Algorithm's Running Time

The running time of the algorithm is as follows. First we build the R table. The size of the table is $O(n|\Sigma|)$ where the direct computation of each element $R[i,j]$ may take $O(n)$ time. However, it is easy to see that $\#_\sigma(P)$ can be computed over all σ in $O(m \log |\Sigma|)$ time by saving the characters in a binary search tree, and associating with each a counter. Moreover, for each σ we can compute $\#_\sigma(T[0,i])$

Algorithm Permuted Scaled Matching

-Build a table R of size $n \times 2|\Sigma| + 1$.
-For every $0 \le i \le n-1$:
    For every $0 \le j \le |\Sigma| - 1$:
$$R(i,j) = \#_{\sigma_j}(T[0,i]) \bmod \#_{\sigma_j}(P).$$
    For every $|\Sigma| \le j \le 2|\Sigma| - 1$:
$$R(i,j) = \left\lfloor \frac{\#_{\sigma_j}(T[0,i])}{\#_{\sigma_j}(P)} \right\rfloor - \left\lfloor \frac{\#_{\sigma_{j+1}}(T[0,i])}{\#_{\sigma_{j+1}}(P)} \right\rfloor.$$
-Each vector is associated with its location i,
  i.e. the vector i is $< R[i,0], \ldots, R[i,2|\Sigma|-1], i >$.
-Sort the vectors using Radix sort.
-Group the vectors into equivalence classes acording to their prefix of length $2|\Sigma|-1$.
-For each equivalence class containing locations $i_1, i_2, \ldots, i_l$ announce appearances $T[i+1,j]$ for each $i,j \in \{i_1, i_2, \ldots, i_l\}$, s.t. $i < j$.

Fig. 1. The permuted scaled matching algorithm

for all i's in $O(n)$ time. We note that we want to only do so for characters that appear in P. So, first we check every character in T whether it exists in P in time $O(n \log |\Sigma|)$ using the binary search tree of the P alphabet. If there is a character in T that is not in P then we partition the text into two separate texts, as no match can occur with this character. At last, we can compute $\#_\sigma(T[0, i])$ for all i's of T in $O(n)$ time for each $\sigma \in \Sigma$. Hence, we can compute all $R[i, j]$ in $O((n + m)|\Sigma|)$ time.

Since we use radix sort on the n vectors of length $2|\Sigma| + 1$ and the values of each location are bounded by n we can implement the radix sort in linear time which is $O(n|\Sigma|)$.

Finally, the running time of the last line announcing appearances is $O(n + occ)$, where occ is the number of permuted scaled appearances of the pattern. In summary,

Theorem 1. *The running time of the permuted scaled matching algorithm is* $O(n|\Sigma| + occ)$.

4.2   Output Representation

The output of the algorithm which we denoted occ may be as large as $O(n^2/m)$. A simple example is text a^n and pattern a^m. In order to reduce this potentially large number of appearances we will redefine the problem to output only the shortest match beginning at any each location i of the text, see e.g. in Figure 2 for two matches starting at the same location. The following claim, for which we omit the straightforward proof, shows that this method does not lose information and is actually just a succinct way of presenting all the appearances.

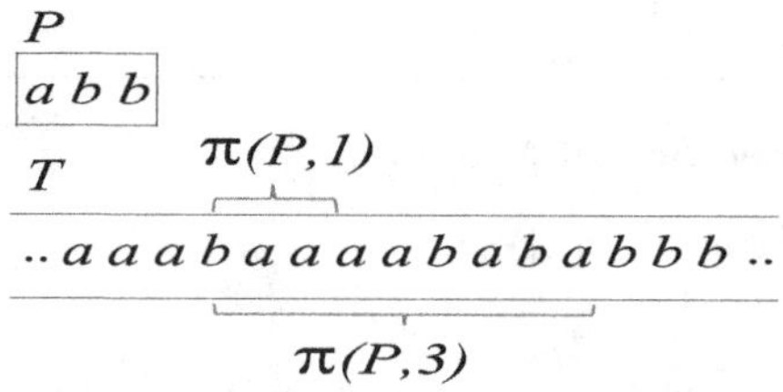

Fig. 2. An example of two scaled permuted matches starting at same location

Claim. Let $T = t_0, t_1, \ldots, t_{n-1}$ be a text and $P = p_0, p_1, \ldots, p_{m-1}$ be a pattern. Let $i < j < h$ be three text locations. Assume $T[i, j]$ is a permuted scaled appearance of P. Then $T[i, h]$ is a permuted scaled appearance of P iff $T[j+1, h]$ is a permuted scaled appearance of P.

To accommodate this we change the last line of the algorithm.

Note that the output is now no more than $O(n)$. Hence, the overall running time of the algorithm is $O(n|\Sigma|)$.

Permuted Scaled Matching Output - Shortest Appearances

-For each entry q' containing linked list $i_1, i_2, \ldots, i_l$ announce appearances $T[i_r + 1, i_{r+1}]$ for each $i_r \in \{i_1, i_2, \ldots, i_l\}$.

Finally we note that the running time can be improved to a deterministic $O(n \log |\Sigma|)$ and to a randomized $O(n)$, the former based on fingerprints and the latter based on hashing of the Rabin-Karp form. We leave the details for the full version of this paper.

Observation 1 *Let Ham denote the Hamming distance function. Then Ham($<$ $R[i, 0], \ldots, R[i, 2|\Sigma| - 1] >, < R[i+1, 0], \ldots, R[i+1, 2|\Sigma| - 1] >) \leq 3$. Moreover, we know the locations in the array that change.*

References

1. Amihood Amir, Alberto Apostolico, Gad M. Landau, and Giorgio Satta. Efficient text fingerprinting via parikh mapping. *J. Discrete Algorithms*, 1(5-6), 2003.
2. Amihood Amir, Ayelet Butman, and Moshe Lewenstein. Real scaled matching. *Inf. Process. Lett.*, 70(4):185–190, 1999.
3. Amihood Amir, Ayelet Butman, Moshe Lewenstein, and Ely Porat. Real two dimensional scaled matching. *Algorithmica*, 53(3):314–336, 2009.
4. Amihood Amir, Ayelet Butman, Moshe Lewenstein, Ely Porat, and Dekel Tsur. Efficient one-dimensional real scaled matching. *J. Discrete Algorithms*, 5(2):205–211, 2007.
5. Amihood Amir and Gruia Călinescu. Alphabet-independent and scaled dictionary matching. *J. Algorithms*, 36(1):34–62, 2000.
6. Amihood Amir and Eran Chencinski. Faster two dimensional scaled matching. *Algorithmica*, 56(2):214–234, 2010.
7. Amihood Amir, Gad M. Landau, and Uzi Vishkin. Efficient pattern matching with scaling. *J. Algorithms*, 13(1):2–32, 1992.
8. Robert S. Boyer and J. Strother Moore. A fast string searching algorithm. *Commun. ACM*, 20(10):762–772, 1977.
9. Peter Burcsi, Ferdinando Cicalese, Gabriele Fici, and Zsuzsanna Lipták. Algorithms for jumbled pattern matching in strings. *Int. J. Found. Comput. Sci.*, 23(2):357–374, 2012.
10. Peter Burcsi, Ferdinando Cicalese, Gabriele Fici, and Zsuzsanna Lipták. On approximate jumbled pattern matching in strings. *Theory Comput. Syst.*, 50(1):35–51, 2012.
11. Ayelet Butman, Revital Eres, and Gad M. Landau. Scaled and permuted string matching. *Inf. Process. Lett.*, 92(6):293–297, 2004.
12. Ferdinando Cicalese, Gabriele Fici, and Zsuzsanna Lipták. Searching for jumbled patterns in strings. In *Stringology*, pages 105–117, 2009.
13. Ferdinando Cicalese, Eduardo Sany Laber, Oren Weimann, and Raphael Yuster. Near linear time construction of an approximate index for all maximum consecutive sub-sums of a sequence. In *CPM*, pages 149–158, 2012.

14. Michael J. Fisher and Michael S. Paterson. String matching and other products. *Complexity of Computation, R.M. Karp (editor), SIAM AMS Proceeding*, 7:113–125, 1974.
15. Zvi Galil and Joel I. Seiferas. Time-space-optimal string matching. *J. Comput. Syst. Sci.*, 26(3):280–294, 1983.
16. G.A.Sathishkumar, S. Ramachandran, and K. Bhoopathy Bagan. Image encryption using random pixel permutation by chaotic mapping. In *IEEE Symposium on Computers and Informatics (ISCI)*, pages 247–251, 2012.
17. Petteri Jokinen, Jorma Tarhio, and Esko Ukkonen. A comparison of approximate string matching algorithms. *Softw., Pract. Exper.*, 26(12):1439–1458, 1996.
18. Richard M. Karp, Raymond E. Miller, and Arnold L. Rosenberg. Rapid identification of repeated patterns in strings, trees and arrays. In *STOC*, pages 125–136, 1972.
19. Richard M. Karp and Michael O. Rabin. Efficient randomized pattern-matching algorithms. *IBM Journal of Research and Development*, 31(2):249–260, 1987.
20. D. E. Knuth, J. H. Morris, and V. B. Pratt. Fast Pattern Matching in Strings. *SIAM Journal on Computing*, 6(2):323–350, 1977.
21. Tanaeem M. Moosa and M. Sohel Rahman. Sub-quadratic time and linear space data structures for permutation matching in binary strings. *J. Discrete Algorithms*, 10:5–9, 2012.
22. V. Patidar, G. Purohit, K.K. Sud, and N.K. Pareek. Image encryption through a novel permutation-substitution scheme based on chaotic standard map. In *International Workshop on Chaos-Fractals Theories and Applications (IWCFTA)*, pages 164–169, 2010.
23. Uzi Vishkin. Deterministic sampling - a new technique for fast pattern matching. *SIAM J. Comput.*, 20(1):22–40, 1991.

The Worst Case Complexity
of Maximum Parsimony

Amir Carmel, Noa Musa-Lempel, Dekel Tsur, and Michal Ziv-Ukelson

Department of Computer Science, Ben-Gurion University of the Negev
{karmela,noamu,dekelts,michaluz}@cs.bgu.ac.il

Abstract. One of the core classical problems in computational biology
is that of constructing the most parsimonious phylogenetic tree interpret-
ing an input set of sequences from the genomes of evolutionarily related
organisms. We re-examine the classical Maximum Parsimony (MP) optimization problem for the general (asymmetric) scoring matrix case, where
rooted phylogenies are implied, and analyze the worst case bounds of three
approaches to MP: The approach of Cavalli-Sforza and Edwards [5], the
approach of Hendy and Penny [12], and a new agglomerative, "bottom-
up" approach we present in this paper. We show that the second and
third approaches are faster than the first by a factor of $\Theta(\sqrt{n})$ and $\Theta(n)$,
respectively.

1 Introduction

Phylogenetics is the study of evolutionary relationships among groups of organisms (e.g. species, populations), which are discovered through molecular sequencing data and morphological data matrices. *Phylogenies* (also called dendograms)
are graph-like structures whose topology describes the inferred evolutionary history among a set of biological entities, such as species or DNA sequences. Phylogenies are classically computationally modeled as either rooted or unrooted
labeled binary trees, where the input entities are assigned to the leaf vertices.
An *unrooted phylogeny* is an acyclic connected labeled graph in which every vertex has degree of either three or one. Each vertex of degree one has a distinct
label. A *rooted phylogeny*, on the other hand, is similar to an unrooted phylogeny,
except that it has one internal vertex of degree two, which is designated as the
root. In a rooted phylogeny the edges are directed from the root towards the
leaves.

The decision of whether to model phylogenies as rooted versus unrooted trees
depends either on the availability of a molecular clock, or on the nucleotide
or amino acid substitution scoring matrix representing the evolutionary mutation events. Modeling phylogenies as unrooted trees requires the assumption of
symmetric scoring matrices. However, when the symmetry restriction on scoring matrices is removed, the tree rooting becomes meaningful. A simple literature review of current biological research shows that the symmetric scoring
matrices, though computationally convenient, do not yield a biologically reliable
model [10, 16, 18, 19]. Various recent biological publications apply asymmetric

A.S. Kulikov, S.O. Kuznetsov, and P. Pevzner (Eds.): CPM 2014, LNCS 8486, pp. 79–88, 2014.

scoring matrices to the alignment of genomic sequences, and many papers can nowadays be found on the construction of asymmetric scoring matrices consisting of nucleotides [3, 20, 21] and amino acids [2, 14]. Thus, in this work we do not assume symmetric scoring matrices and therefore construct general, rooted phylogenies.

Phylogenetic trees among a nontrivial number of input sequences are constructed using computational phylogenetics methods. Methods for phylogeny reconstruction can be classified into *distance-based* versus *character-based* methods. Given a set of input sequences, a distance-based method, such as UPGMA and neighbor joining (NJ), first computes pairwise distances according to some measure [7]. Then, the actual data is discarded and the fixed distances are used in the derivation of trees. In contrast, in character-based methods (such as Maximum Parsimony and Maximum Likelihood) the inference depends upon models describing the evolutionary changes of characters (e.g. nucleotides or amino acids) that led from an original sequence in some common ancestor to the evolution of the observed input sequences. The great advantage of character-based algorithms for phylogenetic reconstruction is that, given a good multiple alignment of the input sequences, they can exploit the potential phylogenetic inferences with great sensitivity. Their weakness, however, is in their computational intensity. In this paper we focus on the classical *Maximum Parsimony* character-based phylogenetic approach.

1.1 Phylogentic Reconstruction Based on Parsimony Maximization

Parsimony Maximization (i.e. preferring the simpler of two otherwise equally adequate theorizations) is one of the classical approaches to computationally reconstruct a phylogeny for a given set of biologically related sequences. When applied to computational phylogenetics, the parsimony maximization approach seeks the phylogenetic tree that supposes the least amount of evolutionary change explaining the observed data [5]. There are two classical problems inferred from phylogenetic parsimony maximization: Small Parsimony (SP) and Maximum Parsimony (MP), explained below.

Problem 1: Small Parsimony (SP). The Small Parsimony problem is to compute, for a proposed phylogeny, a reconstruction of events leading to the input data with as few changes as possible over the whole tree. The input to this problem is a multiple alignment of n input sequences of length m each, and a topology in the form of a rooted phylogenetic tree over n leaves, where each leaf is associated with a distinct sequence from the input set. Based on this input, the objective is to compute a labeling of the internal vertices of the input phylogeny which optimizes some predefined scoring scheme.

Problem 2: Maximum Parsimony (MP). The Maximum Parsimony problem is to seek, among all possible phylogenies over a given set of leaves, the phylogeny that yields the best SP score. Similarly to SP, the input to this problem is a multiple alignment of n input sequences of length m each. However, here the topology is not given. The Maximal Parsimony (MP) problem is NP-Hard [6, 9].

Measuring SP and MP complexity in terms of basic operations. SP and MP algorithms work by computing some information for every internal vertex of the input phylogeny. This information, as well as the complexity of its computation, depend on the scoring scheme employed by the parsimony algorithm. Thus, in what follows, we will use the term *basic operation* to denote the work invested in the computation of the information of a single vertex of a considered phylogeny for a general scoring scheme. For example, in the Fitch SP algorithm [8], which computes a minimal Hamming Distance SP score, an $O(m)$-time basic operation is applied, while in the Sankoff algorithm [17] which optimizes an SP score of minimal weighted edit distance, an $O(m\sigma^2)$-time basic operation is applied, where σ denotes the size of the alphabet spelling the input sequences.

Several constrained variants of classical MP were studied over the years [7]. The most famous among them is the Perfect Phylogeny problem, which is also NP-Hard [4] and for which FPT algorithms were proposed [1,13].

Our Contribution. In this work, we examine the complexity of MP in terms of the total number of basic operations executed throughout the run of the algorithm. Using this measure, we analyze the worst case bounds of the time complexity of three approaches to MP. The first, basic approach, is the one proposed by Cavalli-Sforza and Edwards [5], which performs $(n-1) \cdot (2n-3)!!$ basic operations, where $(2n-3)!! = 1 \times 3 \times 5 \times \cdots \times (2n-3)$. The second approach is based on the Hendy and Penny [12] MP search space, which interleaves the SP computations within the tree-space development flow. This search space was originally proposed for the purpose of a branch-and-bound MP search, and its theoretical worst-case bound was not previously properly bounded. One of our results is Theorem 2.2 in Section 2, in which we analyze the basic operations complexity of the Hendy and Penny approach, and show that it improves the Cavalli-Sforza and Edwards complexity bound by a factor of $\Theta(\sqrt{n})$.

Both Hendy and Penny's work as well as follow-up exact branch and bound extensions [7] still kept the same traditional order of search space development, based on the Cavalli-Sforza tree enumeration order. *In order to further improve MP efficiency, we propose to turn the state development order of the classical MP search tree "upside down", i.e. from the classical top down incremental tree extension by a single edge per each new state, to a bottom-up agglomerative approach which merges two previous subtrees per each state.* This idea is based on the observation that the "top down" approach to MP search goes "against the grain" of the update operations applied by the Hendy and Penny SP subroutines per each new search space node.

In order to avoid this contradiction, we suggest a "bottom up" directed MP search, based on subtree merging, which flows along with the natural "bottom up" direction of Small Parsimony. By design, traversing the search space according to our proposed search tree requires performing exactly one basic operation per node of the search tree (the basic operation is applied on the new root of the two merged trees in the agglomerative step). Thus, the complexity of our new MP algorithm is equal to the size of our proposed search space tree. We show that this size is approximately $e \cdot (2n-3)!!$, where e is the base of the natural logarithm. *Thus, our new approach yields a reduction of the Cavalli-Sforza search-space bound by a factor of $\Theta(n)$, and that of Hendy and Penny by a factor of $\Theta(\sqrt{n})$ (Theorem 5 in Section 4).* We note that the bound obtained by our proposed algorithm is on the same order of the number of topologies considered by MP, and therefore any approach to exact MP that must enumerate all topologies will not improve our result.

We note that the original work of Hendy and Penny assumed a symmetric scoring matrix. Under the assumption of symmetric scoring matrices, it can be shown that rooted MP can be solved via the simpler unrooted MP. However, such is not the case in practice, when the restriction to symmetrical scoring matrices is removed [10, 16, 18, 19], and the tree rooting becomes meaningful. The algorithms we propose and analyze in this paper generalize the previous most efficient exact solutions to this problem and do not assume the use of symmetrical scoring matrices [10, 18, 19]. *Due to space restrictions, additional materials, including omitted proofs and figures, can be found in* http://www.cs.bgu.ac.il/ negevcb/publications.php

2 Analysis of Previous Approaches

Throughout the paper, a *phylogeny* is a rooted binary labeled tree. Each leaf in the tree has a distinct label from $\{1, \ldots, n\}$, where n is the number of leaves. We also define a *forest* to be a collection of rooted binary labeled trees. Each leaf in the forest has a distinct label from $\{1, \ldots, n\}$, where n is the number of leaves in all the trees.

We will later define another type of trees called search space trees. In order to make the text more clear, we will use different terminology for the two types of trees: we will use *vertex* for phylogenies and *node* for search space trees.

2.1   The Algorithm of Cavali-Sforza and Edwards

The algorithm of Cavalli-Sforza and Edwards [5] enumerates all phylogenies with n leaves, and then solves the Small Parsimony problem on each tree. Cavalli-Sforza and Edwards showed that the number of phylogenies with n leaves is $(2n-3)!!$. Moreover, each tree has exactly $n-1$ internal vertices. The following complexity bound is obtained.

Theorem 1 (Cavalli-Sforza and Edwards [5]). *The basic operations complexity of the algorithm of Cavalli-Sforza and Edwards is $(n-1) \cdot (2n-3)!!$.*

The enumeration of all phylogenies can be modeled by a *search space tree*. The search space tree $\mathcal{T}_n^{\text{CSE}}$ consists of n levels (see Figure 1). The nodes of level $i-1$ correspond to all phylogenies with i leaves. For a node v of the search space tree, denote by F_v the phylogeny that corresponds to v. A node v of level $i-1$ has $2i-1$ children defined as follows: For each edge e in F_v, there is a child v_e of v whose corresponding phylogeny F_{v_e} is obtained from F_v by splitting the edge e into two edges connected in a new vertex x. Moreover, the vertex x has an additional child, which is a new leaf with label $i + 1$ that is added to F_{v_e} as a child of x. The node v has an additional child v' whose corresponding phylogeny $F_{v'}$ is obtained from F_v by adding a new root vertex x. The children of x are the root of F_v and a new leaf (with label $i + 1$). From the definition of the search space tree, it is clear that level i of the tree contains $(2i - 3)!!$ nodes, and in particular, there are $(2n - 3)!!$ leaves in the tree. Thus, the number of phylogenies with n leaves is $(2n - 3)!!$. The enumeration of the phylogenies with n leaves is achieved by performing a traversal of the search space tree, and building the phylogeny F_v when reaching each node v.

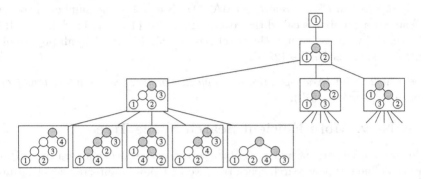

Fig. 1. An example of the search space tree $\mathcal{T}_4^{\mathrm{CSE}}$. Only some of the nodes of the search space tree are shown. Highlighted in gray within the phylogenies in each node of the search tree are the vertices updated by basic operations when applying Hendy and Penny's MP approach on $\mathcal{T}_4^{\mathrm{CSE}}$.

2.2 The Algorithm of Hendy and Penny

Hendy and Penny [12] interleaved the Small Parsimony operations within search space tree traversal. The algorithm traverses the tree $\mathcal{T}_n^{\mathrm{CSE}}$ and when reaching a node v, it solves the Small Parsimony problem for the phylogeny F_v. However, the Small Parsimony algorithm does not need to compute the required information for all internal vertices of F_v since the information computed for the phylogeny of v's parent can be used. More precisely, let u be the parent of v, and let x be the newly added leaf in F_v. If y is an internal vertex of F_v that is not an ancestor of x, then the information for the vertex y in F_v is identical to the information for the vertex y in F_u. Since the latter was already computed, in order to solve the Small Parsimony problem on F_v, only vertex information for the ancestors of x needs to be computed. For a vertex v in $\mathcal{T}_n^{\mathrm{CSE}}$, let NUMANC($v$) denote the number of ancestors of x in F_v, where x is the newly added leaf in F_v. Therefore, the basic operations complexity of the algorithm of Hendy and Penny is $\sum_{v \in \mathcal{T}_n^{\mathrm{CSE}}}$ NUMANC(v) (the summation is performed over nodes from all levels of the search tree). We next give a formula for the summation above.

Definition 1. *Let H_i be the sum of* NumAnc(v) *for all nodes v in level $i+1$ of T_n^{CSE}.*

By definition, $\sum_{v \in T_n^{CSE}}$ NumAnc$(v) = \sum_{i=1}^{n-1} H_i$.

Lemma 1. $H_i = (2i)!! - (2i-1)!!$.

Proof. Let v be a node at level $i+1$ in T_n^{CSE}, and let x be the newly added leaf in F_v. Let F' be the forest obtained from F_v by removing x and its ancestors, and removing all the edges incident on these vertices. Note that every tree in F' is a subtree of a distinct ancestor of x in F_v, and every ancestor of x corresponds to exactly one tree in F'. Therefore, NumAnc(v) is equal to the number of trees in F'. The latter number is called the cover number for $\{1, \ldots, i\}$ in F_v in [15]. It is shown in [15] that the sum of the cover number for $\{1, \ldots, i\}$ in all phylogenies with $i+1$ leaves is $(2i)!! - (2i-1)!!$. □

Theorem 2. *The basic operations complexity of the algorithm of Hendy and Penny is $\Theta(\sqrt{n}(2n-3)!!)$.*

3   A New, More Efficient Search Space Tree

In this section we present a new, "bottom up" search space enumerating MP. In order to define our new search space tree, we first build a directed acyclic graph $\mathcal{G}_n$. We will later transform $\mathcal{G}_n$ into a tree $\mathcal{T}_n$ by removing some of the edges.

We define a *merge operation* on a forest F to be an operation that generates a new forest F' by adding a new vertex to the forest and hanging two trees of F on this vertex. In the graph $\mathcal{G}_n$, every node v corresponds to a forest with n leaves. The graph contains a single node at level 0 whose corresponding forest consists of n singletons. The nodes of level i in the graph correspond to all the forests that can be obtained from the forests that correspond to the nodes of level $i-1$ by single merge operations. There is an edge (u,v) in the graph if and only if F_v is obtained from F_u by a merge operation.

Note that a node in $\mathcal{G}_n$ can have several incoming edges. We next transform $\mathcal{G}_n$ into a tree $\mathcal{T}_n$, by selecting exactly one edge from the edges entering a node, and removing all other edges.

For a tree T in a forest define the *label* of T to be

$$\text{label}(T) = \min\{\text{label}(v) : v \text{ is a leaf in } T\}.$$

The label of a forest F is

$$\text{label}(F) = \min\{\text{label}(T) : T \text{ is a non-singleton tree in } F\}.$$

For a node u in the graph $\mathcal{G}_n$, let T_u denote the tree in F_u for which label$(T_u) = $ label(F_u). Using the definitions above we can define the search space tree $\mathcal{T}_n$ (see Figure 2).

Definition 2. *The search space tree $\mathcal{T}_n$ is a tree whose nodes are the nodes of $\mathcal{G}_n$. A node v is the parent of a node u in $\mathcal{T}_n$ if F_v is obtained from F_u by deleting the root of T_u.*

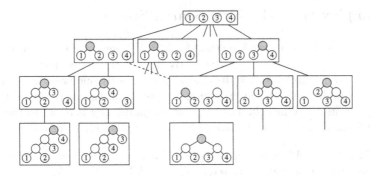

Fig. 2. An example of the search space tree $\mathcal{T}_4$. Only some of the nodes of the tree are shown. The dashed line is an edge of $\mathcal{G}_4$ that is not an edge of $\mathcal{T}_4$. Highlighted in gray within the phylogenies in each node of the search tree are the vertices updated by basic operations when applying our MP approach on $\mathcal{T}_4$.

The definition above gives an implicit characterization for the children of a node. We now show explicit characterization.

Lemma 2. *A node u is a child of a node v in $\mathcal{T}_n$ if and only if F_u is obtained from F_v by merging two trees T_1 and T_2 from F_v with* label$(T_1) <$ label(T_2) *and either*

1. *$T_1 = T_v$ (and in particular, T_1 is not a singleton), or*
2. *T_1 is a singleton and* label$(T_1) <$ label(T_v).

Proof. ($\Rightarrow$) Let u be a child of v in $\mathcal{T}_n$. By definition, deleting the root of T_u gives the forest F_v. In other words, if we denote the two trees formed from T_u by removing its root, then F_u is obtained from F_v by merging T_1 and T_2. Without loss of generality, assume label$(T_1) <$ label(T_2). By definition, label$(T_u) =$ min(label(T_1), label(T_2)) = label(T_1). The tree T_u is the non-singleton tree of F_u with minimum label. It follows that in F_u, the leaves with labels $1, \ldots,$ label$(T_1) - 1$ are each inside a singleton. Since F_u is obtained from F_v by a single merge operation, these leaves are also in singletons in F_v. If T_1 is non-singleton then it is the non-singleton tree in F_v with minimum label$(\cdot)$ value. Thus, $T_1 = T_v$, namely case 1 of the lemma occurs. If T_1 is a singleton then, since the leaves with labels $1, \ldots,$ label(T_1) are in singletons in F_v, we obtain that label$(T_v) >$ label(T_1) and case 2 of the lemma occurs.

($\Leftarrow$) Suppose F_u is obtained by merging two trees T_1 and T_2 from F_v, and let T' denote the tree obtained by this merge. In both cases of the lemma we have that the leaves with labels $1, \ldots,$ label$(T_1) - 1$ are each inside a singleton in F_v, and therefore these leaves are also in singletons in F_u. Therefore, T' is the non-singleton tree in F_u with minimum label, namely $T' = T_u$. Thus, v is the parent of u in $\mathcal{T}_n$. $\square$

4   Complexity of the New Search Space

We have proposed a new search method for MP. We now wish to bound the basic operations complexity of our approach. By design, traversing the search space according to our search tree requires performing exactly one basic operation per node: For a node v, the basic operation is applied on the new root of the two merged trees in F_v. Thus, the basic operation complexity is equal to the size of $\mathcal{T}_n$. We will next show that the size of $\mathcal{T}_n$ is approximately $e \cdot (2n - 3)!!$.

Definition 3. *Let A_i^n denote the number of nodes in level i of $\mathcal{T}_n$, and let $L_{i,k}^n$ denote the number of nodes v in level i for which* label$(F_v) = k$.

Example 1. $A_0^n = 1$, $A_1^n = \binom{n}{2}$, and $A_{n-1}^n = (2n - 3)!!$.

Note that for a node v in level i of $\mathcal{T}_n$, the forest F_v contains $n - i$ trees.

Observation 3. *If v is a node in level i then* label$(F_v) \le n - i$.

Proof. The forest F_v is obtained from the forest of n singletons by a sequence of i merge operations. At least one merge operation must involve a tree containing a leaf with label at most $n - i$. Thus, the label of the resulting tree is at most $n - i$. This tree can participate in other merge operations, which either decrease or do not change the label of the tree. At the end of the merge operations the labels of the tree is at most $n - i$ and thus label$(F_v) \le n - i$.   □

Observation 4. $A_i^n = \sum_{k=1}^{n-i} L_{i,k}^n$.

Lemma 3. $L_{i+1,k}^n = (n - i - k) \sum_{l=k}^{n-i} L_{i,l}^n$.

Proof. First, we count the number of children with label k for a node in level i with label l (we will soon show that this number does not depend on the topology of the corresponding forest). Let v be some node in level i with label l. Note that the forests corresponding to the children of v have labels at most l. For a child u of v for which label$(F_u) = k$, the forest F_u is obtained by merging two trees T_1 and T_2 from F_v such that label$(T_1) = k$ and label$(T_2) > k$ (note that $T_1 = T_v$ if $k = l$, and otherwise T_1 is a singleton). It follows that the number of children of v with label k is the number of trees T in F_v with label$(T) > k$. Since $k \le l$ and the leaves with labels $1, \ldots, l - 1$ are in singletons, it follows that there are exactly k trees with label at most k. The total number of trees in F_v is $n - i$ and therefore F_v contains $n - i - k$ trees with label greater than k. Thus, the number of children of v with label k is $n - i - k$ (not depending on the topology nor on l).

From the above we can deduce a recursive formula for $L_{i+1,k}^n$. A node in level $i + 1$ with label k is the child of a node in level i with label $l \ge k$. The lemma follows.   □

Lemma 4. $L_{i,k}^n = (2i - 1)!!\binom{n-k+i-1}{2i-1}$.

Lemma 5. $A_i^n = (2i - 1)!!\binom{n+i-1}{2i}$.

Let μ_n be the number of nodes in a search space tree $\mathcal{T}_n$. Then, $\mu_n = 1 + \sum_{i=1}^{n-1} A_i^n = \sum_{i=0}^{n-1}(2i - 1)!!\binom{n+i-1}{2i}$. The series (μ_n) has already been studied, and the following result was obtained [11].

$$\lim_{n\to\infty} \frac{\mu_n}{e(2n - 3)!!} = 1 \tag{1}$$

We now compare the size of our search tree to the size of the search tree $\mathcal{T}_n^{CSE}$. Let η_n be the number of nodes in $\mathcal{T}_n^{CSE}$. Then, $\eta_n = \sum_{i=1}^{n}(2i - 3)!!$. It is easy to verify that

$$\lim_{n\to\infty} \frac{\eta_n}{(2n - 3)!!} = 1 \tag{2}$$

From equations (1) and (2) we obtain that $\lim_{n\to\infty} \frac{\mu_n}{\eta_n} = e$.

Finally, since solving MP using our search tree requires performing exactly one basic operation per node of the search tree, we obtain the following theorem.

Theorem 5. *The basic operations complexity for solving MP using the new search space is $(1 + o(1)) \cdot e \cdot (2n - 3)!!$. Moreover, this complexity is $\Theta(\sqrt{n})$ times smaller than the complexity of the algorithm of Hendy and Penny.*

Proof. The theorem follows from Theorem 2 and Equation (1). □

Acknowledgments. The research of A.C, N.M-L and M.Z-U was partially supported by ISF grant 478/10. The research of A.C and D.T was partially supported by ISF grant 981/11. The research of A.C, D.T, N.M-L and M.Z-U was partially supported by the Frankel Center for Computer Science at Ben Gurion University of the Negev.

References

1. Agarwala, R., Fernández-Baca, D.: A polynomial-time algorithm for the perfect phylogeny problem when the number of character states is fixed. SIAM Journal on Computing 23(6), 1216–1224 (1994)
2. Bastien, O., Roy, S., Maréchal, É.: Construction of non-symmetric substitution matrices derived from proteomes with biased amino acid distributions. Comptes Rendus Biologies 328(5), 445–453 (2005)
3. Blouin, M.S., Yowell, C.A., Courtney, C.H., Dame, J.B.: Substitution bias, rapid saturation, and the use of mtDNA for nematode systematics. Molecular Biology and Evolution 15(12), 1719–1727 (1998)
4. Bodlaender, H.L., Fellows, M.R., Warnow, T.J.: Two strikes against perfect phylogeny. Springer (1992)
5. Cavalli-Sforza, L.L., Edwards, A.W.: Phylogenetic analysis. Models and estimation procedures. American Journal of Human Genetics 19(3 pt. 1), 233 (1967)

6. Day, W.H., Sankoff, D.: Computational complexity of inferring phylogenies by compatibility. Systematic Biology 35(2), 224–229 (1986)
7. Felsenstein, J., Felenstein, J.: Inferring phylogenies, vol. 2. Sinauer Associates Sunderland (2004)
8. Fitch, W.M.: Toward defining the course of evolution: minimum change for a specific tree topology. Systematic Biology 20(4), 406–416 (1971)
9. Foulds, L.R., Graham, R.L.: The steiner problem in phylogeny is np-complete. Advances in Applied Mathematics 3(1), 43–49 (1982)
10. Gojobori, T., Ishii, K., Nei, M.: Estimation of average number of nucleotide substitutions when the rate of substitution varies with nucleotide. J. of Molecular Evolution 18(6), 414–422 (1982)
11. Grosswald, E.: Bessel Polynomials (1978)
12. Hendy, M., Penny, D.: Branch and bound algorithms to determine minimal evolutionary trees. Mathematical Biosciences 59(2), 277–290 (1982)
13. McMorris, F.R., Warnow, T.J., Wimer, T.: Triangulating vertex colored graphs. In: Proceedings of the Fourth Annual ACM-SIAM Symposium on Discrete Algorithms, pp. 120–127. Society for Industrial and Applied Mathematics (1993)
14. Müller, T., Rahmann, S., Rehmsmeier, M.: Non-symmetric score matrices and the detection of homologous transmembrane proteins. Bioinformatics 17(suppl. 1), S182–S189 (2001)
15. Ochiumi, N., Kanazawa, F., Yanagida, M., Horibe, Y.: On the average number of nodes covering a given number of leaves in an unordered binary tree. J. of Combinatorial Mathematics and Combinatorial Computing 76, 3 (2011)
16. Rodriguez, F., Oliver, J.L., Marin, A., Medina, J.R.: The general stochastic model of nucleotide substitution. J. of Theoretical Biology 142(4), 485–501 (1990)
17. Sankoff, D.: Minimal mutation trees of sequences. SIAM Journal on Applied Mathematics 28(1), 35–42 (1975)
18. Tajima, F., Nei, M.: Estimation of evolutionary distance between nucleotide sequences. Molecular Biology and Evolution 1(3), 269–285 (1984)
19. Takahata, N., Kimura, M.: A model of evolutionary base substitutions and its application with special reference to rapid change of pseudogenes. Genetics 98(3), 641–657 (1981)
20. Tamura, K.: The rate and pattern of nucleotide substitution in drosophila mitochondrial DNA. Molecular Biology and Evolution 9(5), 814–825 (1992)
21. Tamura, K., Nei, M.: Estimation of the number of nucleotide substitutions in the control region of mitochondrial DNA in humans and chimpanzees. Molecular Biology and Evolution 10(3), 512–526 (1993)

From Indexing Data Structures to de Bruijn Graphs*

Bastien Cazaux[1], Thierry Lecroq[2], and Eric Rivals[1]

[1] L.I.R.M.M. & Institut Biologie Computationnelle, Université de Montpellier II,
CNRS U.M.R. 5506, Montpellier, France
[2] LITIS EA 4108, NormaStic CNRS FR 3638, Université de Rouen, France
{cazaux,rivals}@lirmm.fr, thierry.lecroq@univ-rouen.fr

Abstract. New technologies have tremendously increased sequencing throughput compared to traditional techniques, thereby complicating DNA assembly. Hence, assembly programs resort to de Bruijn graphs (dBG) of k-mers of short reads to compute a set of long contigs, each being a putative segment of the sequenced molecule. Other types of DNA sequence analysis, as well as preprocessing of the reads for assembly, use classical data structures to index all substrings of the reads. It is thus interesting to exhibit algorithms that directly build a dBG of order k from a pre-existing index, and especially a contracted version of the dBG, where non branching paths are condensed into single nodes. Here, we formalise the relationship between suffix trees/arrays and dBGs, and exhibit linear time algorithms for constructing the full or contracted dBGs. Finally, we provide hints explaining why this bridge between indexes and dBGs enables to dynamically update the order k of the graph.

1 Introduction

The de Bruijn graph (dBG) of order k on an alphabet Σ with σ symbols has σ^k vertices corresponding to all the possible distinct strings of length k on the alphabet Σ and there is a directed edge from vertex u to vertex v if the suffix of u of length $k-1$ equals the prefix of v of length $k-1$. De Bruijn graphs have various properties and are more commonly defined on all the k-mers of the strings of a finite set rather than on all the possible strings of length k on the alphabet. When a vertex u has only one outgoing edge to vertex v and when v has only one ingoing edge from vertex u then the two vertices can be merged. By applying this rule whenever possible, one gets a contracted dBG. dBGs occur in different contexts. In bioinformatics they are largely used in *de novo* assembly due to a result of Pevzner *et al* [14]. Indeed recent sequencing technologies allow to obtain hundreds of million of short sequencing reads (about 100 nucleotides long) from one DNA sample. Next step is to reconstruct the genome sequence using assembly algorithms. However, the volume of read data to process has forced

---

* This work is supported by ANR Colib'read (ANR-12-BS02-0008) and Défi MASTODONS SePhHaDe from CNRS and Labex NumEV.

A.S. Kulikov, S.O. Kuznetsov, and P. Pevzner (Eds.): CPM 2014, LNCS 8486, pp. 89–99, 2014.
© Springer International Publishing Switzerland 2014

the shift from the classical overlap graph approach, which requires too much memory, towards a de Bruijn Graph where vertices are k-mers of the reads. In this context, there exist compact exact data structures for storing dBGs [7,3,15,5] and probabilistic data structures such as Bloom filters [12,6,5]. Onodera *et al* propose to add to the succinct dBG representation of [3] a bit vector marking the branching nodes, thereby enabling them to simulate efficiently a contracted dBG, where each simple path is reduced to one edge [11].

Suffix trees are well-known indexing data structures that enable to store and retrieve all the factors of a given string. They can be adapted to a finite set of strings and are then called generalised suffix trees (GSTs). They can be built in linear time and space. They have been widely studied and used in a large number of applications (see [1,9]). In practice, they consume too much space and are often replaced by the more economical suffix arrays [10], which have the same properties.

Read analysis and assembly include preliminary steps like filtering and error correction. To speed up such steps, some algorithms index the substrings, or the k-mers of the reads. Hence, before the assembly starts, the read set has already been indexed and mined. For instance, the error correction software hybrid-shrec builds a GST of all reads [16]. It can thus be efficient to enable the construction of the dBG for the subsequent assembly, directly from the index rather than from scratch. For these reasons, we set out to find algorithms that transform usual indexes into a dBG or a contracted dBG. It is also of theoretical interest to build bridges between well studied indexes and this graph on words. Despite recent results [15,11], formal methods for constructing dBG from suffix trees are an open question. Notably, the String Graph, which is also used for genome assembly, can be constructed from a FM-index [17].

In this article, given a finite collection S of strings and an integer k we formalise the relationship between GSTs and dBGs and show how to linearly build the dBG of order k for S. Next we show how to directly build the contracted dBG of order k for S in linear time and space, without building the dBG. We also show how to perform the same task using suffix arrays. Finally, we give some hints on how to dynamically adapt our dBG construction from order k to $k-1$ or from k to $k+1$.

2   Preliminaries

An *alphabet* Σ is a finite set of *letters*. A finite sequence of elements of Σ is called a *word* or a *string*. The set of all words over Σ is denoted by Σ^*, and ε denotes the empty word. For a word x, $|x|$ denotes the *length* of x. Given two words x and y, we denote by xy the *concatenation* of x and y. For every $1 \leq i \leq j \leq |x|$, $x[i]$ denotes the i-th letter of x, and $x[i .. j]$ denotes the *substring* or *factor* $x[i]x[i+1]\ldots x[j]$. Let k be a positive integer. If $|x| \geq k$, $first_k(x)$ is the *prefix* of length k of x and $last_k(x)$ is the *suffix* of length k of x. A substring of length k of x is called a k-mer of x. For i such that $1 \leq i \leq |x| - k + 1$, $(x)_{k,i}$ is the k-mer of x starting in position i, *i.e.* $(x)_{k,i} = x[i .. i + k - 1]$.

$$
\begin{array}{c c c c c c c c}
 & 1 & 2 & 3 & 4 & 5 & 6 & 7 \\
s_1 & b & a & c & b & a & b & \\
s_2 & c & b & a & b & c & a & a \\
s_3 & b & c & a & a & c & b & \\
s_4 & c & b & a & a & c & & \\
s_5 & b & b & a & c & b & a & a \\
\end{array}
$$

Fig. 1. $Support_S(ba)$ = $\{(1,1),(1,4),(2,2),(4,2),(5,2),(5,5)\}$, $RC_S(ba)$ = $\{\varepsilon, c, cb, cba, cbab, b, bc, bca, bcaa, a, ac, cbaa\}$, $LC_S(ba)$ = $\{\varepsilon, c, ac, bac, b, bbac\}$ and $d_S(ba) = 0$. One has $RC_S(ba) \cap \Sigma = \{a, b, c\}$. Thus, the word ba is not right extensible in S (see Def. 2).

Thus we have $first_k(x) = (x)_{k,1}$ and $last_k(x) = (x)_{k,|x|-k+1}$. We denote by $\sharp(\Lambda)$ the cardinality of any finite set Λ.

Let $S = \{s_1, \ldots, s_n\}$ be a finite set of words. Let us denote the sum of the lengths of the input strings by $\|S\| := \sum_{s_i \in S} |s_i|$. We denote by F_S the set of factors of words of S. For a word w of F_S,

- $Support_S(w)$ is the set of pairs (i, j), where w is the substring $(s_i)_{|w|,j}$. $Support_S(w)$ is called the support of w in S.
- $RC_S(w)$ (resp. $LC_S(w)$) is the set of *right context* (resp. *left context*) of the word w in S, *i.e.* the set of words w' such that $ww' \in F_S$ (resp. $w'w \in F_S$).
- $\lceil w \rceil_S$ is the word ww' where w' is the longest word of $RC_S(w)$ such that $Support_S(w) = Support_S(ww')$. In other words, such that w and ww' have exactly the same support in S.
- $\lfloor w \rfloor_S$ is the word w' where w' is the longest prefix of w such that $Support_S(w') \neq Support_S(w)$.
- $d_S(w) := |\lceil w \rceil_S| - |w|$.

In other words, $\lceil w \rceil_S$ is the longest extension of w having the same support than w in S, while $\lfloor w \rfloor_S$ is the shortest reduction of w with a support different from that of w in S. These definitions are illustrated in a running example, with $S := \{bacbab, cbabcaa, bcaacb, cbaac, bbacbaa\}$, presented in Fig. 1.

We give the definition of a de Bruijn graph for assembly (dBG for short), which differs from the original definition of a complete graph over all possible words of length k stated by de Bruijn [8].

Definition 1. *Let k be a positive integer and $S := \{s_1, \ldots, s_n\}$ be a set of n words. The de Bruijn graph of order k for S, denoted by DBG_k^+, is a directed graph, $DBG_k^+ := (V^+, E^+)$, whose vertices are the k-mers of words of S and where an arc links u to v if and only if u and v are two successive k-mers of a word of S, i.e.: $V^+ := F_S \cap \Sigma^k$ and $E^+ := \{(u, v) \in V^{+2} \mid last_{k-1}(u) = first_{k-1}(v) \text{ and } v[k] \in RC_S(u)\}$.*

Examples of arcs are displayed on Fig. 2.

Let us introduce now the notions of extensibility for a substring of S and that of a Contracted dBG (CdBG for short).

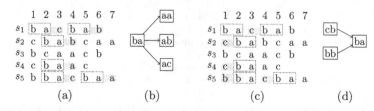

Fig. 2. Examples of arcs from DBG_k^+. (a) letters in the right context of ba, and (b) the successors of node ba in DBG_2^+; one for each letter in $RC_S(w) \cap \Sigma$. (c) letters in the left context of ba, and (d) the predecessors of node ba in DBG_2^+.

Definition 2 (Extensibility). *Let w be a word of F_S, w is* right extensible *in S if and only if $\sharp(RC_S(w) \cap \Sigma) = 1$ and w is* left extensible *in S if and only if $\sharp(LC_S(w) \cap \Sigma) = 1$.*

As S is clear from the context, we simply omit the "in S". Let w be a word of $\Sigma^\star$. The word w is said to be a *unique k'-mer of S* if and only if $k' \geq k$ and for all $i \in [1..k' - k + 1]$, $(w)_{k,i} \in F_S$ and for all $j \in [1..k' - k]$, $(w)_{k,j}$ is right extensible and $(w)_{k,j+1}$ is left extensible.

Definition 3. *A* contracted de Bruijn graph *of order k, denoted by $CDBG_k^+ = (V_c^+, E_c^+)$, is a directed graph where:*
$V_c^+ = \{w \in \Sigma^\star \mid w \text{ is a } k'\text{-mer unique maximal by substring and } k' \geq k\}$ *and*
$E_c^+ = \{(u, v) \in V_c^{+2} \mid last_{k-1}(u) = first_{k-1}(v) \text{ and } v[k] \in RC_S(last_k(u))\}$.

Note that in the previous definition, an element w in V_c^+ does not necessarily belong to F_S, since w may only exist as the substring of the agglomeration of two words of S. Thus, let w be a k'-mer unique maximal by substring with $k' \geq k$: $last_k(w)$ is not right extensible or $RC_S(last_k(w)) \cap \Sigma = \{a\}$ and $last_{k-1}(w) \cdot a$ is not left extensible, $first_k(w)$ is not left extensible or $LC_S(first_k(w)) \cap \Sigma = \{a\}$ and $a \cdot first_{k-1}(w)$ is not right extensible. With this argument, we have both following propositions.

Proposition 1. *Let $(u, v) \in E_c^+$; $(last_k(u), first_k(v)) \in E^+$ and there exists $w \in V^+$ such that $(w, first_k(v)) \in E^+ \backslash \{(last_k(u), first_k(v))\}$ or $(last_k(u), w) \in E^+ \backslash \{(last_k(u), first_k(v))\}$.*

Proposition 2. *Let $(u, v) \in E^+$. If u is right extensible and v is left extensible, then there exists $w \in V_c^+$ such that $uv[k]$ is a substring of w. Otherwise, there exists $(u', v') \in E_c^+$ such that $u = last_k(u')$ and $v = first_k(v')$.*

According to Prop. 1 and 2, $CDBG_k^+$ is the graph DBG_k^+ where the arcs (u, v) are contracted if and only if u is right extensible and v is left extensible.

3   Definition of de Bruijn Graphs with Words

Let k be a positive integer. We define the following three subsets of F_S.

- $InitExact_{S,k} = \{w \in F_S \mid |w| = k \text{ and } d_S(w) = 0\}$
- $Init_{S,k} = \{w \in F_S \mid |w| \geq k \text{ and } d_S(first_k(w)) = |w| - k\}$
- $SubInit_{S,k} = InitExact_{S,k-1}$

A word of $InitExact_{S,k}$ is either only the suffix of some s_i or has at least two right extensions, while the first k-mer of a word in $Init_{S,k} \setminus InitExact_{S,k}$ has only one right extension.

Proposition 3. $InitExact_{S,k} = Init_{S,k} \cap \{w \in F_S \mid |w| = k\}$.

For w an element of $Init_{S,k}$, $first_k(w)$ is a k-mer of S. Given two words w_1 et w_2 of $Init_{S,k}$, $first_k(w_1)$ and $first_k(w_2)$ are distinct k-mers of S. Furthermore for each k-mer w' of S, there exists a word w of $Init_{S,k}$ such that $first_k(w) = w'$. From this, we get the following proposition.

Proposition 4. *There exists a bijection between $Init_{S,k}$ and the set of the k-mers of S.*

According to Def. 1 and Prop. 4, each vertex of DBG_k^+ can be assimilated to a unique element of $Init_{S,k}$. To define the arcs between the words of $Init_{S,k}$, which correspond to arcs of DBG_k^+, we need the following proposition, which states that each single letter that is a right extension of w gives rise to a single arc.

Proposition 5. *For $w \in InitExact_{S,k}$ and $a \in \Sigma \cap RC_S(w)$, there exists a unique $w' \in Init_{S,k}$ such that $last_{k-1}(w)a$ is a prefix of w'.*

The set $Init_{S,k}$ represents the nodes of DBG_k^+. Let us now build the set of arcs that is isomorphic to E^+. Let w be a word of $Init_{S,k}$ and $Succ(w)$ denote the set of successors of $first_k(w)$: $Succ(w) := \{x \in Init_{S,k} \mid (first_k(w), first_k(x)) \in E^+\}$. We know that for each letter a in $RC_S(w)$, there exists an arc from $first_k(w)$ to $first_k(last_{|w|-1}(w)a)$ in DBG_k^+. We consider two cases depending on the length of w:

Case 1 $|w| = k$. According to Prop. 3, $w \in InitExact_{S,k}$ and hence $last_{k-1}(w) \in SubInit_{S,k}$. Therefore, the outgoing arcs of w in DBG_k^+ are the arcs from w to w' satisfying the condition of Prop. 5. Then,
$Succ(w) = \cup_{a \in \Sigma \cap RC_S(w)} \lceil last_{k-1}(w)a \rceil_S$.

Case 2 $|w| > k$. As w is longer than k, it contains the next k-mer; hence $first_k(last_{|w|-1}(w)a) = first_k(last_{|w|-1}(w))$, and there exists a unique outgoing arc of w: that from w to $\lceil w[2..k] \rceil_S$. Indeed, by definition of $Init_{S,k}$, $\lceil w[2..k] \rceil_S \in Init_{S,k}$, and thus $Succ(w) = \{\lceil w[2..k] \rceil_S\}$.

Now, we can build integrally DBG_k^+ or more exactly an isomorphic graph of DBG_k^+. Thus for simplicity, from now on we confound the graph we build with DBG_k^+. To do the same with $CDBG_k^+$, we need to characterise the concepts of right and left extensibility in terms of word properties. By the construction of DBG_k^+, we have the following results.

Proposition 6. *Let w be a word of $Init_{S,k}$. $first_k(w)$ is right extensible if and only if $|w| > k$ or $\sharp(RC_S(w) \cap \Sigma) = 1$.*

Proposition 7. *Let w be a word of $Init_{S,k}$ such that $first_k(w)$ is right extensible. Let the letter a be the unique element of $RC_S(first_k(w)) \cap \Sigma$, then $last_{k-1}(first_k(w))a$ is left extensible if and only if $\sharp(Support_S(first_k(w))) = \sharp(Support_S(last_{k-1}(first_k(w))a) \setminus \{(i,1) \mid 1 \le i \le n\})$.*

We present a generic algorithm to build incrementally $CDBG_k^+$. In the following sections, we exhibit algorithms to compute DBG_k^+ and $CDBG_k^+$ for two important indexing structures.

4   Transition from the Suffix Tree to de Bruijn Graphs

A *generalised* ST (GST) can index the substrings of a set of words. Generally for this sake, all words are concatenated and separated by a special symbol not occurring elsewhere. However, this trick is not compulsory, and an alternative is to keep the indication of a terminating node within each node.

4.1   The Suffix Tree and Its Properties

The *Generalised Suffix Tree* (GST) of a set of words S is the suffix tree of S, where each word of S does not finish necessarily by a letter of unique occurrence. Hence, for each node v of the GST of S, we keep in memory the set, denoted by $Suff_S(v)$, of pairs (i, j) such that the word represented by v is the suffix of s_i starting at position j. Let us denote by T the GST of S (from now on, we simply say the tree) and by V_T its set of nodes. For $v \in V_T$, $Children(v)$ denotes its set of children and $f(v)$ its parent.

Some nodes of T may have just one child. The size of the union of $Suff_S(v)$ for all node v of T equals the number of leaves in the GST when the words end with a terminating symbol. Hence, the space to store T and the sets $Suff_S(.)$ is linear in $\|S\|$. By simplicity, for a node v of T, the word represented by v is confused with v. For each node v of T, $v \in F_S$. As all elements of F_S are not necessarily represented by a node of T, we give the following proposition.

Proposition 8. *The set of nodes of T is exactly the set of words w of F_S such that $d_S(w) = 0$.*

We recall the notion of a suffix link (SL) for any node v of T (leaves included). Let $sl(v)$ denote the node targeted by the suffix link of v, i.e. $sl(v) = v[2 .. |v|]$. By definition of a suffix tree, for all $w \in F_S$, there exists a node v of T such that w is a prefix of v. Let v' the node of minimal length of T such that w is a prefix of v, then $|v'| = |w| + d_S(w)$, and therefore $\lceil w \rceil_S = v'$.

Proposition 9. *Let $w \in F_S$. Then $|\lceil w \rceil_S| \ge |w| > |f(\lceil w \rceil_S)|$, where $f(\lceil w \rceil_S)$ is the parent of $\lceil w \rceil_S$ in T.*

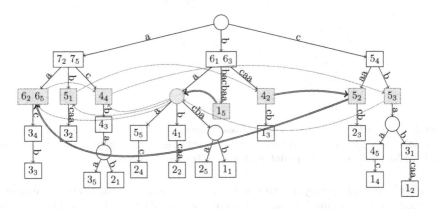

Fig. 3. The GST for our running example and the constructed dBG for $k := 2$. Square nodes represent words that occur as a suffix of some s_i, circle nodes are the other nodes of T. Grey nodes represent the vertices of the dBG. Each square node contains i_j when it represents the suffix of s_j starting at position i. The curved arrows are the edges of the dBG; those in dotted lines correspond to Case 1 and those in doubled lines to Case 2 (see p. 95).

4.2   Construction of DBG_k^+

Let $[x_1..x_m]$ be the set of k-mers of S. According to the definition of $Init_{S,k}$ and to Prop. 4, $Init_{S,k} = [\lceil x_1 \rceil_S .. \lceil x_m \rceil_S]$. Thus, by Prop. 9, $Init_{S,k} = \{v \in V_T \mid |f(v)| < k$ and $|v| \geq k\}$. Similarly, $InitExact_{S,k} = \{v \in V_T \mid |v| = k\}$. Now, it appears clearly that $InitExact_{S,k}$ is a subset of $Init_{S,k}$, since for all $v \in V_T$, $|f(v)| < |v|$.

We consider the same two cases as for the construction of E^+ on p. 93, but in the case of a tree. Let $v \in Init_{S,k}$.

Case 1 $|v| = k$ (Fig. 4a). As $v \in InitExact_{S,k}$, $sl(v) \in SubInit_{S,k}$. Therefore, each child u of $sl(v)$ is an element of $Init_{S,k}$. Thus, the outgoing arcs of v in DBG_k^+ are the arcs from v to the child u of $sl(v)$ where the first letter of the label between $sl(v)$ and u is an element of the right context of v. As the set of the first letters of the label between v and children of v is exactly $RC_S(v) \cap \Sigma$, the number of outgoing arcs of v in DBG_k^+ is the number of children of v. To build the outgoing arcs of v in DBG_k^+, for each child u' of v, we associate v with the node of $Init_{S,k}$ between the root and $sl(u')$, i.e. $\lceil first_k(sl(u')) \rceil_S$.

Case 2 $|v| > k$ (Fig. 4b and 4c). We have that $sl(v)$ is a node of V_T. As $|v| > k$, $|sl(v)| \geq k$. Thus, there exists an element of $Init_{S,k}$ between the root and $sl(v)$. We associate v with this node, i.e. $\lceil first_k(sl(v)) \rceil_S$.

We illustrate these two cases in Fig. 3. For Case 1: v is $\boxed{6_2\ 6_5}$, $sl(v)$ is $\boxed{7_2\ 7_5}$, the unique child u' of v is $\boxed{3_4}$, and $sl(u')$ is $\boxed{4_4}$, which is in $Init_{S,k}$. For Case 2: v is $\boxed{1_5}$, $sl(v)$ is $\boxed{2_5}$, and $\lceil first_k(sl(v)) \rceil_S$ is $\bigcirc$.

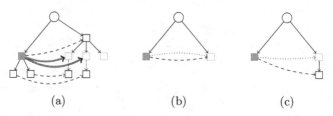

Fig. 4. Figures (a), (b) and (c) show Case 1 and Case 2 for the arcs of DBG_k^+. The grey node is v. The dashed arcs correspond to suffix links. The arcs of the dBG are in bold (a) for the Case 1 and in dotted lines (b) (c) for the Case 2.

In both cases, building the arcs of E^+ requires to follow the SL of some node. The node, say u, pointed at by a SL may not be initial. Hence, the initial node representing the associated first k-mer of u is the only ancestral initial node of u. We equip each such node u with a pointer $p(u)$ that points to the only initial node on its path from the root. In other words, for any $u \notin Init_{S,k}$ such that $|u| > k$, one has $p(u) := \lceil first_k(u) \rceil_S$.

The algorithm to build the DBG_k^+ is as follows. A first depth first traversal of T allows to collect the nodes of $Init_{S,k}$ and for each such node to set the pointer $p(.)$ of all its descendants in the tree. Finally to build E^+, one scans through $Init_{S,k}$ and for each node v one adds $Succ(v)$ to E^+ using the formula given above. Altogether this algorithm takes a time linear in the size of T. Moreover, the number of arcs in E^+ is linear in the total number of children of initial nodes. This gives us the following result.

Theorem 1. *For a set of words S, building the de Bruijn Graph of order k, DBG_k^+ takes linear time and space in $|T|$ or in $\|S\|$.*

4.3   Construction of CDBG$_k^+$

In Section 3, we have seen an algorithm that allows to compute directly CDBG$_k^+$ provided that one can determine if a node v is right extensible and if next(v) is left extensible, where next(v) denotes the only successor of v. Let us see how to compute the extensibility in the case of a Suffix Tree.

By applying Prop. 6 in the case of tree, for an element v of $Init_{S,k}$, $first_k(v)$ is right extensible if and only if $|v| > k$ or $\sharp(Children(v)) = 1$. Thus checking the right extensibility of a node takes constant time.

For the left extensibility of the single successor of a node, one only needs the size of support of some nodes (Prop. 7). Let us see first how to compute $\sharp(Support_S(.))$ on the tree, and then how to apply Prop. 7.

Proposition 10. *Let v be a word of F_S and $V_T(\lceil v \rceil_S)$ denotes the set of nodes of the subtree rooted in $\lceil v \rceil_S$. $Support_S(v) = \cup_{v' \in V_T(\lceil v \rceil_S)} Suff_S(v')$.*

Along a traversal of the tree, we compute and store $\sharp(Support_S(v))$ and $\sharp(Support_S(v) \cap \{(i,1) \mid 1 \leq i \leq n\})$ for each node v in linear time in $|T|$.

Let v be a word of $Init_{S,k}$ such that $first_k(v)$ is right extensible.

Case 1 If $|v| = k$ then $first_k(v) = v$ and $\sharp(Children(v)) = 1$. Let u be the only child of v. Thus, $|u| > k$, $\sharp(RC_S(v) \cap \Sigma) = \{u[k+1]\}$, and $last_{k-1}(v)u[k+1] = first_k(sl(u))$. Hence, $\sharp(Support_S(v)) = \sharp(Support_S(first_k(sl(u))) \setminus \{(i,1) \mid 1 \leq i \leq n\})$ and by Prop. 7, $first_k(sl(u))$ is left extensible.

Case 2 If $|v| > k$ then $\sharp(RC_S(first_k(v)) \cap \Sigma) = \{v[k+1]\}$ and $last_{k-1}(first_k(v)) \cdot v[k+1] = last_k(first_{k+1}(v)) = first_k(sl(v))$. By Prop. 7, $first_k(sl(v))$ is left extensible if and only if
$$\sharp(Support_S(first_k(v))) = \sharp(Support_S(first_k(sl(v))) \setminus \{(i,1) \mid 1 \leq i \leq n\})$$

As $\sharp(Support_S(first_k(v))) = \sharp(Support_S(\lceil first_k(v) \rceil_S))$ and $\sharp(Support_S(v) \setminus \{(i,1) \mid 1 \leq i \leq n\}) = \sharp(Support_S(v)) - \sharp(Support_S(v) \cap \{(i,1) \mid 1 \leq i \leq n\})$, determining if $next(v)$ is left extensible takes constant time. To conclude, as for any initial node v, we can compute in $O(1)$ its set of successors $Succ(v)$, its right extensibility, and the left extensibility of its single successor, we obtain a complexity that is linear in the size of DBG_k^+, since each successor is accessed only once. This yields Theorem 2.

Theorem 2. *For a set of words S, building the Contracted de Bruijn Graph of order k, $CDBG_k^+$ takes linear time and space in $|T|$ or in $\|S\|$.*

5   dBG and CdBG from Suffix Array

For lack of space the reader is referred to [4] for the full details.

Theorem 3. *The dBG of order k, $CDBG_k^+$, for a set of words S can be built in a time and space that are linear in $\|S\|$ using the generalised suffix array of S.*

6   Dynamically Updating the Order of DBG^+

Genome assembly from short reads requires to test multiple values of k for the dBG. Indeed, the presence of genomic repeats, makes some order k appropriate to assemble non repetitive regions, and larger orders necessary to disentangle (at least some) repeated regions. Combining assemblies obtained from DBG_k^+ for successive values of k is the key of IDBA assembler, but the dBG is rebuilt for each value [13]. Other tools also exploit this idea [2]. It is thus interesting to dynamically change the order of the dBG. Here, we argue that starting the construction from an index instead of the raw sequences ease the update. On page 94, we mention which information are needed in general to build DBG_k^+. Assume the words are indexed in a suffix tree T (as in Section 4.2). Consider changing k to $k-1$. First, only the nodes of $Init_{S,k}$ whose parent represents a word of length $k-1$ are substituted by their parent in DBG_{k-1}^+, all other nodes remain unchanged. Thus, any arc of order k either stays as such or has some of its endpoints shifted toward the parent node in T. In any case, updating an arc depends only on the nature of its nodes in DBG_{k-1}^+ (whether they belong to $Init_{S,k-1}$ or $InitExact_{S,k-1}$), and can be computed in constant time.

The same situation arises when changing k to $k + 1$. First, only nodes of $InitExact_{S,k}$ change in DBG^+_{k+1}: they are substituted by their children. Updating an arc also depends on the nature of its nodes: it can create a fork towards the children of the destination node if the latter changes, or it can be multiplied and join each children of the source to one children of the destination if both nodes change. Then, the label of the children in T indicate which children to connect to. It can be seen that updating from DBG^+_k to DBG^+_{k+1} in either direction takes linear time in the size of T. Moreover, as updating the support of nodes in T is straightforward, we can readily apply the contraction algorithm to obtain $CDBG^+_{k+1}$ (see Section 4.3).

References

1. Apostolico, A.: The myriad virtues of suffix trees. In: Apostolico, A., Galil, Z. (eds.) Combinatorial Algorithms on Words. NATO Advanced Science Institutes, Series F, vol. 12, pp. 85–96. Springer (1985)
2. Bankevich, A., Nurk, S., Antipov, D., Gurevich, A.A., et al.: SPAdes: a new genome assembly algorithm and its applications to single-cell sequencing. Journal of Computational Biology 19(5), 455–477 (2012)
3. Bowe, A., Onodera, T., Sadakane, K., Shibuya, T.: Succinct de Bruijn Graphs. In: Raphael, B., Tang, J. (eds.) WABI 2012. LNCS, vol. 7534, pp. 225–235. Springer, Heidelberg (2012)
4. Cazaux, B., Lecroq, T., Rivals, E.: From Indexing Data Structures to de Bruijn Graphs. Technical report, lirmm-00950983 (February 2014)
5. Chikhi, R., Limasset, A., Jackman, S., Simpson, J., Medvedev, P.: On the representation of de Bruijn graphs. ArXiv e-prints (January 2014)
6. Chikhi, R., Rizk, G.: Space-efficient and exact de Bruijn graph representation based on a Bloom filter. Algorithms for Molecular Biology 8, 22 (2013)
7. Conway, T.C., Bromage, A.J.: Succinct data structures for assembling large genomes. Bioinformatics 27(4), 479–486 (2011)
8. de Bruijn, N.: On bases for the set of integers. Publ. Math. Debrecen 1, 232–242 (1950)
9. Gusfield, D.: Algorithms on strings, trees and sequences: computer science and computational biology. Cambridge University Press, Cambridge (1997)
10. Manber, U., Myers, G.: Suffix arrays: a new method for on-line string searches. SIAM J. Comput. 22(5), 935–948 (1993)
11. Onodera, T., Sadakane, K., Shibuya, T.: Detecting superbubbles in assembly graphs. In: Darling, A., Stoye, J. (eds.) WABI 2013. LNCS, vol. 8126, pp. 338–348. Springer, Heidelberg (2013)
12. Pell, J., Hintze, A., Canino-Koning, R., Howe, A., Tiedje, J., Brown, C.: Scaling metagenome sequence assembly with probabilistic de Bruijn graphs. Proc. Natl Acad. Sci. USA 109(33), 13272–13277 (2012)
13. Peng, Y., Leung, H.C.M., Yiu, S.M., Chin, F.Y.L.: IDBA – A Practical Iterative de Bruijn Graph De Novo Assembler. In: Berger, B. (ed.) RECOMB 2010. LNCS, vol. 6044, pp. 426–440. Springer, Heidelberg (2010)
14. Pevzner, P., Tang, H., Waterman, M.: An Eulerian path approach to DNA fragment assembly. Proc. Natl Acad. Sci. USA 98(17), 9748–9753 (2001)

15. Rødland, E.A.: Compact representation of k-mer de Bruijn graphs for genome read assembly. BMC Bioinformatics 14, 313 (2013)
16. Salmela, L.: Correction of sequencing errors in a mixed set of reads. Bioinformatics 26(10), 1284–1290 (2010)
17. Simpson, J.T., Durbin, R.: Efficient construction of an assembly string graph using the FM-index. Bioinformatics 26(12), i367–i373 (2010)

Randomized and Parameterized Algorithms for the Closest String Problem

Zhi-Zhong Chen[1], Bin Ma[2], and Lusheng Wang[3]

[1] Division of Information System Design, Tokyo Denki University, Hatoyama,
Saitama 350-0394, Japan
zzchen@mail.dendai.ac.jp
[2] School of Computer Science, University of Waterloo, 200 University Ave. W,
Waterloo, ON, Canada N2L3G1
binma@uwaterloo.ca
[3] Department of Computer Science, City University of Hong Kong, Tat Chee
Avenue, Kowloon, Hong Kong SAR
cswangl@cityu.edu.hk

Abstract. Given a set $S = \{s_1, s_2, \ldots, s_n\}$ of strings of equal length L
and an integer d, the *closest string problem* (CSP) requires the computation of a string s of length L such that $d(s, s_i) \leq d$ for each $s_i \in S$,
where $d(s, s_i)$ is the Hamming distance between s and s_i. The problem
is NP-hard and has been extensively studied in the context of approximation algorithms and parameterized algorithms. Parameterized algorithms provide the most practical solutions to its real-life applications
in bioinformatics. In this paper we develop the first randomized parameterized algorithms for CSP. Not only are the randomized algorithms
much simpler than their deterministic counterparts, their expected-time
complexities are also significantly better than the previously best known
(deterministic) algorithms.

1 Introduction

Given a set $S = \{s_1, s_2, \ldots, s_n\}$ of strings of equal length L and an integer d
(called *radius*), the *closest string problem* (CSP) requires the computation of a
string s of length L such that $d(s, s_i) \leq d$ for each $s_i \in S$, where $d(s, s_i)$ is the
Hamming distance between s and s_i. Such a string s is referred to as a *center
string* of S with radius d.

CSP has attracted great attention in recently years due to its important applications in bioinformatics [18]. For example, one needs to solve numerous CSP instances over a binary alphabet in order to find the approximate gene clusters using
the Center Gene Cluster model [1,15]. Degenerated Primer Design [30] also involves
to solve CSP instances over the DNA alphabet. Other applications include universal PCR primer design [19,17,8,26,13,30], genetic probe design [17], antisense drug
design [17,7], finding unbiased consensus of a protein family [3], and gene regulatory
motif finding [17,13,28,6,10], etc. Consequently, CSP has been extensively studied
in computational biology [17,18,21,14,24,13,23,16,9,12,27,7,25,28]. In particular,
CSP has been proved to be NP-hard [11,17].

A.S. Kulikov, S.O. Kuznetsov, and P. Pevzner (Eds.): CPM 2014, LNCS 8486, pp. 100–109, 2014.

One approach to CSP is to design approximation algorithms. Along this line, Lancto *et al.* [17] presented the first non-trivial approximation algorithm for CSP, which achieves a ratio of $\frac{4}{3}$. Li *et al.* [18] designed the first polynomial-time approximation scheme (PTAS) for CSP. Subsequently, the time complexity of the PTAS was improved in [21,20]. However, the best-known PTAS in [20] has time complexity $\mathcal{O}(mn^{\mathcal{O}(\epsilon^{-2})})$ which is prohibitive for even a moderately small $\epsilon > 0$.

A more practical approach to CSP is via parameterized algorithms. Parameterized algorithms for CSP are based on the observation that the radius d in a practical instance of CSP is usually reasonably small and hence an algorithm with time complexity $\mathcal{O}(f(d) \times poly(n))$ for a polynomial function $poly(n)$ and exponential function $f(d)$ may still be acceptable. Along this line, Stojanovic *et al.* [27] designed a linear-time algorithm for the special case of CSP where d is fixed to 1. Gramm *et al.* [14] proposed the first parameterized algorithm for CSP, which runs in $\mathcal{O}(nL + nd \cdot (d + 1)^d)$ time. Ma and Sun [20] designed an algorithm that runs in $\mathcal{O}(nL + nd \cdot (16(|\Sigma| - 1))^d)$ time. This algorithm is the first polynomial-time algorithm for the special case of CSP where d is logarithmic in the input size and the alphabet size $|\Sigma|$ is a constant. Improved algorithms for CSP along this line were given in [29,5,31,4]. Among them, the algorithm with the best theoretical time complexity for general alphabets is given in [5]. For small alphabets, the best time complexity is achieved by the algorithm in [4]. In particular, this algorithm runs in $\mathcal{O}(nL + nd^3 \cdot 6.731^d)$ time for binary strings, while runs in $\mathcal{O}(nL + nd \cdot 13.183^d)$ time for DNA strings. Noticeably, in order to achieve better time complexity, these best-performing algorithms combined multiple techniques, which made the algorithms rather complicated.

Randomization has been widely employed to design parameterized algorithms for many NP-hard problems [22]. However, randomization has not been used to design parameterized algorithms for CSP, and it is unclear if randomization will be of any benefit at all to solving CSP exactly. The only randomized algorithm that we are aware of is a randomized heuristic algorithm for the binary case of CSP proposed by Boucher and Brown [2]. With large synthetic as well as real-genomic data, they demonstrated the heuristic algorithm could detect motifs efficiently. However, no theoretical bounds on the running time or the success probability were provided.

In this paper, we demonstrate that randomization indeed helps design much simpler and more efficient parameterized algorithms for CSP. Several randomized algorithms are proposed. The first algorithm is presented in Section 3 and is for the binary case of CSP. The algorithm is as simple as the following: It starts with a string t that is initialized to s_1. At each iteration it selects an s_i with $d(t, s_i) > d$ and randomly flips one bit of t where s_i disagrees with t. If a center string is not found within d iterations, the algorithm starts over again. This algorithm for binary case uses very similar heuristic as in [2]. However, the procedure to apply the heuristic, as well as the start and end conditions are changed in order to achieve the theoretical bounds proved in this paper. Through rigorous analysis, we show that for any given binary CSP instance, this surprisingly simple algorithm finds a correct center string in $\mathcal{O}(nL + n\sqrt{d} \cdot (2e)^d) \approx \mathcal{O}(nL + n\sqrt{d} \cdot 5.437^d)$ expected time, where e is the base of the natural logarithm. This time bound is significantly better than the bound $\mathcal{O}(nL + nd^3 \cdot 6.731^d)$ achieved by the previously best known (deterministic) algorithm for CSP [4].

The algorithm is then extended in the rest of the paper to solve the general-alphabet case and to provide better time complexity. More specifically, the algorithm is slightly changed to solve the general-alphabet case in Section 4: Instead of flipping the bit at a randomly selected position of t, the letter at that position is changed to a letter randomly selected from the alphabet according to a carefully designed probability distribution. As a result, we show that CSP can be solved in $\mathcal{O}(nL + n\sqrt{d} \cdot (e\sigma)^d)$ expected time, where σ is the size of the given alphabet. For DNA strings where $\sigma = 4$, this new bound is $\mathcal{O}(nL + n\sqrt{d} \cdot (10.874)^d)$, which is significantly better than $\mathcal{O}(nL + nd \cdot 13.183^d)$ achieved by the best known (deterministic) algorithm [4].

In Section 5, the algorithm is further improved by a simple strategy that avoids repeated changes at the same position of the candidate center string t. Also, in each iteration the selection of s_i maximizes $d(t, s_i)$. We show that with these small changes, the expected time complexity of the algorithm is reduced to $\mathcal{O}(nL + n\sqrt{d} \cdot (2.5\sigma)^d)$. Therefore, for binary and DNA strings, the bounds are $\mathcal{O}(nL + n\sqrt{d} \cdot 5^d)$ and $\mathcal{O}(nL + n\sqrt{d} \cdot 10^d)$, respectively.

Noticing that the time complexity $\mathcal{O}(nL + n\sqrt{d} \cdot (2.5\sigma)^d)$ is not better than the algorithm in [5] for large σ, two different strategies are introduced in Sections 6 and 7 to specifically deal with large alphabets. The algorithm in Section 6 runs in $\mathcal{O}(nL + n\sqrt{d} \cdot (2\sigma + 4)^d)$ expected time. This provides better time complexity than the previously best algorithm in [5] for large σ. For example, for protein strings ($\sigma = 20$), the new algorithm runs in $\mathcal{O}(nL + n\sqrt{d} \cdot 44^d)$ expected time while the algorithm in [5] runs in $\mathcal{O}(nL + nd \cdot 47.21^d)$ time. The algorithm in Section 7 has even better time complexity for large σ. However, the resulting time bound of our analysis is not a closed-form expression. Via numerical computation, we show that the algorithm runs in $\mathcal{O}(nL + nd^2 \cdot 9.81^d)$ and $\mathcal{O}(nL + nd^2 \cdot 40.1^d)$ expected time for DNA and protein strings, respectively.

2   Notations

In this paper, a string is over an alphabet with a fixed size $\sigma \geq 2$. For each positive integer k, $[1..k]$ denotes the set $\{1, 2, \ldots, k\}$. For a string s, $|s|$ denotes the length of s. For each $i \in [1..|s|]$, $s[i]$ denotes the letter of s at its i-th position. Thus, $s = s[1]s[2] \ldots s[|s|]$. A position set of a string s is a subset of $[1..|s|]$. For two strings s and t of the same length, $d(s, t)$ denotes their Hamming distance.

Two strings s and t of the same length L *agree* (respectively, *differ*) at a *position* $i \in [1..L]$ if $s[i] = t[i]$ (respectively, $s[i] \neq t[i]$). The *position set where s and t agree* (respectively, *differ*) is the set of all positions $i \in [1..L]$ where s and t agree (respectively, differ). The following special notations will be very useful. For two strings s_1, s_2 of the same length, $\{s_1 \equiv s_2\}$ (respectively, $\{s_1 \not\equiv s_2\}$) denotes the set of all positions where s_1 and s_2 agree (respectively, differ). Moreover, for three strings s_1, s_2, s_3 of the same length, $\{s_1 \not\equiv s_2 \not\equiv s_3\}$ denotes the position set $\{s_1 \not\equiv s_2\} \cap \{s_1 \not\equiv s_3\} \cap \{s_2 \not\equiv s_3\}$, while $\{s_1 \equiv s_2 \not\equiv s_3\}$ denotes the position set $\{s_1 \equiv s_2\} \cap \{s_2 \not\equiv s_3\}$.

3 Randomized Algorithm for Binary Alphabets

To get familiar with the techniques used in this paper, we start with the binary case of the problem in this section.

Most parameterized algorithms for CSP are based on the bounded search tree method. These algorithms start with a suboptimal solution t, and gradually change t to the center string by altering the letters at certain positions of t. At each step, if $d(t, s_i) > d$ for an input string s_i, then at least one of the positions in $\{t \not\equiv s_i\}$ needs to be changed. Different strategies for choosing the position (or positions) to change lead to very different time complexities. Gramm et al. [14] exhaustively tried every position in a size-$(d+1)$ subset of $\{t \not\equiv s_i\}$, resulting in the first parameterized algorithm for CSP with time complexity $\mathcal{O}\left(nL + nd \cdot (d+1)^d\right)$. Ma and Sun [20] instead tried every subset of $\{t \not\equiv s_i\}$ with bounded size and changed the positions in a subset all at once, which led to an $\mathcal{O}\left(nL + nd \cdot (16\sigma - 1)^d\right)$ algorithm with a very nontrivial proof. Boucher and Brown [2] proposed a seemingly simple strategy to choose a position from $\{t \not\equiv s_i\}$ uniformly at random. The resulting heuristic algorithm in their paper deviates from the bounded search tree scheme since it can alter the original t more than d times. Although there was no theoretical proof about the better performance of the heuristic algorithm, it worked efficiently on simulated data. Here, we apply the same strategy under the bounded search tree method in Algorithm 1, and show that such a randomized strategy is not only simpler, but also provides a cleaner proof and improves the time complexity.

Algorithm 1. BoundedGuess-Binary

Input: Strings $s_1, \ldots, s_n$ each of length L, and a nonnegative integer d.
Output: A center string of $\{s_1, \ldots, s_n\}$ with radius d if there is any, and "no" otherwise.
1 Let $t = s_1$ and $\Delta = d$.
2 While $\Delta > 0$, perform the following steps:
  2.1 If $d(t, s_i) \leq d$ for all $1 \leq i \leq n$, output t and halt.
  2.2 Find a string s_i among the input strings such that $d(t, s_i) > d$.
  2.3 Select a position p in $\{t \not\equiv s_i\}$ uniformly at random and flip the bit of t at position p.
  2.4 Decrease Δ by 1.
3 Output "no" and halt.

By analyzing BoundedGuess-Binary, we can prove that the algorithm outputs a correct center string with probability at least $\frac{d!}{(2d)^d} \approx \sqrt{2\pi d}\left(\frac{1}{2e}\right)^d$. Thus, by repeating the algorithm $\mathcal{O}\left(\frac{(2e)^d}{\sqrt{d}}\right)$ times, the center string can be computed with high probability. Thus, the following theorem can be proved. The details of the proof are given in the full version of this paper.

Theorem 1. *The binary case of the closest string problem can be solved in* $\mathcal{O}\left(nL + n\sqrt{d} \cdot (2e)^d\right) = \mathcal{O}\left(nL + n\sqrt{d} \cdot 5.437^d\right)$ *expected time.*

The time bound in Theorem 1 is better than $\mathcal{O}\left(nL + nd^3 \cdot 6.731^d\right)$, which is the best time bound achieved by the fastest deterministic algorithm for the binary case [4].

4   Randomized Algorithm for General Alphabets

In this section, we extend the algorithm in Section 3 to the general case. In Step 2.3 of Algorithm 1, a randomly selected $p \in \{t \not\equiv s_i\}$ has a good chance to be such that $t[p]$ is different from the center string. Consequently, in the binary case, changing $t[p]$ to $s_i[p]$ will make t one step closer to the center string. However, when the alphabet size is greater than 2, this is not the optimal strategy any more, because $s_i[p]$ can be either equal to or different from the center string. The algorithm has to carefully bet on one of the two cases during the search. Thus, we modify Step 2.3 in Algorithm 1 as follows:

Algorithm 2. BoundedGuess

2.3 Select a position p in $\{t \not\equiv s_i\}$ uniformly at random, change $t[p]$ to $s_i[p]$ with probability $\frac{2}{\sigma}$ while to each of the other $\sigma - 2$ letters with probability $\frac{1}{\sigma}$.

By analyzing BoundedGuess, we can prove the next theorem. The details are given in the full version of this paper.

Theorem 2. *The closest string problem can be solved in* $\mathcal{O}\left(nL + n\sqrt{d} \cdot (\sigma e)^d\right)$ *expected time.*

For DNA strings ($\sigma = 4$), the time bound $\mathcal{O}\left(nL + n\sqrt{d} \cdot (\sigma e)^d\right)$ becomes $\mathcal{O}\left(nL + n\sqrt{d} \cdot 10.874^d\right)$ and is better than $\mathcal{O}\left(nL + nd \cdot 13.183^d\right)$, which is the best bound achieved by the fastest deterministic algorithm for the problem [4]. However, for large σ (say, $\sigma = 20$), the time bound in [5] is better.

5   An $\mathcal{O}(nL + n\sqrt{d} \cdot (2.5\sigma)^d)$ Time Algorithm

In this section, we obtain a more efficient algorithm by refining BoundedGuess in Section 4.

First, a close inspection reveals that we do not need to modify a position of t again once it has been modified. Thus, if we record the already modified positions with a set F, and avoid those positions in later steps, the algorithm's time complexity may be reduced. More specifically, we can augment Step 1 by also initializing $F = \emptyset$, modify Step 2.3 by selecting p from $\{t \not\equiv s_i\} \setminus F$ instead of from $\{t \not\equiv s_i\}$, and augment Step 2.4 by also adding p to F. A crucial observation is that if we find an s_i in Step 2.2 so that $d(t, s_i)$ is maximized, then we can prove that $\{t \not\equiv s_i\} \setminus F$ is significantly smaller than $\{t \not\equiv s_i\}$. Consequently, the probability of selecting a correct position in Step 2.3 is increased. This will be proved in the full version of this paper.

Secondly, if $d(t, s_i) - d \geq 2$ for the string s_i found in Step 2.2, then we still have $d(t, s_i) > d$ after modifying only one position of t in Step 2.3 and hence the same s_i can be used in Step 2.3 during the next iteration of the while-loop. More generally, if $d(t, s_i) - d = \ell$ for the string s_i found in Step 2.2, then we can use s_i in Step 2.3 during ℓ consecutive iterations of the while-loop.

Based on the above observations, we now obtain a new algorithm as follows:

Algorithm 3. NonRedundantGuess

1 Let $t = s_1$, $\Delta = d$, and $F = \emptyset$.
2 While $\Delta > 0$, perform the following steps:
    2.1 If $d(t, s_i) \leq d$ for every input string s_i, then output t and halt.
    2.2 Find a string s_i among the input strings such that $d(t, s_i)$ is maximized.
    2.3 Compute $\ell = d(t, s_i) - d$.
    2.4 Select a set R of ℓ positions in $\{t \not\equiv s_i\} \setminus F$ uniformly at random.
    2.5 For each $p \in R$, change $t[p]$ to $s_i[p]$ with probability $\frac{2}{\sigma}$; and to each of the other $\sigma - 2$ letters with probability $\frac{1}{\sigma}$.
    2.6 Decrease Δ by ℓ and add the positions in R to F.
3 Output "no" and halt.

By analyzing NonRedundantGuess, we can prove the next theorem and corollary. The details are given in the full version of this paper.

Theorem 3. *The closest string problem can be solved in expected time*

$$\mathcal{O}\left(nL + n\sqrt{d} \cdot \left(\frac{2^{1+\epsilon}\sigma}{(1+\epsilon)^{\frac{1+\epsilon}{2}}(1-\epsilon)^{\frac{1-\epsilon}{2}}} \right)^d \right), \text{ where } \epsilon = \ell_1/d \text{ and } \ell_1 \text{ is the integer } \ell$$

obtained in Step 2.3 during the 1st iteration of the while-loop.

Corollary 1. *The closest string problem can be solved in* $\mathcal{O}\left(nL + n\sqrt{d} \cdot (2.5\sigma)^d \right)$ *expected time.*

The time bound in Corollary 1 is better than that in Theorem 2.

6 More Efficient Algorithm for Large Alphabets

When $\sigma < 16$, the time complexity in Corollary 1 is better than the previously best known algorithms [5,4]. However, for large σ (such as $\sigma = 20$ for protein

sequences), the algorithm in [5] is still better. In this section, we refine NonRedundantGuess to obtain a new algorithm (named *LargeAlphabet1*) that has a
time complexity $\mathcal{O}(nL + n\sqrt{d} \cdot (2\sigma + 4)^d)$. This time complexity is better than
the one of Corollary 1 for $\sigma \geq 9$. Also, the new time complexity is faster than
the algorithm in [5] for reasonably large alphabet sizes (such as $\sigma = 20$).

When the alphabet size is large, the most expensive factor in the time complexity in Corollary 1 is the σ^d. This factor arises from the fact that for each
position that needs to be changed in Step 2.5 of NonRedundantGuess, we need
to guess the letter from $\sigma - 1$ possibilities. And in total there can be as many
as d such guessing events. However, in the first iteration of the while-loop of
NonRedundantGuess, there can be a large number of positions $p \in \{t \not\equiv s_i\}$
such that $t[p]$ needs to be changed to $s_i[p]$. Denote the set of these positions by
B. For the positions in B, the algorithm actually does not need to guess the
letters. Moreover, by Lemma 3.1 in [5], $|B| \geq \ell_1$. Based on this fact, we obtain
LargeAlphabet1 by modifying Step 2.5 of NonRedundantGuess as follows:

Algorithm 4. LargeAlphabet1

2.5 If $\Delta = d$, then change $t[p]$ to $s_i[p]$ for each $p \in R$. Otherwise, for each
$p \in R$, change $t[p]$ to $s_i[p]$ with probability $\frac{2}{\sigma}$; and to each of the other
$\sigma - 2$ letters with probability $\frac{1}{\sigma}$.

By analyzing LargeAlphabet1, we can prove the next theorem and corollary.
The details are given in the full version of this paper.

Theorem 4. *The closest string problem can be solved in expected time*
$$\mathcal{O}\left(nL + n \cdot \frac{1}{\sqrt{\epsilon(1-\epsilon)}} \cdot \left(\frac{2^{1+\epsilon}\sigma^{1-\epsilon}}{\epsilon^\epsilon(1-\epsilon)^{1-\epsilon}}\right)^d\right), \text{ if } 0 < \epsilon = \ell_1/d < 1.$$

Corollary 2. *The closest string problem can be solved in* $\mathcal{O}\left(nL + n\sqrt{d} \cdot (2\sigma + 4)^d\right)$
expected time.

Obviously, when $\sigma \geq 9$, the time bound in Corollary 2 is better than that in
Corollary 1. In particular, for protein strings (i.e., when $\sigma = 20$), the time
bound in Corollary 2 becomes $\mathcal{O}\left(nL + n\sqrt{d} \cdot 44^d\right)$, which is better than the
best time bound $\mathcal{O}\left(nL + nd \cdot 47.21^d\right)$ achieved by a deterministic algorithm for
the problem [5].

Since the value of $\epsilon = \ell_1/d$ in Theorems 3 and 4 is known at the very beginning
of NonRedundantGuess and LargeAlphabet1, one can choose between the two
algorithms depending on which of the two bounds in Theorems 3 and 4 is smaller.
For $\sigma \geq 5$, numerical computation shows that this combined algorithm has
a better time complexity than the bounds proved in Corollaries 1 and 2. For
example, when $\sigma = 20$, we can show that the combined algorithm has time
complexity $\mathcal{O}\left(nL + n\sqrt{d} \cdot 43.54^d\right)$.

7 More Efficient Algorithm for Nonbinary Alphabets

In this section, we refine NonRedundantGuess in a different way to obtain a new algorithm (named *LargeAlphabet2*). The improvement has two consequences. First, LargeAlphabet2 runs faster than the algorithm in [5] for every $\sigma \geq 2$; whereas LargeAlaphabet1 developed in the previous section has a higher complexity than the algorithm in [5] for very large alphabets ($\sigma > 36$). Secondly, we will show that for $\sigma \geq 3$, the combination of NonRedundantGuess and LargeAlphabet2 has a better time bound than all other bounds obtained in this paper and the algorithm in [5]. However, the only drawback of LargeAlphabet2 is that we are unable to obtain a closed-form expression for its time complexity.

We inherit the notations from Section 5. Recall that the idea behind LargeAlphabet1 is as follows: In the first iteration of the while-loop, we first randomly guess ℓ_1 positions from $\{t \not\equiv s_i\}$ and then change the letter of t at each guessed position p to $s_i[p]$. The idea behind LargeAlphabet2 is to randomly guess more in the first iteration. More specifically, in the first iteration, we guess $B = \{s_i \equiv s \not\equiv t\}$, $H = \{t \not\equiv s_i \not\equiv s\}$, and $s[p]$ for each $p \in H$. The crucial point is that after we change $t[p]$ to $s[p]$ for each $p \in B \cup H$ in the first iteration, we can decrease Δ down to $\min\{d - |B| - |H|, |B| - \ell_1\}$ (instead of down to $d - |B| - |H|$). This follows from Lemma 3.1 in [5].

Based on the discussion in the last paragraph, we obtain LargeAlphabet2 from NonRedundantGuess by replacing Steps 2.4 through 2.6 with the next step:

Algorithm 5. LargeAlphabet2

2.4 If $\Delta < d$, then perform Steps 2.4, 2.5, and 2.6 of NonRedundantGuess. Otherwise, perform the following three steps:

2.4.1 Select an integer $h \in \{0, 1, \ldots, d - \ell\}$, a subset H of $\{t \not\equiv s_i\}$ with $|H| = h$, and a subset B of $\{t \not\equiv s_i\} \setminus H$ with $\ell \leq |B| \leq d - h$ all uniformly at random.

2.4.2 Let $R = B \cup H$. For each $p \in R$, change $t[p]$ to $s_i[p]$ if $p \in B$, while change $t[p]$ to one of the $\sigma - 2$ letters not equal to $s_i[p]$ uniformly at random if $p \in H$.

2.4.3 Set $\Delta = \min\{d - |R|, |B| - \ell\}$ and $F = R$. [1]

By analyzing LargeAlphabet2, we can prove the next theorem. The details are given in the full version of this paper.

Theorem 5. *The closest string problem can be solved in* $\mathcal{O}\left(nL + nd^2 \cdot \left(\frac{1}{\gamma}\right)^d\right)$ *expected time, where* $\gamma = \dfrac{(1-\epsilon)^{\frac{1-\epsilon}{2}}(1+\epsilon)^{1+\epsilon}}{(3+\epsilon)^{\frac{3+\epsilon}{2}}}$ *if* $\sigma = 2$; *otherwise*

---

[1] Instead of setting $F = R$, we can set $F = \{t \not\equiv s_i\}$. This change can only speed up the algorithm. The reason for setting $F = R$ is for the clarity of the proof.

$$\gamma = \min_{0 \le \alpha \le 1-\epsilon} \frac{1}{(\sigma-2)^\alpha} \left(\frac{1-\epsilon-\alpha}{\sigma} \right)^{\frac{1-\epsilon-\alpha}{2}} \frac{2^{\frac{1-\epsilon+\alpha}{2}} \alpha^\alpha (1+\epsilon-\alpha)^{(1+\epsilon-\alpha)}}{(3+\epsilon-\alpha)^{\frac{3+\epsilon-\alpha}{2}}}.$$

Although it is difficult to find a closed-form expression for the value of $\frac{1}{\gamma}$, we can perform numerical computation to obtain an approximate upper bound on $\frac{1}{\gamma}$ for any given σ. Note that the time bound in Theorem 5 is not as good as those in Corollaries 1 and 2 for small σ (such as $\sigma = 4$) but is significantly better for large σ (say, $\sigma = 20$ or 50). Furthermore, numerical computation also shows that for every $\sigma \ge 2$, the bound in Theorem 5 is better than the time bound in [5].

For $\sigma \ge 3$, numerical computation shows that the smaller bound between the two in Theorems 3 and 5 is better than the bounds in Corollaries 1 and 2 and Theorem 5. For example, when $\sigma = 4$ (respectively, $\sigma = 20$), we can show that the smaller bound between the two in Theorems 3 and 5 is $\mathcal{O}\left(nL + nd^2 \cdot 9.81^d\right)$ (respectively, $\mathcal{O}\left(nL + nd^2 \cdot 40.1^d\right)$).

References

1. Böcker, S., Jahn, K., Mixtacki, J., Stoye, J.: Computation of median gene clusters. Journal of Computational Biology 16(8), 1085–1099 (2009)
2. Boucher, C., Brown, D.G.: Detecting motifs in a large data set: Applying probabilistic insights to motif finding. In: Rajasekaran, S. (ed.) BICoB 2009. LNCS, vol. 5462, pp. 139–150. Springer, Heidelberg (2009)
3. Ben-Dor, A., Lancia, G., Perone, J., Ravi, R.: Banishing bias from consensus sequences. In: Hein, J., Apostolico, A. (eds.) CPM 1997. LNCS, vol. 1264, pp. 247–261. Springer, Heidelberg (1997)
4. Chen, Z.-Z., Ma, B., Wang, L.: A three-string approach to the closest string problem. Journal of Computer and System Sciences 78, 164–178 (2012)
5. Chen, Z.-Z., Wang, L.: Fast exact algorithms for the closest string and substring problems with application to the planted (ℓ, d)-motif model. IEEE/ACM Transactions on Computational Biology and Bioinformatics 8(5), 1400–1410 (2011)
6. Davila, J., Balla, S., Rajasekaran, S.: Space and time efficient algorithms for planted motif search. In: Proc. of the International Conference on Computational Science, pp. 822–829 (2006)
7. Deng, X., Li, G., Li, Z., Ma, B., Wang, L.: Genetic design of drugs without side-effects. SIAM J. Comput. 32(4), 1073–1090 (2003)
8. Dopazo, J., Rodríguez, A., Sáiz, J.C., Sobrino, F.: Design of primers for PCR amplification of highly variable genomes. CABIOS 9, 123–125 (1993)
9. Evans, P.A., Smith, A.D.: Complexity of approximating closest substring problems. In: Proc. of the 14th International Symposium on Foundations of Complexity Theory, pp. 210–221 (2003)
10. Fellows, M.R., Gramm, J., Niedermeier, R.: On the parameterized intractability of motif search problems. Combinatorica 26(2), 141–167 (2006)
11. Frances, M., Litman, A.: On covering problems of codes. Theoret. Comput. Sci. 30, 113–119 (1997)
12. Gramm, J., Guo, J., Niedermeier, R.: On exact and approximation algorithms for distinguishing substring selection. In: Lingas, A., Nilsson, B.J. (eds.) FCT 2003. LNCS, vol. 2751, pp. 195–209. Springer, Heidelberg (2003)

13. Gramm, J., Hüffner, F., Niedermeier, R.: Closest strings, primer design, and motif search. In: Florea, L., et al (eds.), Currents in Computational Molecular Biology. Poster Abstracts of RECOMB 2002, pp. 74–75 (2002)

14. Gramm, J., Niedermeier, R., Rossmanith, P.: Fixed-parameter algorithms for closest string and related problems. Algorithmica 37, 25–42 (2003)

15. Hufsky, F., Kuchenbecker, L., Jahn, K., Stoye, J., Böcker, S.: Swiftly computing center strings. In: Moulton, V., Singh, M. (eds.) WABI 2010. LNCS, vol. 6293, pp. 325–336. Springer, Heidelberg (2010)

16. Jiao, Y., Xu, J., Li, M.: On the k-closest substring and k-consensus pattern problems. In: Sahinalp, S.C., Muthukrishnan, S.M., Dogrusoz, U. (eds.) CPM 2004. LNCS, vol. 3109, pp. 130–144. Springer, Heidelberg (2004)

17. Lanctot, K., Li, M., Ma, B., Wang, S., Zhang, L.: Distinguishing string search problems. Inform. and Comput. 185, 41–55 (2003)

18. Li, M., Ma, B., Wang, L.: On the closest string and substring problems. J. ACM 49(2), 157–171 (2002)

19. Lucas, K., Busch, M., Mösinger, S., Thompson, J.A.: An improved microcomputer program for finding gene- or gene family-specific oligonucleotides suitable as primers for polymerase chain reactions or as probes. CABIOS 7, 525–529 (1991)

20. Ma, B., Sun, X.: More efficient algorithms for closest string and substring problems. SIAM J. Comput. 39(4), 1432–1443 (2010)

21. Marx, D.: Closest substring problems with small distances. SIAM J. Comput. 38(4), 1382–1410 (2008)

22. Marx, D.: Randomized techniques for parameterized algorithms. In: Thilikos, D.M., Woeginger, G.J. (eds.) IPEC 2012. LNCS, vol. 7535, p. 2. Springer, Heidelberg (2012)

23. Mauch, H., Melzer, M.J., Hu, J.S.: Genetic algorithm approach for the closest string problem. In: Proc. of the 2nd IEEE Computer Society Bioinformatics Conference (CSB), pp. 560–561 (2003)

24. Meneses, C.N., Lu, Z., Oliveira, C.A.S., Pardalos, P.M.: Optimal solutions for the closest-string problem via integer programming. INFORMS J. Comput. (2004)

25. Nicolas, F., Rivals, E.: Complexities of the centre and median string problems. In: Baeza-Yates, R., Chávez, E., Crochemore, M. (eds.) CPM 2003. LNCS, vol. 2676, pp. 315–327. Springer, Heidelberg (2003)

26. Proutski, V., Holme, E.C.: Primer master: A new program for the design and analysis of PCR primers. CABIOS 12, 253–255 (1996)

27. Stojanovic, N., Berman, P., Gumucio, D., Hardison, R., Miller, W.: A linear-time algorithm for the 1-mismatch problem. In: Rau-Chaplin, A., Dehne, F., Sack, J.-R., Tamassia, R. (eds.) WADS 1997. LNCS, vol. 1272, pp. 126–135. Springer, Heidelberg (1997)

28. Wang, L., Dong, L.: Randomized algorithms for motif detection. J. Bioinform. Comput. Biol. 3(5), 1039–1052 (2005)

29. Wang, L., Zhu, B.: Efficient algorithms for the closest string and distinguishing string selection problems. In: Deng, X., Hopcroft, J.E., Xue, J. (eds.) FAW 2009. LNCS, vol. 5598, pp. 261–270. Springer, Heidelberg (2009)

30. Wang, Y., Chen, W., Li, X., Cheng, B.: Degenerated primer design to amplify the heavy chain variable region from immunoglobulin cDNA. BMC Bioinform. 7(suppl. 4), S9 (2006)

31. Zhao, R., Zhang, N.: A more efficient closest string algorithm. In: Proc. of the 2nd International Conference on Bioinformatics and Computational Biology (2010)

Indexed Geometric Jumbled Pattern Matching

Stephane Durocher[1,*], Robert Fraser[1], Travis Gagie[2,**],
Debajyoti Mondal[1], Matthew Skala[1], and Sharma V. Thankachan[3]

[1] Department of Computer Science, University of Manitoba, Canada
{durocher, fraser, jyoti, mskala}@cs.umanitoba.ca
[2] Department of Computer Science, University of Helsinki, Finland
gagie@cs.helsinki.fi
[3] Cheriton School of Computer Science, University of Waterloo, Canada
thanks@uwaterloo.ca

Abstract. We consider how to preprocess n colored points in the plane
such that later, given a multiset of colors, we can quickly find an axis-
aligned rectangle containing a subset of the points with exactly those
colors, if one exists. We first give an index that uses $o(n^4)$ space and
$o(n)$ query time when there are $\mathcal{O}(1)$ distinct colors. We then restrict
our attention to the case in which there are only two distinct colors. We
give an index that uses $\mathcal{O}(n)$ bits and $\mathcal{O}(1)$ query time to detect whether
there exists a matching rectangle. Finally, we give a $\mathcal{O}(n)$-space index
that returns a matching rectangle, if one exists, in $\mathcal{O}(\lg^2 n/\lg\lg n)$ time.

1 Introduction

Over the past ten years, researchers have studied online jumbled pattern matching for strings, graphs and, most recently, point sets. Butman et al. [6] showed
how, given a string and multiset of characters, in linear time we can find all
the substrings consisting of a permutation of those characters. The multiset is
usually represented as a Parikh vector of characters, i.e., a list of the distinct
characters' frequencies. Lacroix et al. [16] showed how, given a node-colored tree
and a Parikh vector of colors with a constant number of non-zero entries, in
polynomial time we can find all the connected subgraphs whose nodes have exactly the prescribed colors. Fellows et al. [11] extended this result to graphs with
bounded tree-width and, implicitly, to all graphs when the Parikh vector's L1
norm (i.e., the multiset's size) is at most logarithmic in the size of the graph [3].
Barba et al. [2] showed how, given a set of colored points in the plane and a
Parikh vector of colors, in cubic time we can find all the minimal axis-aligned
rectangles containing subsets of points with exactly the prescribed colors.

Over the past five years, researchers have also studied indexed jumbled pattern
matching for strings and graphs. Cicalese et al. [8] showed how, given a binary
string, in quadratic time we can build a linear-space index such that later, given
frequencies of 0s and 1s, we can detect in constant time whether there exists a

* Supported by the Natural Sciences and Engineering Research Council of Canada.
** Supported by the Academy of Finland.

A.S. Kulikov, S.O. Kuznetsov, and P. Pevzner (Eds.): CPM 2014, LNCS 8486, pp. 110–119, 2014.

matching substring. Subsequent authors [1,5,9,14,17] have reduced the construction time slightly for the general case, to $\mathcal{O}(n^2/\lg^2 n)$, and more significantly for special cases or relaxations. Gagie et al. [13] reduced the space from a linear number of words to a linear number of bits and extended the result to trees whose nodes each have one of two colors, still with $\mathcal{O}(n^2/\lg^2 n)$ construction time.

In this paper we study indexed jumbled pattern matching in point sets. In Section 2 we generalize Kociumaka et al.'s [15] recent proof that, given a string of length n over a $\mathcal{O}(1)$-size alphabet, we can build a $o(n^2)$-space index such that later, given a Parikh vector of characters, in $o(n)$ time we can find a matching substring if one exists. We show that, given a set of n points in the plane each of which is one of $\mathcal{O}(1)$ distinct colors, we can build a $o(n^4)$-space index such that later, given a Parikh vector of colors, in $o(n)$ time we can find one instance of a matching axis-aligned rectangle if one exists. Since these indexes provide an example match, we call them *witnessing* indexes. Notice that, when dealing with trees, graphs or point sets, the total number of matches can be superlinear, so returning only one instance seems reasonable. We will show in the full version of this paper how to derandomize our Las Vegas construction.

In Section 3 we generalize Cicalese et al.'s detection index for binary strings. We show how, given n points in the plane in general position (i.e., with each x- and y-coordinate unique), each of which is one of two distinct colors, in $\mathcal{O}(n^3 \lg n)$ time we can build an index that occupies $\mathcal{O}(n)$ bits such that later, given a Parikh vector of those two colors, in $\mathcal{O}(1)$ time we can detect whether there exists a matching axis-aligned rectangle. As part of our construction, we use a new dynamic detection index for jumbled pattern matching in binary strings, with $\mathcal{O}(n \lg n)$ update time and $\mathcal{O}(1)$ query time. Although $\mathcal{O}(n \lg n)$-time updates may seem slow, we note that $o(n/\lg^2 n)$-time updates would imply a faster way to build static indexes for binary strings than that which is known. Due to space constraints, we leave the description of this dynamic index to full version of this paper, where we will also show that $\mathcal{O}(n)$-bit, $\mathcal{O}(1)$-time detection indexes exist not only for rectangles but for any shape that contains a point from which the entire interior is visible (i.e., star shapes).

In Section 4 we apply a recent technique by Cicalese et al. [10] to turn our detection index for bichromatic point sets into a witnessing index. We show how, given n points in the plane in general position, each of which is one of two distinct colors, in $\mathcal{O}(n^3 \lg n)$ time we can build a $\mathcal{O}(n)$-space index such that later, given a Parikh vector of those two colors, in $\mathcal{O}(\lg^2 n/\lg\lg n)$ time we can return a matching axis-aligned rectangle, if one exists.

2   A Witnessing Index for $\mathcal{O}(1)$ Colors

It is possible to build a $\mathcal{O}(n^2)$-space index with $\mathcal{O}(1)$ query time for jumbled pattern matching in a string over a $\mathcal{O}(1)$-size alphabet, by building a perfect hash table of all its substrings' Parikh vectors; or to store nothing but the string itself and search it in $\mathcal{O}(n)$ time with Butman et al.'s algorithm for each query.

Nevertheless, Kociumaka et al.'s was the first (and, so far as we know, still the only) index for jumbled pattern matching in strings over ternary or larger alphabets, to use simultaneously $o(n^2)$ space and $o(n)$ query time. Specifically, for any alphabet size $\sigma = \mathcal{O}(1)$ and any positive $\epsilon < 1$, we can set their index to use $\mathcal{O}(n^{2-\epsilon})$ space and $\mathcal{O}(m^{(2\sigma-1)\epsilon})$ query time, where $m \leq n$ is the L1 norm of the Parikh vector in the query. Thus, choosing $\epsilon < 1/(2\sigma - 1)$ means we use $o(n^2)$ space and $o(m) = o(n)$ query time.

Similarly, it is possible to build a $\mathcal{O}(n^4)$-space index with $\mathcal{O}(1)$ query time for jumbled pattern matching in a point set, by building a perfect hash table of all the Parikh vectors of subsets that can be enclosed in axis-aligned rectangles; or to store nothing but the point set itself and search it in $\mathcal{O}(n^3)$ time with Barba et al.'s algorithm for each query. Therefore, an obvious starting point is to describe a $o(n^4)$-space index with $o(n^3)$ query time, analogous to Kociumaka et al.'s. This makes sense only for $\sigma \geq 4$ because, for $\sigma = 3$, we can store in $\mathcal{O}(n^3)$ space a perfect hash table of all Parikh vectors with L1 norm at most n.

In fact, for any positive $\epsilon < 1$, we can use $\mathcal{O}(n^{4-\epsilon})$ space and the same query time as Kociumaka et al.'s, $\mathcal{O}(m^{(2\sigma-1)\epsilon})$. Choosing again $\epsilon < 1/(2\sigma - 1)$, therefore, means we use $o(n^4)$ space and $o(n)$ query time.

Theorem 1. *Given a set of n points in the plane each of which is one of a constant number σ of distinct colors, we can store them in $\mathcal{O}(n^{4-\epsilon})$ space for any given positive $\epsilon < 1$ such that later, given a vector $C = (c_1, \ldots, c_\sigma)$ with L1 norm m, in $\mathcal{O}(m^{(2\sigma-1)\epsilon})$ time we can return an axis-aligned rectangle containing exactly c_i points of the ith color, for $1 \leq i \leq \sigma$, if such a rectangle exists.*

Proof. Without loss of generality, assume we are working in rank space (i.e., on an $n \times n$ grid with one point in each row and each column). For the moment, assume we know in advance values b and h such that $b \leq m \leq b + h$. For each axis-aligned rectangle containing between b and $b + h$ points, we call the set of the lowest b points the rectangle's *body* and we call the remaining points its *head*.

Consider all $\mathcal{O}(n^3)$ axis-aligned (not necessarily minimal) rectangles whose bottoms and tops hit points and that contain exactly b points each. There are $\mathcal{O}(n^3)$ such rectangles because, once we have chosen the bottom point and left and right sides of such a rectangle, the fact it contains b points determines which point its top must hit. We say a vector with L1 norm b is *light* if there are between 1 and t such rectangles that match that vector, where t is a threshold we specify later; otherwise, we say the vector is *heavy*.

We store a perfect hash table of the light vectors and with each of those vectors, we store a list of the locations of the $\mathcal{O}(t)$ rectangles matching that vector. This takes a total of $\mathcal{O}(n^3)$ space regardless of t. We also store a $\mathcal{O}(n^2)$-space data structure that, given a 4-sided range, in $\mathcal{O}(1)$ time tells us how many points of each color lie in that range. We store another perfect hash table of all the distinct vectors for rectangles containing between b and $b + h$ points whose bodies have heavy vectors; with each one, we store the location of one rectangle matching that distinct vector. Since there are $\mathcal{O}(n^3/t)$ heavy vectors and there are $\mathcal{O}(h^\sigma)$ possible distinct vectors for heads, this takes a total of $\mathcal{O}(n^3 h^\sigma/t)$ space.

Given C, we consider all $\mathcal{O}(h^{\sigma-1})$ ways to choose two vectors B and H with non-negative entries such that $C = B + H$, B has L1 norm b and H has L1 norm at most h. For each choice of B and H, we check whether B is light and, if so, we run through the list of the locations of the $\mathcal{O}(t)$ rectangles matching B. For each such rectangle, we check whether the $|H|$ points immediately above the rectangle match H; if so, we return the location of the rectangle extended to include those H points, then stop. This takes $\mathcal{O}(t)$ time for each choice of B and H. If B is not light then we search for C in our second perfect hash table, which takes $\mathcal{O}(1)$ time.

Overall, we use $\mathcal{O}(n^3 + n^3 h^\sigma/t)$ space and $\mathcal{O}(h^{\sigma-1}t)$ query time. Setting $t = h^\sigma$, our space becomes $\mathcal{O}(n^3)$ and our query time becomes $\mathcal{O}(h^{2\sigma-1})$. For any upper bound M on m, if we repeat this construction for $b = 0, M^\epsilon, 2M^\epsilon, \ldots, M$ and $h = M^\epsilon$, then we use $\mathcal{O}(n^3 M^{1-\epsilon})$ space and we can answer any query with $m \leq M$ in $\mathcal{O}(M^{(2\sigma-1)\epsilon})$ time. Storing data structures for $M = 1, 2, 4, \ldots, n$ takes a total of $\mathcal{O}(n^{4-\epsilon})$ space and lets us answer any query in $\mathcal{O}(m^{(2\sigma-1)\epsilon})$ time. $\square$

3   A Detection Index for Two Colors

Suppose a binary string contains one substring of length m including a copies of 1, and another of length m including c copies of 1. Then by sliding a window of length m between those two positions, we can always find a substring of length m including b copies of 1 for any b between a and c. Cicalese et al.'s [8] index for jumbled pattern matching in binary strings makes use of that fact. It stores, for each substring length m, the minimum and maximum numbers of 1s to occur in length-m substrings. This $\mathcal{O}(n)$-space data structure answers detection queries in $\mathcal{O}(1)$ time by checking whether the desired number of 1s is between the stored bounds. If it also records where the minimum and maximum counts occur, then it can do a binary search to answer witnessing queries in $\mathcal{O}(\lg n)$ time.

Fici and Lipták [12] observed that the minimum or maximum number of 1s can only stay the same or increment when we increment m. Gagie et al. [13] pointed out that this means we can store the lists of minima and maxima as bitvectors, with 1s indicating increments, and use rank queries to support $\mathcal{O}(1)$ time access, shrinking the detection index to $\mathcal{O}(n)$ bits (rather than words) of space while retaining $\mathcal{O}(1)$ query time.

Gagie et al. also noted that this idea can be generalized to connected graphs, by generalizing the discrete continuity argument used for strings. If in a connected graph with nodes colored black and white there exists a connected subgraph with m nodes of which a are white, and another connected subgraph with m nodes of which c are white, then there must be a connected subgraph with m nodes of which b are white for each b between a and c. A sequence of connected subgraphs each with m nodes serves the same purpose as the sliding window in the string case. Therefore, there exist $\mathcal{O}(n)$-bit, $\mathcal{O}(1)$-query-time detection indexes also for jumbled pattern matching in connected graphs with two colors. Building such indexes is NP-hard in general, because it is NP-complete to determine whether there is a connected subgraph with m nodes of which a

given number are white, but it takes polynomial time for graphs with bounded tree-width.

Even more generally, suppose we have a hypergraph on n nodes in which each node is black or white and, for each pair of hyperedges e and e' with m nodes each, there is a sequence of hyperedges with m nodes each that starts with e and ends with e' and in which each consecutive pair differs on two nodes. Then there exists a $\mathcal{O}(n)$-bit, $\mathcal{O}(1)$-time detection index for jumbled pattern matching in this hypergraph, although it may not be feasible to build it. The construction time depends on how quickly we can determine the minimum and maximum numbers of white nodes in hyperedges of each size.

In the following lemmas we apply that argument to rectangles on the plane. Given a set of n black and white points in general position on the plane, let them be the nodes of a hypergraph whose hyperedges are the subsets of points that can be contained by axis-aligned rectangles. We first show the existence of a sequence of hyperedges with the necessary continuity property from the minimum to maximum number of white points for each size m; that implies the existence of a detection index. We then show how to construct the index efficiently.

Lemma 1. *Given a set of n black and white points in general position in the plane and a pair of axis-aligned rectangles R and R' each containing m points, there exists a sequence of axis-aligned rectangles containing m points each that starts with R and ends with R' and in which each consecutive pair differs on two points.*

Proof. We denote an axis-aligned rectangle by an ordered quadruple (a, b, c, d) where (a, b) are the coordinates of the lower left corner and (c, d) are the coordinates of the upper right corner. Consider a set of n black and white points in general position in the plane and a pair of axis-aligned rectangles $R = (x_1, y_1, x_2, y_2)$ and $R' = (x_1', y_1', x_2', y_2')$ each containing m points. Unless R and R' contain exactly the same subset of points (in which case the lemma holds trivially), neither R nor R' can completely contain the other. Therefore, each rectangle has at least one edge completely outside the other. Up to symmetry, this leaves two cases for us to consider: $x_1 \leq x_1'$ and $y_1' \leq y_1$; or $x_1 \leq x_1'$ and $x_2 \leq x_2'$ and $y_1 \leq y_1' \leq y_2' \leq y_2$.

$\mathbf{x_1 \leq x_1'}$ **and** $\mathbf{y_1' \leq y_1}$**:** Since $R \subseteq (x_1, y_1', x_2, y_2)$, there exists a rectangle (x_1, y_1', x_2, y_2'') with $y_2'' \leq y_2$ that contains exactly m points. Similarly, since $R' \subseteq (x_1, y_1', x_2', y_2')$, there exists a rectangle (x_1, y_1', x_2'', y_2') with $x_2'' \leq x_2'$ that contains exactly m points. Figure 1a shows an example.

If we can construct three sequences of axis-aligned rectangles containing m points each such that each consecutive pair of rectangles differs on two points and

- one sequence S_1 starts with $R = (x_1, y_1, x_2, y_2)$ and ends with (x_1, y_1', x_2, y_2''),
- one sequence S_2 starts with (x_1, y_1', x_2, y_2'') and ends with (x_1, y_1', x_2'', y_2'),
- one sequence S_3 starts with (x_1, y_1', x_2'', y_2') and ends with $R' = (x_1', y_1', x_2', y_2')$,

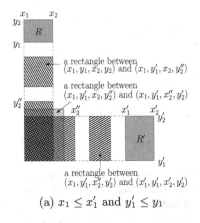

(a) $x_1 \leq x_1'$ and $y_1' \leq y_1$

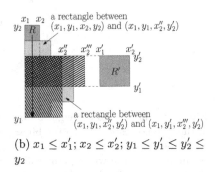

(b) $x_1 \leq x_1'$; $x_2 \leq x_2'$; $y_1 \leq y_1' \leq y_2' \leq y_2$

Fig. 1. Suppose there is a pair of axis-aligned rectangles $R = (x_1, y_1, x_2, y_2)$ and $R' = (x_1', y_1', x_2', y_2')$ each containing m points. We claim there is a sequence of axis-aligned rectangles containing m points each that starts with R and ends with R' and in which each consecutive pair differs on two points.

then by concatenation we can construct another such sequence that starts with R and ends with R'.

To construct S_1, we start with $R = (x_1, y_1, x_2, y_2)$ and alternately decrease the first y-coordinate until it equals y_1' or a point enters the rectangle, then decrease the second y-coordinate until it equals y_2'' or a point leaves the rectangle, and repeat. This way, we keep the number of points within the rectangle the same; since the points are in general position, they enter and leave the rectangle one by one. We construct S_3 similarly.

To construct S_2, we start with (x_1, y_1', x_2, y_2''). Assume $x_2 \leq x_2''$ and $y_2' \leq y_2''$; the other cases are symmetric. We alternately increase the second x-coordinate until it equals x_2'' or a point enters the rectangle, then decrease the second y-coordinate until it equals y_2' or a point leaves the rectangle, and repeat.

$\mathbf{x_1 \leq x_1'}$; $\mathbf{x_2 \leq x_2'}$; $\mathbf{y_1 \leq y_1' \leq y_2' \leq y_2}$: For the sake of brevity, we leave some of the details of this case to the full version of this paper, but Figure 1b shows an illustration. There exist axis-aligned rectangles (x_1, y_1, x_2'', y_2') and $(x_1, y_1', x_2''', y_2')$ with $x_2 \leq x_2'' \leq x_2''' \leq x_2'$ that each contain exactly m points.

If we can construct three sequences of axis-aligned rectangles containing m points each such that each consecutive pair of rectangles differ on two points and

- one sequence S_4 starts with $R = (x_1, y_1, x_2, y_2)$ and ends with (x_1, y_1, x_2'', y_2'),
- one sequence S_5 starts with (x_1, y_1, x_2'', y_2') and ends with $(x_1, y_1', x_2''', y_2')$,
- one sequence S_6 starts with $(x_1, y_1', x_2''', y_2')$ and ends with $R' = (x_1', y_1', x_2', y_2')$,

then by concatenation we can construct another such sequence that starts with R and ends with R'.

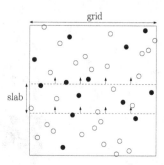

Fig. 2. Sliding a slab of height 9 up a 40×40 grid. In the previous step, the contents of the slab corresponded to the binary string 110011010. One black point left the slab and another entered it, and the contents of the slab currently correspond to 101011010. In the next step, a white point will leave the slab and a black point will enter it, and the contents of the slab will correspond to 101001010.

We construct S_4 and S_5 similarly to how we construct sequence S_2 as described above. We construct S_6 similarly to how we construct S_1 and S_3. Concatenating all these sequences, the lemma follows.   □

Lemma 1 implies the existence of a $\mathcal{O}(n)$-bit space, $\mathcal{O}(1)$-time, detection index for jumbled pattern matching on axis-aligned rectangles with bichromatic points. The index stores the minimum and maximum number of white points using succinct bitvectors, much as in the substring problem. It remains to actually construct that index as quickly as possible. Construction time of $\mathcal{O}(n^4)$ is sufficient by reducing the point set to rank space, building one data structure for $\mathcal{O}(1)$-time range counting for the black points and another for the white points (using $\mathcal{O}(n^2)$ time and space), and checking the number of white and black points in each of the $\mathcal{O}(n^4)$ possible rectangles. We show a tighter bound.

Barba et al. [2] enumerate all possible rectangles by considering slabs of rows in the grid of each height from 1 to n, sweeping each of them up from the bottom to the top of the grid. We avoid the need to consider all rectangles individually by maintaining a dynamic list of the points in the current slab and solving a one-dimensional version of the problem within the slab. Figure 2 shows an example.

Dynamically maintaining the point set in the current slab such that we can find the minimum and maximum numbers of white points in rectangles that cover the vertical extent of the slab, is equivalent to maintaining lists of the minimum and maximum numbers of 1s in substrings of each size in a dynamic binary string subject to insertions and deletions. We use the following lemma, whose proof we defer to the full version.

Lemma 2. *We can maintain a dynamic binary string of up to n bits in $\mathcal{O}(n)$ space such that inserting or deleting a bit takes $\mathcal{O}(n \lg n)$ time and, given m, determining the minimum and maximum numbers of 1s in a substring of length m takes $\mathcal{O}(1)$ time.*

Combining Lemmas 1 and 2 gives the construction time of the detection data structure.

Theorem 2. *Given a set of n points in the plane colored black and white, in $\mathcal{O}(n^3 \lg n)$ time we can build a $\mathcal{O}(n)$-bit index such that later, given a vector $C = (c_1, c_2)$, in $\mathcal{O}(1)$ time we can determine whether there exists an axis-aligned rectangle containing exactly c_1 white points and c_2 black points.*

Proof. By Lemma 1, there exists an axis-aligned rectangle containing exactly c_1 points of the first color and exactly c_2 points of the second color if and only if c_1 is between the minimum and maximum number of white points in all axis-aligned rectangles containing $m = c_1 + c_2$ points. As in the substring data structure, we need only store the minimum and maximum values of c_1 for each value of m, and we can do that in the stated space and time bounds.

To construct the data structure, we consider for each possible coordinate for the bottom of a rectangle, every possible coordinate for the top of the rectangle. That gives us a sequence of $\mathcal{O}(n^2)$ slabs, each differing from the previous one by insertion or deletion of one point. Maintaining the data structure of 2 on the string of bits indicating the colors of points in the current slab, we can find the global minimum and maximum numbers of white points for all rectangles in $\mathcal{O}(n^3 \lg n)$ time, and the result follows.   □

4   A Witnessing Index for Two Colors

The witnessing index of Cicalese et al. [10] for jumbled pattern matching in a binary string is similar to their detection index, but also stores, for every substring length m, the positions of two substrings that witness the minimum and maximum number of 1s, as well as storing the string itself as a bitvector. Then a binary search between the stored extrema gives $\mathcal{O}(\lg n)$ query time on the $\mathcal{O}(n)$-space data structure. At each step of the search they perform $\mathcal{O}(1)$ rank queries on the bit vector to determine the number of 1s in the length-m substring at the current position.

The sequence of rectangles described in Lemma 1 makes possible the same kind of binary search on axis-aligned rectangles in bichromatic point sets. Instead of using a bitvector, however, we use a $\mathcal{O}(n)$-space data structure by Brodal et al. [4] that supports two-dimensional three-sided range-selection queries in $\mathcal{O}(\lg n / \lg \lg n)$ time. Given a three-sided range and an integer k, this data structure returns the kth point in that range, counting away from the middle side. We can use instances of the same data structure to support range counting of the black and white points, with the same query time (although there are faster alternatives [7]).

Theorem 3. *Given a set of n black and white points in the plane, in $\mathcal{O}(n^3 \lg n)$ time we can build in $\mathcal{O}(n^3 \lg n)$ time a $\mathcal{O}(n)$-space index such that later, given a vector $C = (c_1, c_2)$, in $\mathcal{O}(\lg^2 n / \lg \lg n)$ time we can return an axis-aligned rectangle containing exactly c_1 white points and exactly c_2 black points, if one exists.*

Proof. We use the same quadruple notation for rectangles as in the proof of Lemma 1, and as in Theorem 2, we build the data structure in $\mathcal{O}(n^3 \lg n)$ time by examining horizontal slabs to find the global maximum and minimum number of white points for rectangles with each value of $m = c_1 + c_2$. We record the coordinates of the extremal rectangles as well as their white point counts.

We do not include a figure specifically for this proof; instead, we refer the reader back to Figure 1a.

For each m, we record whether $R = (x_1, y_1, x_2, y_2)$ and $R' = (x_1', y_1', x_2', y_2')$ are symmetric to the first case we considered in Lemma 1 (i.e., $x_1 \leq x_1'$ and $y_1' \leq y_1$) or symmetric to the second case (i.e., $x_1 \leq x_1'$ and $x_2 \leq x_2'$ and $y_1 \leq y_1' \leq y_2' \leq y_2$). In this version of this paper, we describe how to deal only with the first case; the second is similar.

Suppose $x_1 \leq x_1'$ and $y_1' \leq y_1$. Then we also record the positions of the intermediate rectangles (x_1, y_1', x_2, y_2'') and (x_1, y_1', x_2'', y_2') described for the this case in Lemma 1. Finally, we record the number of white points in each of R, (x_1, y_1', x_2, y_2''), (x_1, y_1', x_2'', y_2') and R'.

Let S_1, S_2 and S_3 again be the sequences of axis-aligned rectangles containing m points each that we described for this case. We can determine from the number of white points in R, (x_1, y_1', x_2, y_2''), (x_1, y_1', x_2'', y_2') and R' whether we should binary search in S_1, S_2 or S_3. Searching in S_3 is analogous to searching in S_1.

To perform binary search in S_1—i.e., in the sequence of axis-aligned rectangles between $R = (x_1, y_1, x_2, y_2)$ and (x_1, y_1', x_2, y_2'') each containing m points, with $y_1' \leq y_1$—we perform a binary search in the range $[y_1', y_1]$. For the sake of simplicity, we assume we reduce all coordinates to rank space when we build the index, so y_1' and y_1 are integers that differ by at most n.

At each step of the binary search, we choose some integer y_1''' between the endpoints of our current range of integers, as the first y-coordinate of the rectangle we will test. We use a range-selection query to find the mth point from the bottom of the range $(x_1, y_1''', x_2, \infty)$, which tells us the second y-coordinate y_2''' of that test rectangle. We then count the number of white points in $(x_1, y_1''', x_2, y_2''')$, which tells us on which half of the current range we should recurse. We use $\mathcal{O}(\lg n / \lg \lg n)$ time for each step of the binary search, so $\mathcal{O}(\lg^2 n / \lg \lg n)$ time overall.

To perform binary search in S_2—i.e., in the sequence of axis-aligned rectangles between (x_1, y_1', x_2, y_2'') and (x_1, y_1', x_2'', y_2') each containing m points—we perform binary search in the range $[x_2, x_2'']$. Again, we assume $x_2 \leq x_2''$ and $y_2' \leq y_2''$, because the other cases are symmetric.

At each step of the binary search, we choose some integer x_2''' between the endpoints of our current range of integers, as the second x-coordinate of the rectangle we will test. We use a range-selection query on to find the mth point from the bottom of the range $(x_1, y_1', x_2''', \infty)$, which tells us the second y-coordinate y_2''' of that test rectangle. We then count the number of white points in $(x_1, y_1', x_2''', y_2''')$, which tells us on which half of the current range we should recurse. Again, we use $\mathcal{O}(\lg n / \lg \lg n)$ time for each step of the binary search, so $\mathcal{O}(\lg^2 n / \lg \lg n)$ time overall. $\square$

Acknowledgments. Many thanks to the organizers and participants of CCCG 2013 and Stringmasters 2013, especially L. Barba, F. Cicalese, S. Denzumi, M. He, J. Holub, J. Kärkkäinen, A. Kawamura, D. Kempa, T. Kociumaka, Z. Lipták, J. Tarhio and G. Zhou; and to E. Giaquinta, M. Lewenstein, R. Rizzi and A. I. Tomescu.

References

1. Badkobeh, G., Fici, G., Kroon, S., Lipták, Z.: Binary jumbled string matching for highly run-length compressible texts. IPL 113, 604–608 (2013)
2. Barba, L., et al.: On k-enclosing objects in a coloured point set. In: Proc. CCCG, 229–234 (2013)
3. Björklund, A., Kaski, P., Kowalik, L.: Probably optimal graph motifs. In: Proc. STACS, pp. 20–31 (2013)
4. Brodal, G.S., Gfeller, B., Jørgensen, A.G., Sanders, P.: Towards optimal range medians. TCS 412, 2588–2601 (2011)
5. Burcsi, P., Cicalese, F., Fici, G., Lipták, Z.: Algorithms for jumbled pattern matching in strings. IJFCS 23, 357–374 (2012)
6. Butman, A., Eres, R., Landau, G.M.: Scaled and permuted string matching. IPL 92, 293–297 (2004)
7. Chan, T.M., Wilkinson, B.T.: Adaptive and approximate orthogonal range counting. In: Proc. SODA, pp. 241–251 (2013)
8. Cicalese, F., Fici, G., Lipták, Z.: Searching for jumbled patterns in strings. In: Proc. PSC, pp. 105–117 (2009)
9. Cicalese, F., Laber, E., Weimann, O., Yuster, R.: Near linear time construction of an approximate index for all maximum consecutive sub-sums of a sequence. In: Kärkkäinen, J., Stoye, J. (eds.) CPM 2012. LNCS, vol. 7354, pp. 149–158. Springer, Heidelberg (2012)
10. Cicalese, F., Gagie, T., Giaquinta, E., Laber, E.S., Lipták, Z., Rizzi, R., Tomescu, A.I.: Indexes for jumbled pattern matching in strings, trees and graphs. In: Kurland, O., Lewenstein, M., Porat, E. (eds.) SPIRE 2013. LNCS, vol. 8214, pp. 56–63. Springer, Heidelberg (2013)
11. Fellows, M.R., Fertin, G., Hermelin, D., Vialette, S.: Upper and lower bounds for finding connected motifs in vertex-colored graphs. JCSS 77, 799–811 (2011)
12. Fici, G., Lipták, Z.: On prefix normal words. In: Mauri, G., Leporati, A. (eds.) DLT 2011. LNCS, vol. 6795, pp. 228–238. Springer, Heidelberg (2011)
13. Gagie, T., Hermelin, D., Landau, G.M., Weimann, O.: Binary jumbled pattern matching on trees and tree-like structures. In: Bodlaender, H.L., Italiano, G.F. (eds.) ESA 2013. LNCS, vol. 8125, pp. 517–528. Springer, Heidelberg (2013)
14. Giaquinta, E., Grabowski, S.: New algorithms for binary jumbled pattern matching. IPL 113, 538–542 (2013)
15. Kociumaka, T., Radoszewski, J., Rytter, W.: Efficient indexes for jumbled pattern matching with constant-sized alphabet. In: Bodlaender, H.L., Italiano, G.F. (eds.) ESA 2013. LNCS, vol. 8125, pp. 625–636. Springer, Heidelberg (2013)
16. Lacroix, V., Fernandes, C.G., Sagot, M.-F.: Motif search in graphs: Application to metabolic networks. TCBB 3, 360–368 (2006)
17. Moosa, T.M., Rahman, M.S.: Sub-quadratic time and linear space data structures for permutation matching in binary strings. JDA 10, 5–9 (2012)

An Improved Query Time
for Succinct Dynamic Dictionary Matching

Guy Feigenblat[1,2], Ely Porat[1], and Ariel Shiftan[1,*]

[1] Department of Computer Science, Bar-Ilan University, Ramat Gan 52900, Israel
[2] IBM Haifa Research Lab, Haifa University Campus, Haifa 31905, Israel
{feigeng,porately,shiftaa}@cs.biu.ac.il

Abstract. In this work, we focus on building an efficient succinct dynamic dictionary that significantly improves the query time of the current best known results. The algorithm that we propose suffers from only a $O((\log \log n)^2)$ multiplicative slowdown in its query time and a $O(\frac{1}{\epsilon} \log n)$ slowdown for insertion and deletion operations, where n is the sum of all of the patterns' lengths, the size of the alphabet is $polylog(n)$ and $\epsilon \in (0, 1)$. For general alphabet the query time is $O((\log \log n) \log \sigma)$, where σ is the size of the alphabet.

A byproduct of this paper is an Aho-Corasick automaton that can be constructed with only a compact working space, which is the first of its type to the best of our knowledge.

1 Introduction

Traditional pattern matching and dictionary matching are fundamental research domains in computer science. In the pattern matching scenario, a text $T = t_1, ..., t_m$ and a pattern $P = p_1, ..., p_t$ are given, and the text is being searched for all of the occurrences of P in T. In the dictionary matching scenario, alternatively, a set of patterns $P_1, P_2, ..., P_d$ are indexed and given a text T, it is searched for all of the occurrences of the patterns in it. In the static case, the set of patterns is known in advance and cannot be changed, as opposed to the dynamic case, in which patterns can be added, deleted or updated.

Traditional dictionary matching algorithms, both static and dynamic [1–4], requires space, in words, that is proportional to the total length of the patterns stored in the structure. Assuming that the total length of the patterns is n, the space used by these algorithms is $O(n \log n)$; depending on the alphabet, this amount could be up to $O(\log n)$ times more than for the optimal algorithm, if we count the space used in bits rather than words. As the technology evolves and the amount of data that must be indexed becomes larger, researchers become interested in data structures that occupy space that is close to the information-theoretic lower bound without sacrificing the running time drastically. Grossi and Vitter [5] and Ferragina and Manzini [6] were pioneers in this field, and the field has became widely explored since their discoveries.

* This research was supported by the Kabarnit Cyber consortium funded by the Chief Scientist in the Israeli Ministry of Economy under the Magnet Program.

A.S. Kulikov, S.O. Kuznetsov, and P. Pevzner (Eds.): CPM 2014, LNCS 8486, pp. 120–129, 2014.

Suppose that Z is the information-theoretic optimal number of bits that are needed to maintain some data D, and let σ be the size of the alphabet. An index for D is said to be succinct if it requires $Z + o(z)$ bits. It is said to be compact if it requires $O(Z)$ bits and compressed if it requires space proportional to the space of a compressed representation of D, as measured by its zeroth-order entropy, $H_0(T)$, or its k-th-order entropy, $H_k(T)$.

In this work, we focus on building an efficient succinct dynamic dictionary matching algorithm. Our succinct structure suffers from only a $O((\log \log n)^2)$ multiplicative slowdown in the query time which is much better than previous best known results of [7], where n is the sum of the lengths of all of the patterns and the alphabet size is poly-logarithmic in n. For insertion and deletion operations, which are less frequent, the bound becomes a bit worse, specifically $O(\frac{1}{\epsilon} \log n)$ multiplicative slowdown in time, for $\epsilon \in (0, 1)$. As a byproduct, we show how an Aho-Corasick automaton can be constructed with only a compact working space, $O(n \log \sigma + d \log n)$ bits, which is the first of its type to the best of our knowledge. The construction is later being used as one of the building blocks of the dictionary matching algorithm.

For a comparison to previous results, refer to table 1.

Table 1. Comparison of dynamic dictionary matching algorithms, where n is the total length of d patterns, σ is the alphabet size and $\epsilon \in (0, 1)$. Space is given in bits.

	query time	update time	space
[8]	$O((|T| + occ) \log^2 n)$	$O(|P| \log^2 n)$	$O(n\sigma)$
[7]	$O(|T| \log n + occ)$	$O(|P| \log \sigma + \log n)$	$O(n \log \sigma)$
[7]	$O(|T| \log n + occ)$	$O(|P| \log \sigma + \log n)$	$(1 + o(1))n \log \sigma + O(d \log n)$
This paper	$O(|T|(\log \log n) \log \sigma + occ)$	$O(\frac{1}{\epsilon}|P| \log n)$	$(1 + \epsilon)(n \log \sigma + 6n) +$ $o(n) + O(d \log n)$
This paper, for $\sigma = polylog(n)$, using $\epsilon = \frac{1}{\log \log n}$	$O(|T|(\log \log n)^2 + occ)$	$O(|P| \log n \log \log n)$	$(1 + o(1))n \log \sigma + O(d \log n)$

1.1   Preliminaries

Let P be a set of d patterns $P = \{P_1, ..., P_d\}$ of cumulative size $n = \sum_{i=1}^{d} |P_i|$ over an alphabet Σ of size σ, and let T be a text of m characters. Furthermore, denote by occ the number of occurrences of the patterns from the set P in a given text T. For any two patterns, $P_i < P_j$ denotes that P_i is lexicographically smaller than P_j. Additionally, $(P_i)^r$ is the reversal of P_i. We assume a unit-cost RAM with word size $O(log\sigma)$ bits, in which standard arithmetic operations on word-sized operands can be performed in constant time.

1.2   Related Work

The research of dictionary matching algorithms goes back to the known Aho-Corasick automaton from 1975 [3]. Their algorithm is a generalization of the KMP automaton [9], which enables us to solve the static dictionary matching problem optimally using an $O(((|T| + n) \log \sigma + occ)$[1] time algorithm. Amir et al. in [1] proposed the first dynamic algorithm for that problem. Their algorithm is based on generalized suffix trees, and they utilized the McCreight [10] construction algorithm to insert and remove patterns. Shortly afterward Amir et al. [4] introduced an alternative faster algorithm. In the core of this algorithm, there is a reduction between the dynamic marked ancestors problem to the balanced parentheses problem, which provides an efficient way to mark and un-mark nodes in a suffix tree. Assuming that P is a pattern that is removed or added to the dictionary, a dictionary update costs $O(P \frac{\log n}{\log \log n} + \log \sigma)$ which is $O(\log \log n)$ faster than the previous solution [1].

Sahinalp and Vishkin in [2] closed the gap on that problem and proposed the first optimal dynamic algorithm. Their algorithm uses a deterministic labeling technique called the LCP (Local Consistent Parsing) procedure and they show how to assign labels for the patterns in a way that allows a fast comparison of sub-strings. As a result, querying for a pattern's existence becomes $O(|T| + occ)$ (assuming $|P| \leq |T|$), and the insertion or deletion of pattern P costs $O(|P|)$.

All of the above algorithms use space in words, which is linear in the total length of the patterns, namely $O(n \log n)$ bits. Chan et al. [8] were the first to propose a compact (for constant alphabet size), $O(n\sigma)$ bit, data structure for a dynamic dictionary that is based on suffix trees. Their algorithm builds a dynamic suffix tree that uses $O(n\sigma)$ bits of space and supports the updating of a pattern P in $O(|P| \log^2 n)$ time and a query in $O((|T| + occ) \log^2 n)$ time. Their structure utilizes the FM index of [6] and the compressed suffix arrays of [5].

Hon et al. in [7] proposed the first succinct dynamic index for dictionary matching. Their solution is much simpler than [8], and it is based on a orderly sampling technique. Their key observation was that it is possible to reduce the space by grouping characters of a pattern into a single meta character. The space used is $(1 + o(1))n \log \sigma + O(d \log n)$ bits, the query time is $O(|T| \log n + occ)$ and the update time of a pattern P is $O(|P| \log \sigma + \log n)$. Their dynamic solution was preceded by a compressed static construction for dictionary matching based on a similar approach [11].

Belazzougui in [12] showed how the Aho-Corasick automaton can be converted to a static succinct structure that occupies $n(\log \sigma + 3.443 + o(1)) + d \log(\frac{n}{d})$ bits. His data structure provides the optimal query time of $O(|T| + occ)$, can be built in linear time, but not with a compact working space (the working space required is $O(n \log n)$ bits). The index will be described in the following sections since here we use similar states' representation. Hon et al. [13] showed how Belazzougui's index can be slightly modified to be stored in $nH_k(D) + O(n)$ bits, while the query time remains optimal.

---

[1] $O(|T| + n + occ)$ when using hashing.

1.3 Outline

The rest of the paper is organized as follows. In section 2 we present our algorithm for succinct dynamic dictionary matching. The algorithm utilizes the Aho-Corasick automaton construction that is sketched in section 3. Due to lack of space, the construction's details are left to the full paper.

2 Succinct Dynamic Dictionary Matching Algorithm

In this section, we propose a new succinct data-structure for the dynamic dictionary matching problem. This data-structure utilizes the compact construction of the succinct Aho-Corasick automaton described in section 3. The main contribution of this section is the reduction in query time compared to the best previous result of [7]. This reduction in query time incurs a slight increase in update (insertion or removal of patterns) time, however in most practical usages query time is more prominent. Specifically, the query time of the proposed algorithm is $O(|T| \log \log n \log \sigma + occ)$, which is $O(|T|(\log \log n)^2 + occ)$ for $\sigma = polylog(n)$, while the insertion or removal of a pattern P costs $O(\frac{1}{\epsilon}|P| \log n)$, for $\epsilon \in (0,1)$. For a detailed comparison to previous results, refer to table 1.

2.1 Overview of the Technique

Our technique achieves the reduction in query time by using a specific division of the patterns' domain into groups based on length, and handling each group differently with a designated data-strcture. For simplicity, assume that the total length n is known and thus the division is well defined, later on we will describe how to cope with the changing of n. The hierarchy we define is divided into four groups, which are defined as follows:

1. Group XL (Extra Large): for patterns that are longer than or equal to $\frac{n}{\log \log n}$
2. Group L (Large): for patterns that are longer than or equal to $\frac{n}{\log n \log \log n}$ and shorter than $\frac{n}{\log \log n}$. This group will be further divided into $O(\log \log n)$ levels, based on patterns' length, in the following section.
3. Group M (Medium): for patterns that are longer than $0.5 \log_\sigma n$ and shorter than $\frac{n}{\log n \log \log n}$. When this group becomes full, in case the total size of the patterns in this group exceeds $\frac{n}{\log n \log \log n}$, we move them all together into group L.
4. Group S (Small): for patterns that are shorter than or equal to $0.5 \log_\sigma n$.

Notice that as our goal is to build a succinct dynamic data-structure, we have to make certain that the total size of the whole data structure is succinct. However, due to our hierarchy we do have some flexibility to maintain certain groups unsuccinctly relatively to the total size of the patterns in this group. Specifically, the total length of the patterns in the groups "Small" and "Medium" is relatively small, hence we are not restricted to use only succinct data-structures for maintaining the patterns, as opposed to the upper levels.

We use the above hierarchy as follows. On query, upon given a text T, we query each of the designated data-structures of these groups separately. On insertion, upon given a pattern P, we first make certain that the pattern does not exist in the whole data-structure, by running a query procedure for $T = P$; then, if it does not exist, the pattern is inserted directly to the data-structure of the group that matches its length. Similarly, on deletion, we first query to find the pattern among the groups, and then if it exists delete it from there. The exception for the deletion case is the group "Large", in which we don't remove the pattern immediately, but only mark it as removed, as will be described later.

In order to cope with the variability of n, we count the total length of the inserted and removed patterns and rebuild the whole data-structure once this size exceeds $\epsilon \cdot n$, for $\epsilon \in (0, 1)$; this step is to ensure that the patterns remain at the appropriate group despite the change in n. Also, this ensures that the extra space in the "Large" group, caused by patterns that are pending for deletion, does not exceed ϵ factor of the size of the structure. Next, we explain these procedures in detail for each group.

2.2   Extra Large Group

In this group, by definition, there can be at most $\log\log n$ patterns, which we maintain explicitly. We utilize the constant space KMP of [14] for each pattern separately. On query, we iterate over all the patterns in this group, hence the query time is $O(|T|\log\log n + occ)$. Insertion or deletion of a pattern P are trivial and take $O(|P|)$ time.

2.3   Large Group

Recall that patterns in this group are longer than or equal to $\frac{n}{\log n \log\log n}$ and shorter than $\frac{n}{\log\log n}$. We further divide the patterns in this group into $\log\log n$ levels $t_1, t_2, \ldots t_{\log\log n}$, with exponentially decreasing size. Denote the total length of the patterns in level t_i by $|t_i|$. The patterns are divided into levels such that for level t_1, $\frac{n}{2\log\log n} \leq |t_1| < \frac{n}{\log\log n}$; for the second level $\frac{n}{4\log\log n} \leq |t_2| < \frac{n}{2\log\log n}$, and in general, $\frac{n}{2^{i+1}\log\log n} \leq |t_i| < \frac{n}{2^i\log\log n}$. A pattern in the i-th level can either be: (a) A pattern that was originally inserted to this level because its length was between $\frac{n}{2^{i+1}\log\log n}$ and $\frac{n}{2^i\log\log n}$; (b) A pattern that was moved into this level from the $(i+1)$-th level, when it was full; (c) Applies only to level $t_{\log\log n}$, a pattern that was moved into this level from group "Medium" when it became full.

The patterns in each level are maintained in a static Aho-Corasick automatons of theorem 3.1. In each level except for t_1 there are at most two such structures, and t_1 contains at most $\log\log n$ such structures. We call a level from $t_2, \ldots t_{\log\log n}$ full when it contains two automatons. But, we merge and move them to an upper level only when attempting to insert a third one, as described below.

Insert. When a new pattern P is inserted, we construct an Aho-Corasick automaton for that pattern, using theorem 3.1, and we insert it into a level chosen according to its size, i.e., to the level that contains patterns of total length in the same order of its size. If now there are three structures in this level, then we merge two of them by building a new Aho-Corasick automaton that contains all of the two automatons' patterns and we put the resulting structure into an upper level. We keep merging and moving in this fashion until there are less than three structures in the same level or we get to level t_1. Notice that in level t_1 we allow up to $\log \log n$ structures because it may happen that all the patterns reach this level (by merges or direct insertions).

Lemma 1. *Insertion of a pattern P into the "Large" group is done in $O(|P| \log n)$ amortized time, with an extra $o(n \log \sigma) + O(d \log n)$ bits of working space needed just for constructing the automatons.*

Proof. The given pattern P can potentially climbs to at most $\log \log n$ levels and therefore can be part of at most $\log \log n$ constructions of the Aho-Corasick automaton of theorem 3.1. Thus, the amortized insert cost is $O(|P| \log n)$ time. Assuming the total length of the merged patterns is n', according to theorem 3.1, the construction of such structure uses $O(n' \log \sigma) + O(d \log n')$ extra working space. In our case, we can bound n' with $\frac{n}{\log \log n}$, which is the total length of patterns in each structure of level t_1, therefore the extra working space needed during each construction is $o(n \log \sigma) + O(d \log n)$ bits. $\qquad\square$

Query. On query, we iterate over all of the levels and search for the pattern. The total time is $O(T \log \log n \log \sigma + occ)$, because we have $\log \log n$ levels, and querying the Aho-Corasick automatons takes $O(T \log \sigma + occ)$ time according to theorem 3.1. For cases where $\sigma = polylog(n)$ the total time is $O(|T|(\log \log n)^2 + occ)$.

Delete. For deletion, we first run query operation in order to locate the level that contains the pattern; then, we "remove" the pattern from the Aho-Corasick automaton by un-marking it in $O(1)$ time (according to theorem 3.1). We defer the actual deletion to a later stage, ignoring the un-marked patterns when we rebuild the whole data structure.

The above results are summarized in following lemma:

Lemma 2. *A pattern P can be inserted to the "Large" group in $O(\frac{1}{\epsilon}|P| \log n)$ amortized time and marked for deletion in $O(1)$ amortized time. Querying the "Large" group for a text T takes $O(T \log \log n \log \sigma + occ)$ time, which is $O(|T|(\log \log n)^2 + occ)$ for cases where $\sigma = polylog(n)$. The space needed to accommodate the patterns in this group is at most $(1 + \epsilon)(n \log \sigma + 6n) + o(n) + O(d \log n)$, for $\epsilon \in (0, 1)$.*

Proof. The running time for Insert, Delete and Query directly follows the details described above and in lemma 1. As we do not completely delete patterns from this group, only un-mark them, we need to make certain that the extra space does not grow too much. Hence, we count the total length of the inserted and removed patterns and rebuild the group once it exceeds $\epsilon \cdot n$, for $\epsilon \in (0, 1)$; this ensures that we do not have more than $1 + \epsilon$ factor in space. We use the structures from theorem 3.1 in each level, and each consumes $n' \log \sigma + 6n' + o(n') + O(d \log n')$ bits of space, where n' is the total length of the patters in it. In the "worst case" all the dictionary's patterns are located it this group, and therefore, by linearity, the total space consumption is at most $(1 + \epsilon)(n \log \sigma + 6n) + o(n) + O(d \log n)$. □

2.4   Medium Group

We maintain patterns in this group in the dynamic dictionary by Sahinalp and Vishkin of [2][2]. The query time for a pattern is $O(|T| + occ)$, while the time for insertion and deletion is $O(|P|)$. Following the groups definition above, when the total size of the patterns in this group exceeds $\frac{n}{\log n \log \log n}$, we move them all together into group "Large". Hence, the space consumption is
$$O(\frac{n}{\log n \log \log n} \log \frac{n}{\log n \log \log n}) = O(\frac{n}{\log \log n}) = o(n).$$

2.5   Small Group

Because the total length of the patterns in this group is very small, the patterns can be stored permanently in a dictionary structure of [2], without the need to move up the hierarchy. Similarly to the analysis in [7], the overall space requires to store all the patterns is $o(n)$ because there can be at most $\Theta(\sqrt{n})$ such patterns, with the total length of $\Theta(\sqrt{n} \log_\sigma n)$ and total size of $\Theta(\sqrt{n}(\log_\sigma n)^2) = o(n)$ bits. The query, insert and delete operations are optimal in the size of the query or the pattern, respectively.

2.6   Putting it all Together

The construction is summarized in the following theorem:

Theorem 2.1. *There is a data-structure for the dynamic dictionary matching problem of d patterns with total length n, over alphabet of size σ, for $\epsilon \in (0, 1)$, that uses $(1+\epsilon)(n \log \sigma + 6n) + o(n) + O(d \log n)$ bits of space. It supports querying all occurrences of the d patterns in a text T in $O(|T|(\log \log n) \log \sigma + occ)$ time, and insertion or deletion of a pattern P in $O(\frac{1}{\epsilon}|P| \log n)$ amortized time.*

Proof. Query, insertion and deletion are done on each group separately as was described above and in lemma 2. It takes optimal time to execute these operations on the "Small" and "Medium" groups, while on the "Extra Large" group there is a $\log \log n$ slowdown factor. Hence, the overall time bounds are dominated by the operations on the "Large" group. As for the space consumption, it is also dominated by the "Large" group. In addition, as we do not completely delete patterns from this group, we need to make certain that the extra space does not grow too much. Hence, we count the total length of the inserted and

---

[2] Any other non succinct algorithm that requires up to $O(\log \log n)$ query and update time can be used.

removed patterns and rebuild the whole data-structure once this size exceeds $\epsilon \cdot n$, for $\epsilon \in (0,1)$; this ensures that we do not have more than $1 + \epsilon$ factor in space, and that the patterns remain at the appropriate group despite the change in n. Notice that we do not need to rebuild the "Large" group separately, since the whole data-structure rebuild described here is sufficient. □

Theorem 2.2. *There is a data-structure for the dynamic dictionary problem of d patterns with total length n, over alphabet of size $\sigma = polylog(n)$, that uses $(1 + o(1))n \log \sigma + O(d \log n)$ bits of space. It supports querying all occurrences of the d patterns in a text T in $O(|T|(\log \log n)^2 + occ)$ time, and insertion or deletion of a pattern P in $O(|P| \log n \log \log n)$ amortized time.*

Proof. Directly Follows from theorem 2.1, and choosing $\epsilon = \frac{1}{\log \log n}$. □

3   Compact Construction of Succinct Aho-Corasick

As was described in section 2, the algorithm uses static Aho-Corasick automatons as building blocks, that have to be merged from time to time. The requirement for keeping the merge operation succinct is that the construction of the merged automaton is done by using only a compact working space, relative to the total size of the merged patterns. In this section we sketch such a construction and its main contribution is that it uses only a compact working space, which is the first of its type to the best of our knowledge. We further extend the automaton to support un-marking of patterns, i.e., terminal states can be removed from the report tree. The construction is summarized in theorem 3.1.

Theorem 3.1. *The Aho-Corasick automaton of d patterns, over an alphabet of size σ, that support un-marking of a pattern, can be constructed using $O(n \log \sigma) + O(d \log n)$ bits of space and can be compressed to consume $n \log \sigma + 6n + o(n) + O(d \log n)$ bits for maintaining it, where n is the number of states during the construction (before any deletion occurs). The total time for the construction is $O(\frac{n \log n}{\log \log n})$. Furthermore, this approach supports querying all of the occurrences of the d patterns in a text T in $O(|T| \log \sigma + occ)$ and un-marking a pattern in $O(1)$ amortized time.*

The data structure uses similar states' representation to as described by Belazzougui in [12], who showed how to construct static, succinct Aho-Corasick data-structure in linear time, that supports optimal query time. However, his construction uses more than compact working space, i.e. $O(n \log n)$ bits. As in Belazzougui's construction, we maintain three data-structures, namely, an indexable dictionary for the next transitions and two trees for the failure and report links. We modified the internal structures and the way they are built, and show how an Aho-Corasick automaton can be constructed using only a compact working space. In the rest of this section we first describe Belazzougui's idea, and then we sketch our construction. Due to lack of space the details are left to the full paper.

3.1   Belazzougui's Succinct Static Aho-Corasick Automaton

The Aho-Corasick automaton is a generalized KMP structure, which is essentially a trie with three types of links between internal states (nodes): next transitions, failure and report links. The next transitions links are the regular links of the trie. The failure link points to the longest suffix of the current pattern which is also a prefix of some other pattern in the trie. Similarly, the report link points to the longest suffix of the current pattern that is marked (the terminal state). These links allow a smooth transition between states without the need for backtracking.

Belazzougui showed how to represent each of the Aho-Corasick components using succinct data structures. His main observation was to name the states according to the suffix-lexicographic order of the concatenation of the labels on the edges of the trie. Specifically, the name of a state x, $state(x) \in [0, n-1]$, is the rank of the state in that order, where n is the number of states in the automaton. This arrangement has enabled him to maintain a failure tree that is ordered in states with increasing order. Assume that p_i is the i-th prefix in suffix lexicographic order, and let c_i be the label on the incoming edge in the trie that represents state p_i, s.t. $p_i = p'_i c_i$; then, the next transition is encoded using an indexable dictionary that maintains tuples of the form $(c_i, state(p'_i))$. In addition, the failure and report transitions are maintained in a succinct tree, such that a DFS traversal of the tree will enumerate the states in increasing order. The data-structure is encoded using $n(\log \sigma + 3.443 + o(1)) + d\log(\frac{n}{d})$ bits, where σ is the size of the alphabet and d is the number of patterns, and the query time is $O(|T| + occ)$. For cases where $\sigma < (n)^\epsilon$ for any $0 < \epsilon < 1$, the construction can be compressed to $n(H_0 + 3.443 + o(1)) + d(3\log(\frac{n}{d}) + O(1))$ bits.

3.2   Sketch of the Construction

The main challenge was to build the tuples of the next transition dictionary in compact working space. Specifically, how to name each node in the trie of all patterns by its suffix lexicographic order, without maintaining a map of at least $n \log n$ bits. To overcome it, we use compressed suffix arrays that enable us to sort the suffixes efficiently utilizing a compact working space. To build the next transition dictionary, we start by sorting the patterns lexicographically, then we build a trie of the sorted patterns. Finally, we sort the prefixes in suffix lexicographic order by sorting the reversals of the patterns, and name each state in the trie according to this order. As for the failure links, we build them iteratively, in BFS order fashion, using the next transition dictionary. For each node, we use a standard technique to find its failure, as follows: visit the failure of its parent and check whether there is an outgoing edge with the same character. If so, this node is the failure; otherwise, continue recursively to the parent of the parent. We use the dynamic succinct tree of [15] to maintain the failure tree structure during the build. Eventually, we convert the failure links tree into a report tree, choosing the nearest marked ancestor (end of some pattern) as the direct parent of each node or the root in case there is no such node.

References

1. Amir, A., Farach, M., Galil, Z., Giancarlo, R., Park, K.: Dynamic dictionary matching. J. Comput. Syst. Sci. 49(2), 208–222 (1994)
2. Sahinalp, S.C., Vishkin, U.: Efficient approximate and dynamic matching of patterns using a labeling paradigm. In: FOCS, pp. 320–328. IEEE Computer Society (1996)
3. Aho, A.V., Corasick, M.J.: Efficient string matching: an aid to bibliographic search. Commun. ACM 18(6), 333–340 (1975)
4. Amir, A., Farach, M., Idury, R.M., La Poutré, J.A., Schäffer, A.A.: Improved dynamic dictionary matching. In: Proceedings of the Fourth Annual ACM-SIAM Symposium on Discrete Algorithms, SODA 1993, pp. 392–401. Society for Industrial and Applied Mathematics, Philadelphia (1993)
5. Grossi, R., Vitter, J.S.: Compressed suffix arrays and suffix trees with applications to text indexing and string matching (extended abstract). In: Proceedings of the Thirty-Second Annual ACM Symposium on Theory of Computing, STOC 2000, pp. 397–406. ACM, New York (2000)
6. Ferragina, P., Manzini, G.: Indexing compressed text. J. ACM 52(4), 552–581 (2005)
7. Hon, W.-K., Lam, T.-W., Shah, R., Tam, S.-L., Vitter, J.S.: Succinct index for dynamic dictionary matching. In: Dong, Y., Du, D.-Z., Ibarra, O. (eds.) ISAAC 2009. LNCS, vol. 5878, pp. 1034–1043. Springer, Heidelberg (2009)
8. Chan, H.L., Hon, W.K., Lam, T.W., Sadakane, K.: Dynamic dictionary matching and compressed suffix trees. In: Proceedings of the Sixteenth Annual ACM-SIAM Symposium on Discrete Algorithms, SODA 2005, pp. 13–22. Society for Industrial and Applied Mathematics, Philadelphia (2005)
9. Karp, R.M., Miller, R.E., Rosenberg, A.L.: Rapid identification of repeated patterns in strings, trees and arrays. In: Proceedings of the Fourth Annual ACM Symposium on Theory of Computing, STOC 1972, pp. 125–136. ACM, New York (1972)
10. McCreight, E.M.: A space-economical suffix tree construction algorithm. J. ACM 23(2), 262–272 (1976)
11. Hon, W.K., Lam, T.W., Shah, R., Tam, S.L., Vitter, J.S.: Compressed index for dictionary matching. In: Proceedings of the Data Compression Conference, DCC 2008, pp. 23–32. IEEE Computer Society, Washington, DC (2008)
12. Belazzougui, D.: Succinct dictionary matching with no slowdown. In: Amir, A., Parida, L. (eds.) CPM 2010. LNCS, vol. 6129, pp. 88–100. Springer, Heidelberg (2010)
13. Hon, W.-K., Ku, T.-H., Shah, R., Thankachan, S.V., Vitter, J.S.: Faster compressed dictionary matching. In: Chavez, E., Lonardi, S. (eds.) SPIRE 2010. LNCS, vol. 6393, pp. 191–200. Springer, Heidelberg (2010)
14. Rytter, W.: On maximal suffices and constant-space linear-time versions of kmp algorithm. In: Rajsbaum, S. (ed.) LATIN 2002. LNCS, vol. 2286, pp. 196–208. Springer, Heidelberg (2002)
15. Sadakane, K., Navarro, G.: Fully-functional succinct trees. In: Proceedings of the Twenty-First Annual ACM-SIAM Symposium on Discrete Algorithms, SODA 2010, pp. 134–149. Society for Industrial and Applied Mathematics, Philadelphia (2010)

Order-Preserving Pattern Matching
with k Mismatches

Paweł Gawrychowski[1] and Przemysław Uznański[2,*]

[1] Max-Planck-Institut für Informatik, Saarbrücken, Germany
[2] LIF, CNRS and Aix-Marseille Université, Marseille, France

Abstract. We study a generalization of the order-preserving pattern matching recently introduced by Kubica et al. (Inf. Process. Let., 2013) and Kim et al. (submitted to Theor. Comp. Sci.), where instead of looking for an exact copy of the pattern, we only require that the relative order between the elements is the same. In our variant, we additionally allow up to k mismatches between the pattern of length m and the text of length n, and the goal is to construct an efficient algorithm for small values of k. Our solution detects an order-preserving occurrence with up to k mismatches in $\mathcal{O}(n(\log \log m + k \log \log k))$ time.

1 Introduction

Order-preserving pattern matching, recently introduced in [9] and [10], and further considered in [4], is a variant of the well-known pattern matching problem, where instead of looking for a fragment of the text which is identical to the given pattern, we are interested in locating a fragment which is order-isomorphic with the pattern. Two sequences over integer alphabet are *order-isomorphic* if the relative order between any two elements at the same positions in both sequences is the same. Similar problems have been extensively studied in a slightly different setting, where instead of a fragment, we are interested in a (not necessarily contiguous) subsequence. For instance, pattern avoidance in permutations was of much interest.

For the order-preserving pattern matching, both [9] and [10] present an $\mathcal{O}(n + m \log m)$ time algorithm, where n is the length of the text, and m is the length of the pattern. Actually, the solution given by [10] works in $\mathcal{O}(n + \text{sort}(m))$ time, where $\text{sort}(m)$ is the time required to sort a sequence of m numbers. Furthermore, efficient algorithms for the version with multiple patterns are known [4]. Also, a generalization of suffix trees in the order-preserving setting was recently considered [4], and the question of constructing a forward automaton allowing efficient pattern matching and developing an average-case optimal pattern matching algorithm was studied [1].

Given that the complexity of the exact order-preserving pattern matching seems to be already settled, a natural direction is to consider its approximate version.

---

[*] This work was started while the author was a PhD student at Inria Bordeaux Sud-Ouest, France.

A.S. Kulikov, S.O. Kuznetsov, and P. Pevzner (Eds.): CPM 2014, LNCS 8486, pp. 130–139, 2014.

Such direction was successfully investigated for the related case of *parametrized pattern matching* in [6], where an $\mathcal{O}(nk^{1.5} + mk \log m)$ time algorithm was given for parametrized matching with k mismatches.

We consider a relaxation of order-preserving pattern matching, which we call *order-preserving pattern matching with k mismatches*. Instead of requiring that the fragment we seek is order-isomorphic with the pattern, we are allowed to first remove k elements at the corresponding positions from the fragment and the pattern, and then check if the remaining two sequences are order-isomorphic. In such setting, it is relatively straightforward to achieve running time of $\mathcal{O}(nm \log \log m)$, where n is the length of the text, and m is the length of the pattern. Such complexity might be unacceptable for long patterns, though, and we aim to achieve complexity of the form $\mathcal{O}(nf(k))$. In other words, we would like our running time to be close to linear if the bound on the number of mismatches is very small. We construct a deterministic algorithm with $\mathcal{O}(n(\log \log m + k \log \log k))$ running time. At a very high level, our solution is similar to the one given in [6]. We show how to filter the possible starting positions so that a position is either eliminated in $\mathcal{O}(f(k))$ time, or the structure of the fragment starting there is simple, and we can verify the occurrence in $\mathcal{O}(f(k))$ time. The details are quite different in our setting, though.

A different variant of approximate order-preserving pattern matching could be that we allow to remove k elements from the fragment, and k elements from the pattern, but don't require that they are at the same positions. Then we get order-preserving pattern matching with k errors. Unfortunately, such modification seems difficult to solve in polynomial time: even if the only allowed operation is removing k elements from the fragment, the problem becomes NP-complete [3].

2   Overview of the Algorithm

Given a text $(t_1, \ldots, t_n)$ and a pattern $(p_1, \ldots, p_m)$, we want to locate an order-preserving occurrence with at most k mismatches of the pattern in the text. Such occurrence is a fragment $(t_i, \ldots, t_{i+m-1})$ with the property that if we ignore the elements at some up to corresponding k positions in the fragment and the pattern, the relative order of the remaining elements is the same in both of them.

The above definition of the problem is not very convenient to work with, hence we start with characterising k-isomorphic sequences using the language of subsequences in Lemma 1. This will be useful in some of the further proofs and also gives us a polynomial time solution for the problem, which simply considers every possible i separately. To improve on this naive solution, we need a way of quickly eliminating some of these starting positions. For this we define the signature $S(a_1, \ldots, a_m)$ of a sequence $(a_1, \ldots, a_m)$, and show in Lemma 3 that the Hamming distance between the signatures of two k-isomorphic sequences cannot be too large. Hence such distance between $S(t_i, \ldots, t_{i+m-1})$ and $S(p_1, \ldots, p_m)$ can be used to filter some starting positions where a match cannot happen.

In order to make the filtering efficient, we need to maintain $S(t_i, \ldots, t_{i+m-1})$ as we increase i, i.e., move a window of length m through the text. For this we first provide in Lemma 4 a data structure which, for a fixed word, allows us to maintain a word of a similar length under changing the letters, so that we can quickly generate the first k mismatches between subwords of the current and the fixed word. The structure is based on representing the current word as a concatenation of subwords of the fixed word. Then we observe that increasing i by one changes the current signature only slightly, which allows us to apply the aforementioned structure to maintain $S(t_i, \ldots, t_{i+m-1})$ as shown in Lemma 5. Therefore we can efficiently eliminate all starting positions for which the Hamming distance between the signatures is too large.

For all the remaining starting positions, we reduce the problem to computing the heaviest increasing subsequence, which is a weighted version of the well-known longest increasing subsequence, in Lemma 6. The time taken by the reduction depends on the Hamming distance, which is small as otherwise the position would be eliminated in the previous step. Finally, such weighted version of the longest increasing subsequence can be solved efficiently as shown in Lemma 7. Altogether these results give an algorithm for order-preserving pattern matching with k with the cost of processing a single i depending mostly on k.

An implicit assumption in this solution is that there are no repeated elements in neither the text nor the pattern. The assumption can be removed without increasing the time complexity by carefully modifying all parts of the algorithm. Some of these changes are not completely trivial, for example the definition of a signature becomes more involved, which in turn makes the proofs more complicated. Because of the limited space we defer the details (and a few proofs) to the full version.

3   Preliminaries

We consider strings over an integer alphabet, or in other words sequences of integers. Two such sequences are *order-isomorphic* (or simply *isomorphic*), denoted by $(a_1, \ldots, a_m) \sim (b_1, \ldots, b_m)$, when $a_i \leq a_j$ iff $b_i \leq b_j$ for all i, j. We will also use the usual equality of strings. Whenever we are talking about sequences, we are interested in the relative order between their elements, and whenever we are talking about strings consisting of characters, the equality of elements is of interest to us. For two strings s and t, their *Hamming distance* $H(s, t)$ is simply the number of positions where the corresponding characters differ.

Given a text $(t_1, \ldots, t_n)$ and a pattern $(p_1, \ldots, p_m)$, the *order-preserving pattern matching* problem is to find i such that $(t_i, \ldots, t_{i+m-1}) \sim (p_1, \ldots, p_m)$. We consider its approximate version, i.e., order-preserving pattern matching with k mismatches. We define two sequences *order-isomorphic with k mismatches*, denoted by $(a_1, \ldots, a_m) \overset{k}{\sim} (b_1, \ldots, b_m)$, when we can select (up to) k indices $1 \leq i_1 < \ldots < i_k \leq m$, and remove the corresponding elements from both sequences so that the resulting two new sequences are isomorphic, i.e., $a_j \leq a_{j'}$ iff $b_j \leq b_{j'}$ for any $j, j' \notin \{i_1, \ldots, i_k\}$. In *order-preserving pattern matching with k mismatches* we want i such that $(t_i, \ldots, t_{i+m-1}) \overset{k}{\sim} (p_1, \ldots, p_m)$, see Fig. 1.

Our solution works in the word RAM model, where n integers can be sorted in $\mathcal{O}(n \log \log n)$ time [5], and we can implement dynamic dictionaries using

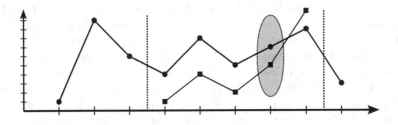

Fig. 1. $[1, 4, 2, 5, 11]$ occurs in $[1, 10, 6, 4, 8, 5, 7, 9, 3]$ (at position 4) with 1 mismatch

van Emde Boas trees. In the restricted comparison model, where we can only compare the integers, all $\log \log$ in our complexities increase to $\log$.

We assume that integers in any sequence are all distinct. Such assumption was already made in one of the papers introducing the problem [9], with a justification that we can always perturb the input to ensure this (or, more precisely, we can consider pairs consisting of a number and its position). In some cases this can change the answer, though[1]. Nevertheless, using a more complicated argument, given in the full version, we can generalize our solution to allow the numbers to repeat. Another simplifying assumption that we make in designing our algorithm is that $n \leq 2m$. We can do so using a standard trick of cutting the text into overlapping fragments of length $2m$ and running the algorithm on each such fragment separately, which preserves all possible occurrences.

4   The Algorithm

First we translate k-isomorphism into the language of subsequences.

Lemma 1. $(a_1, \ldots, a_m) \overset{k}{\approx} (b_1, \ldots, b_m)$ iff there exist $i_1, \ldots, i_{m-k}$ such that $a_{i_1} < \ldots < a_{i_{m-k}}$ and $b_{i_1} < \ldots < b_{i_{m-k}}$.

The above lemma implies an inductive interpretation of k-isomorphism useful in further proofs and a fast method for testing k-isomorphism.

Proposition 1. If $(a_1, \ldots, a_m) \overset{k+1}{\sim} (b_1, \ldots, b_m)$ then there exists $(a'_1, \ldots, a'_m)$ such that $(a_1, \ldots, a_m) \overset{1}{\sim} (a'_1, \ldots, a'_m)$ and $(a'_1, \ldots, a'_m) \overset{k}{\approx} (b_1, \ldots, b_m)$.

Lemma 2. $(a_1, \ldots, a_m) \overset{k}{\approx} (b_1, \ldots, b_m)$ can be checked in time $\mathcal{O}(m \log \log m)$.

Proof. Let π be the sorting permutation of $(a_1, \ldots, a_m)$. Such permutation can be found in time $\mathcal{O}(m \log \log m)$. Let $(b'_1, \ldots, b'_m)$ be a sequence defined by setting $b'_i := b_{\pi(i)}$. Then, by Lemma 1, $(a_1, \ldots, a_m) \overset{k}{\approx} (b_1, \ldots, b_m)$ iff there exists an increasing subsequence of b' of length $m - k$. Existence of such a subsequence can be checked in time $\mathcal{O}(m \log \log m)$ using a van Emde Boas tree [7].   □

---

[1] More precisely, it might make two non-isomorphic sequences isomorphic, but not the other way around.

By applying the above lemma to each of the possible occurrences separately, we can already solve order-preserving pattern matching with k mismatches in time $\mathcal{O}(nm \log \log m)$. However, our goal is to develop a faster $\mathcal{O}(nf(k))$ time algorithm. For this we cannot afford to verify every possible position using Lemma 2, and we need a closer look into the structure of the problem.

The first step is to define the *signature* of a sequence $(a_1, \ldots, a_m)$. Let $\mathrm{pred}(i)$ be the position where the predecessor of a_i among $\{a_1, \ldots, a_m\}$ occurs in the sequence (or 0, if a_i is the smallest element). Then the signature $S(a_1, \ldots, a_m)$ is a new sequence $(1 - \mathrm{pred}(1), \ldots, m - \mathrm{pred}(m))$ (a simpler version, where the new sequence is $(\mathrm{pred}(1), \ldots, \mathrm{pred}(m))$, was already used to solve the exact version). The signature clearly can be computed in time $\mathcal{O}(m \log \log m)$ by sorting. While looking at the signatures is not enough to determine if two sequences are k-isomorphic, in some cases it is enough to detect that they are not.

Lemma 3. *If* $(a_1, \ldots, a_m) \not\approx^k (b_1, \ldots, b_m)$, *then the Hamming distance between* $S(a_1, \ldots, a_m)$ *and* $S(b_1, \ldots, b_m)$ *is at most* $3k$.

Proof. We apply induction on the number of mismatches k.

For $k = 0$, $(a_1, \ldots, a_m) \sim (b_1, \ldots, b_m)$ iff $S(a_1, \ldots, a_m) = S(b_1, \ldots, b_m)$, so the Hamming distance is clearly zero.

Now we proceed to the inductive step. If $(a_1, \ldots, a_m) \overset{k+1}{\sim} (b_1, \ldots, b_m)$, then due to Proposition 1, there exists $(a'_1, \ldots, a'_m)$, such that $(a'_1, \ldots, a'_m) \not\approx^k (b_1, \ldots, b_m)$ and $(a_1, \ldots, a_m) \overset{1}{\sim} (a'_1, \ldots, a'_m)$. Second constraint is equivalent (by application of Lemma 1) to existence of such i, that $(a_1, \ldots, a_{i-1}, a_{i+1}, \ldots, a_m) \sim (a'_1, \ldots, a'_{i-1}, a'_{i+1}, \ldots, a'_m)$.

We want to upperbound the Hamming distance between $S(a_1, \ldots, a_m)$ and $S(a'_1, \ldots, a'_m)$. Let j, j' be indices such that a_j is the direct predecessor of a_i and $a_{j'}$ is the direct successor of a_i, both taken from the set $\{a_1, \ldots, a_m\}$. Similarly, let ℓ, ℓ' be such indices, that a'_ℓ is the direct predecessor, and $a'_{\ell'}$ is the direct successor of a'_i, both taken from the set $\{a'_1, \ldots, a'_m\}$. That is,

$$\ldots < a_j < a_i < a_{j'} < \ldots$$

is the sorted version of $(a_1, \ldots, a_m)$, and

$$\ldots < a'_\ell < a'_i < a'_{\ell'} < \ldots$$

is the sorted version of $(a'_1, \ldots, a'_m)$. The signatures $S(a_1, \ldots, a_m)$ and $S(a'_1, \ldots, a'_m)$ differ on at most 3 positions: j', ℓ', and i. Thus $\mathrm{H}(S(a_1, \ldots, a_m), S(b_1, \ldots, b_m))$ can be upperbounded by

$$\mathrm{H}(S(a_1, \ldots, a_m), S(a'_1, \ldots, a'_m)) + \mathrm{H}(S(a'_1, \ldots, a'_m), S(b_1, \ldots, b_m)) \leq 3k + 3,$$

which ends the inductive step. □

Our algorithm iterates through $i = 1, 2, 3, \ldots$ maintaining the signature of the current $(t_i, \ldots, t_{i+m-1})$. Hence the second step is that we develop in the next two lemmas a data structure, which allows us to store $S(t_i, \ldots, t_{i+m-1})$, update it efficiently after increasing i by one, and compute its Hamming distance to $S(p_1, \ldots, p_m)$.

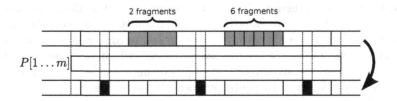

Fig. 2. Updating the representation. Black boxes represent mismatches, gray areas are full fragments between mismatches. Fragments are either left untouched (on the left), or compressed into a single new one (on the right).

Lemma 4. *Given a string $S^P[1..m]$, we can maintain a string $S^T[1..2m]$ and perform the following operations:*

1. *replacing any character $S^T[x]$ in amortized time $\mathcal{O}(\log \log m)$,*
2. *generating the first $3k$ mismatches between $S^T[i..(i + m - 1)]$ and $S^P[1..m]$ in amortized time $\mathcal{O}(k + \log \log m)$.*

The structure is initialized in time $\mathcal{O}(m \log \log m)$.

Proof. We represent the current $S^T[1..2m]$ as a concatenation of a number of fragments. Each fragment is a subword of S^P (possibly single letter) or a special character $\$$ not occurring in S^P. The starting positions of the fragments are kept in a van Emde Boas tree, and additionally each fragment knows its successor and predecessor. In order to bound the amortized complexity of each operation, we maintain an invariant that every element of the tree has 2 credits available, with one credit being worth $\mathcal{O}(\log \log m)$ time. We assume that given any two substrings of S^P, we can compute their longest common prefix in $\mathcal{O}(1)$ time. This is possible after $\mathcal{O}(m)$ preprocessing [2,8].

We initialize the structure by partitioning S^T into $2m$ single characters. The cost of initialization, including allocating the credits, is $\mathcal{O}(m \log \log m)$.

Replacing $S^T[x]$ with a new character c starts with locating the fragment w containing the position i using the tree. If w is a single character, we replace it with the new one. If w is a longer subword $w[i..j]$ of S^P, and we need to replace its ℓ-th character, we first split w into three fragments $w[i..(i + \ell - 1)]$, $w[i + \ell]$, $w[(i + \ell + 1)..j]$. In both cases we spend $\mathcal{O}(\log \log m)$ time, including the cost of inserting the new elements and allocating their credits.

Generating the mismatches begins with locating the fragment corresponding to the position i. Then we scan the representation from left to right starting from there. Locating the fragment takes $\mathcal{O}(\log \log m)$ time, but traversing can be done in $\mathcal{O}(1)$ time per each step, as we can use the information about the successor of each fragment. We will match S^P with the representation of S^T while scanning. This is done using constant time longest common prefix queries. Each such query allows us to either detect a mismatch, or move to the next fragment. Whenever we find a mismatch, if the part of the text between the previous mismatch (or the beginning of the window) and the current mismatch contains at least 3 full fragments, we replace them with a single fragment, which is the corresponding subword of S^P. If there are less than 3 full fragments, we keep the current representation intact, see Fig. 2. We stop the scanning after reaching $(3k+1)$-th mismatch, or after the whole window was processed, whichever comes first. By a standard argument, the amortized cost of processing a single mismatch is $\mathcal{O}(1)$, so we need $\mathcal{O}(k + \log\log m)$ time to generate all the mismatches.   □

Lemma 5. *Given a pattern $(p_1, \ldots, p_m)$ and a text $(t_1, \ldots, t_{2m})$, we can maintain an implicit representation of the current signature $S(t_i, \ldots, t_{i+m-1})$ and perform the following operations:*

1. *increasing i by one in amortized time $\mathcal{O}(\log\log m)$,*
2. *generating the first $3k$ mismatches between $S(p_1, \ldots, p_m)$ and $S(t_i, \ldots, t_{i+m-1})$ in time $\mathcal{O}(k + \log\log m)$.*

The structure is initialized in time $\mathcal{O}(m\log\log m)$.

Proof. First we construct $S(p_1, \ldots, p_m)$ in time $\mathcal{O}(m\log\log m)$ by sorting. Whenever we increase i by one, just a few characters of $S(t_i, \ldots, t_{i+m-1}) = (s_1, \ldots, s_m)$ need to be modified. The new signature can be created by first removing the first character s_1, appending a new character s_{m+1}, and then modifying the characters corresponding to the successors of t_i and t_{i+m}. By maintaining all $t_i, \ldots, t_{i+m-1}$ in a van Emde Boas tree (we can rename the elements so that $t_i \in \{1, \ldots, 2m\}$ by sorting) we can calculate both s_{m+1} and the characters which needs to be modified in $\mathcal{O}(\log\log m)$ time. Current $S(t_i, \ldots, t_{i+m-1})$ is stored using Lemma 4. We initialize $S^T[1..2m]$ to be $S(t_1, \ldots, t_m)$ concatenated with m copies of, say, 0. After increasing i by one, we replace $S^T[i]$, $S^T[i+m]$ and possibly two more characters in $\mathcal{O}(\log\log m)$ time. Generating the mismatches is straightforward using Lemma 4.   □

Now our algorithm first uses Lemma 5 to quickly eliminate the starting positions i such that the Hamming distance between the corresponding signatures is large. For the remaining starting positions, we reduce checking if $(t_i, \ldots, t_{i+m-1}) \overset{k}{\approx} (p_1, \ldots, p_m)$ to a weighted version of the well-known longest increasing subsequence problem on at most $3(k+1)$ elements. In the weighted variant, which we call *heaviest increasing subsequence*, the input is a sequence $(a_1, \ldots, a_\ell)$ and weight w_i of each element a_i, and we look for an increasing subsequence with the largest total weight, i.e., for $1 \le i_1 < \ldots < i_s \le \ell$ such that $a_{i_1} < \ldots < a_{i_s}$ and $\sum_j w_{i_j}$ is maximized.

Lemma 6. *Assuming random access to $(a_1, \ldots, a_m)$, the sorting permutation π_b of $(b_1, \ldots, b_m)$, and the rank of every b_i in $\{b_1, \ldots, b_m\}$, and given ℓ positions where $S(a_1, \ldots, a_m)$ and $S(b_1, \ldots, b_m)$ differ, we can reduce in $\mathcal{O}(\ell\log\log \ell)$ time checking if $(a_1, \ldots, a_m) \overset{k}{\approx} (b_1, \ldots, b_m)$ to computing the heaviest increasing subsequence on at most $\ell + 1$ elements.*

Proof. Let $d_1, \ldots, d_\ell$ be the positions where $S(a_1, \ldots, a_m)$ and $S(b_1, \ldots, b_m)$ differ. From the definition of a signature, for any other position i the predecessors

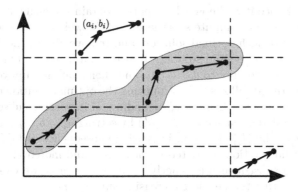

Fig. 3. Partition into maximal paths. The heaviest increasing subsequence is marked.

of a_i and b_i in their respective sequences are at the same position j, which we denote by $j \to i$. For any given i, $j \to i$ for at most one j. Similarly, for any given j, $j \to i$ for at most one i, because the only such i corresponds to the successor of, say, a_j in its sequence. Consider a partition of the set of all positions into maximal *paths* of the form $j_1 \to \ldots \to j_k$ (see Fig. 3). Such partition is clearly unique, and furthermore the first element of every path is one of the positions where the signatures differ (except one possible path starting with the position corresponding to the smallest element). Hence there are at most $\ell + 1$ paths, and we denote by I_j the path starting with d_j. If the smallest element occurs at the same position in both sequences, we additionally denote this position by d_0, and call the path starting there I_0 (we will assume that this is always the case, which can be ensured by appending $-\infty$ to both sequences).

Recall that our goal is to check if $(a_1, \ldots, a_m) \overset{k}{\approx} (b_1, \ldots, b_m)$. For this we need to check if there exist $i_1, \ldots, i_{m-k}$ such that $a_{i_1} < \ldots < a_{i_{m-k}}$ and $b_{i_1} < \ldots < b_{i_{m-k}}$. Alternatively, we could compute the largest s for which there exist a solution $i_1, \ldots, i_s$ such that $a_{i_1} < \ldots < a_{i_s}$ and $b_{i_1} < \ldots < b_{i_s}$. We claim that one can assume that for each path I either none of its elements are among $i_1, \ldots, i_s$, or all of its elements are there. We prove this in two steps.

1. If $i_k \in I$ and $i_k \to j$, then without losing the generality $i_{k+1} = j$. Assume otherwise, so $i_{k+1} \neq j$ or $k = s$. Recall that it means that a_j is the successor of a_{i_k} and b_j is the successor of b_{i_k}. Hence $a_{i_k} < a_j$ and $b_{i_k} < b_j$. If $k = s$ we can extend the current solution by appending j. Otherwise $a_j < a_{i_{k+1}}$ and $b_j < b_{i_{k+1}}$, so we can extend the solution by inserting j between i_k and i_{k+1}.
2. If $i_k \in I$ and $j \to i_k$, then without losing the generality $i_{k-1} = j$. Assume otherwise, so $i_{k-1} \neq j$ or $k = 1$. Similarly as in the previous case, a_j is the predecessor of a_{i_k} and b_j is the predecessor of b_{i_k}. Hence $a_j < a_{i_k}$ and $b_j < b_{i_k}$. If $k = 1$ we can extend the current solution by prepending j. Otherwise $a_{i_{k+1}} < a_j$ and $b_{i_{k+1}} < b_j$, so we can insert j between i_{k-1} and i_k.

Now let the weight of a path I be its length $|I|$. From the above reasoning we know that the optimal solution contains either no elements from a path, or all of its elements. Hence if we know which paths contain the elements used in the optimal solution, we can compute s as the sum of the weights of these paths. Additionally, if we take such optimal solution, and remove all but the first element from every path, we get a valid solution. Hence s can be computed by choosing some solution restricted only to $d_0, \ldots, d_\ell$, and then summing up weights of the corresponding paths. It follows that computing the optimal solution can be done, similarly as in the proof of Lemma 2, by finding an increasing subsequence. We define a new weighted sequence $(a'_0, \ldots, a'_\ell)$ by setting $a'_j = b_{\pi_b(d_j)}$ and choosing the weight of a'_j to be $|I_j|$. Then an increasing subsequence of $(a'_0, \ldots, a'_\ell)$ corresponds to a valid solution restricted to $d_0, \ldots, d_\ell$, and moreover the weight of the heaviest such subsequence is exactly s. In other words, we can reduce our question to computing the heaviest increasing subsequence.

Finally, we need to analyze the complexity of our reduction. Assuming random access to both $(a_1, \ldots, a_m)$ and π_b, we can construct $(a'_0, \ldots, a'_\ell)$ in time $\mathcal{O}(\ell)$. Computing the weight of every a'_j is more complicated. We need to find every $|I_j|$ without explicitly constructing the paths. For every d_j we can retrieve the rank r_j of its corresponding element in $\{b_1, \ldots, b_m\}$. Then I_j contains d_j and all i such that the predecessor of b_i among $\{b_{d_0}, \ldots, b_{d_\ell}\}$ is b_{d_j}. Hence $|I_j|$ can be computed by counting such i. This can be done by locating the successor $b_{d_{j'}}$ of b_{d_j} in $\{b_{d_0}, \ldots, b_{d_\ell}\}$ and returning $r_{d_{j'}} - r_{d_j} - 1$ (if the successor does not exist, $m - r_{d_j}$). To find all these successors, we only need to sort $\{b_{d_0}, \ldots, b_{d_\ell}\}$, which can, again, be done in time $\mathcal{O}(\ell \log \log \ell)$. $\qquad \square$

Lemma 7. *Given a sequence of ℓ weighted elements, we can compute its heaviest increasing subsequence in time $\mathcal{O}(\ell \log \log \ell)$.*

Proof. Let the sequence be $(a_1, \ldots, a_\ell)$, and denote the weight of a_i by w_i. We will describe how to compute the weight of the heaviest increasing subsequence, reconstructing the subsequence itself will be straightforward. At a high level, for each i we want to compute the weight r_i of the heaviest increasing subsequence ending at a_i. Observe that $r_i = w_i + \max\{r_j : j < i \text{ and } a_j < a_i\}$, where we assume that $a_0 = -\infty$ and $r_0 = 0$. We process $i = 1, \ldots, \ell$, so we need a dynamic structure where we could store all already computed results r_j so that we can select the appropriate one efficiently. To simplify the implementation of this structure, we rename the elements in the sequence so that $a_i \in \{1, \ldots, \ell\}$. This can be done in $\mathcal{O}(\ell \log \log \ell)$ time by sorting. Then the dynamic structure needs to store n values $v_1, \ldots, v_n$, all initialized to $-\infty$ in the beginning, and implement two operations:

1. increase any v_k,
2. given k, return the maximum among $v_1, \ldots, v_k$.

Then to compute r_i we first find the maximum among $v_1, \ldots, v_{a_i-1}$, and afterwards update v_{a_i} to be r_i. Both operations can be implemented in amortized time $\mathcal{O}(\log \log \ell)$ using a van Emde Boas tree. $\qquad \square$

By combining the above ingredients (and, as mentioned before, cutting the input into overlapping fragments of length $2m$) we obtain the following result.

Theorem 1. *Order-preserving pattern matching with k mismatches can be solved in time $\mathcal{O}(n(\log\log m + k\log\log k))$, where n is the length of the text and m is the length of the pattern.*

5 Conclusions

Recall that the complexity of our solution is $\mathcal{O}(n(\log\log m + k\log\log k))$. Given that it is straightforward to prove a lower bound of $\Omega(n + m\log m)$ in the comparison model, and that for $k = 0$ one can achieve $\mathcal{O}(n + \mathrm{sort}(m))$ time [10], a natural question is whether achieving $\mathcal{O}(nf(k)) + \mathcal{O}(m\,\mathrm{polylog}(m))$ time is possible. Finally, even though the version with k errors seems hard (see the introduction), there might be an $\mathcal{O}(nf(k))$ time algorithm, with $f(k)$ being an exponential function.

References

1. Belazzougui, D., Pierrot, A., Raffinot, M., Vialette, S.: Single and multiple consecutive permutation motif search. In: Cai, L., Cheng, S.-W., Lam, T.-W. (eds.) ISAAC 2013. LNCS, vol. 8283, pp. 66–77. Springer, Heidelberg (2013)
2. Bender, M.A., Farach-Colton, M.: The LCA problem revisited. In: Gonnet, G.H., Viola, A. (eds.) LATIN 2000. LNCS, vol. 1776, pp. 88–94. Springer, Heidelberg (2000)
3. Bose, P., Buss, J.F., Lubiw, A.: Pattern matching for permutations. Inf. Process. Lett. 65(5), 277–283 (1998)
4. Crochemore, M., et al.: Order-preserving incomplete suffix trees and order-preserving indexes. In: Kurland, O., Lewenstein, M., Porat, E. (eds.) SPIRE 2013. LNCS, vol. 8214, pp. 84–95. Springer, Heidelberg (2013)
5. Han, Y.: Deterministic sorting in $O(n\log\log n)$ time and linear space. In: Proceedings of the Thiry-fourth Annual ACM Symposium on Theory of Computing, STOC 2002, pp. 602–608. ACM, New York (2002)
6. Hazay, C., Lewenstein, M., Sokol, D.: Approximate parameterized matching. ACM Transactions on Algorithms 3(3) (2007)
7. Hunt, J.W., Szymanski, T.G.: A fast algorithm for computing longest common subsequences. Commun. ACM 20(5), 350–353 (1977)
8. Kärkkäinen, J., Sanders, P., Burkhardt, S.: Linear work suffix array construction. J. ACM 53(6), 918–936 (2006)
9. Kim, J., Eades, P., Fleischer, R., Hong, S.H., Iliopoulos, C.S., Park, K., Puglisi, S.J., Tokuyama, T.: Order preserving matching. CoRR abs/1302.4064 (2013)
10. Kubica, M., Kulczyński, T., Radoszewski, J., Rytter, W., Waleń, T.: A linear time algorithm for consecutive permutation pattern matching. Inf. Process. Lett. 113(12), 430–433 (2013)

Parameterized Complexity Analysis for the Closest String with Wildcards Problem

Danny Hermelin[1,*] and Liat Rozenberg[2]

[1] Ben-Gurion University
hermelin@bgu.ac.il
[2] University of Haifa
liatle@gmail.com

Abstract. The Closest String problem asks to find a string s which is not too far from each string in a set of m input strings, where the distance is taken as the Hamming distance. This well-studied problem has various applications in computational biology and drug design. In this paper, we introduce a new variant of Closest String where the input strings can contain wildcards that can match any letter in the alphabet, and the goal is to find a solution string without wildcards. We call this problem the Closest String with Wildcards problem, and we analyze it in the framework of parameterized complexity. Our study determines for each natural parameterization whether this parameterization yields a fixed-parameter algorithm, or whether such an algorithm is highly unlikely to exist.

More specifically, let m denote the number of input strings, each of length n, and let d be the given distance bound for the solution string. Furthermore, let k denote the minimum number of wildcards in any input string. We present fixed-parameter algorithms for the parameters m, n, and $k+d$, respectively. On the other hand, we then show that such results are unlikely to exist when k and d are taken as single parameters. This is done by showing that the problem is NP-hard already for $k = 0$ and $d \geq 2$. Finally, to complement the latter result, we present a polynomial-time algorithm for the case of $d = 1$. Apart from this last result, all other results hold even when the strings are over a general alphabet.

1 Introduction

Let Σ be an arbitrary alphabet set, and let $d(s, s')$ denote the Hamming distance between two strings $s, s' \in \Sigma^n$. In the *Closest String* (CS) problem, we are given a set of m strings, $S = \{s_1, s_2, ..., s_m\} \subseteq \Sigma^n$, and a positive integer d, and the goal is to find (or to determine whether there exists) a string s such that $d(s, S) \leq d$. That is, a string s with $d(s, s_i) \leq d$ for all $i = 1, ..., m$. This problem (which is also known as the *consensus* or *center* string problem)

* The research leading to these results has received funding from the People Programme (Marie Curie Actions) of the European Union's Seventh Framework Programme (FP7/2007-2013) under REA grant agreement number 631163.11

A.S. Kulikov, S.O. Kuznetsov, and P. Pevzner (Eds.): CPM 2014, LNCS 8486, pp. 140–149, 2014.

is one of the most important multiple string comparison problems, with numerous bioinformatic applications such as motif finding [9,12,15], PCR primer design [6,12,15,21], genetic probe design [15], and antisense drug design [15].

In this paper, we address a new variant of the Closest String problem which we call the *Closest String with Wildcards* (CSW) problem. In this variant, the given input strings may include any number of occurrences of the *wildcard* character '*' that matches all the characters in the alphabet Σ. Thus, for two strings $s, s' \in (\Sigma \cup \{*\})^n$, the Hamming distance between these two strings is now defined as $d(s, s') := |\{i : s[i] \neq s'[i], s[i] \neq *, s'[i] \neq *, 1 \leq i \leq n\}|$. However, the solution is required not to contain any wildcards at all (otherwise the all-wildcard string is always a solution).

Definition 1 (*Closest String with Wildcards (CSW)*). *Given m strings $S = \{s_1, s_2, ..., s_m\} \subseteq (\Sigma \cup \{*\})^n$, and a positive integer d, find a string $s \in \Sigma^n$ such that $d(s, S) \leq d$.*

The CSW problem is important not only as a generalization of the well known CSP, but also since it represents more realistic biological sequences that have been created by alignment generating technologies. The development in the next generation sequencing technologies allows generating large amounts of genomic data. As described above, many biological applications ask for a representative string that satisfy a distance constrained from all of the generated sequences. The input strings of CSP have been assumed to represent a gapless sequences. However, biological sequence data is subject to errors in specific fragments, or may include corrupted reads. The missing data is represented as gaps, which we model as wildcard characters that have no influence on the Hamming distance between the rest of the sequence and the closest string.

Related work: The CS problem was proved to be NP-hard, even for binary strings [11]. Therefore, practical solutions for this problem mainly focused on polynomial time approximation scheme (PTAS) algorithms [17,18,19,1,5], and exact algorithm which are based on fixed parameter solutions [13,18,22]. We note that in the context of approximation algorithms, the optimization version of CS asks to find a string s that minimize the Hamming distance from s to the strings in S.

The first approximation algorithm was presented by Lanctot *et al.* [15] and gave a $4/3 + o(1)$ ratio. Li *et al.* [17] gave the first PTAS algorithm that was later improved in [18,19]. The time complexity of the algorithm presented in [18] for $(1 + \epsilon)$ ratio is $O(n^{O(\epsilon^{-2})})$. Stojanovic *et al.* [22] proposed a linear time algorithm for the case of $d = 1$. Gramm *et al.* [13] showed that CSP is FPT when d is a parameter and presented an algorithm that runs in $O(mn + md \cdot d^d)$-time. They also showed that the problem is FPT when the number of strings m is fixed by using an integer linear programming formulation. Ma and Sun [18] gave an algorithm that runs in $O(m|\Sigma|^{O(d)})$-time, which is polynomial time if $d = \log m$ and Σ has constant size. Their result was further improved in [3,23,24]. Chen *et al.* [4] propose a new parameterized algorithm that improved previous results especially for small alphabet.

To the best of our knowledge, there are no published results on the CSW problem variation. The goal of this paper is to initiate a systematic study of this problem. In particular, we will analyze the problem using the framework of parameterized complexity [7].

Parameterized complexity: In parameterized complexity, problem instances are ordered pairs $(x, \kappa) \in \{0,1\}^* \times \mathbb{N}$, where x is a binary string that encodes the combinatorial input of the problem (in our case, a set of strings and a few integers), and κ is a non-negative integer which is referred to as the *parameter*. The parameter κ in most cases is independent of the input length, and is meant to model problem-specific parameters which tend to be small in practice. The central notion of parameterized complexity is that of fixed-parameter tractability: A problem is said to be *fixed-parameter tractable (FPT)* if there is an algorithm that solves the problem in $f(\kappa) \cdot |x|^{O(1)}$ time, where $f()$ is allowed to be an arbitrary (typically super-polynomial) function, and the exponent of $|x|^{O(1)}$ is independent of κ. Thus, for example, a running-time of $O(2^{\kappa^3} \cdot |x|^4)$ is considered feasible in parameterized complexity, while $O(|x|^{\kappa/2})$ is not. We refer the reader to [7,10,20] for more background information on parameterized complexity.

Our results: In this paper, the parameters we will examine for CSW are:

- The length of the input strings n.
- The number of input strings m.
- The distance bound d required of the solution string.
- The maximum number of wildcards k in any given input string.

For these four parameters, and their possible aggregations, we provide a complete picture regarding whether or not these yield fixed-parameter algorithms. In particular, Section 2 shows how to adapt the algorithm of Gramm *et al.* [13] for CS to the case where the input strings have wildcards, yielding a fixed-parameter algorithm with respect to m. In Section 3, we prove that CSW is fixed-parameter tractable when parameterized by n, and then use this algorithm in Section 4 to present an algorithm with respect to parameter $k + d$. Section 5 then shows that unless P=NP, no such algorithm exists when either k or d are taken as single parameters. Finally, in Section 6 we consider the case where the strings are binary and $d = 1$, and present a polynomial time algorithm for this case using a reduction to the 2-SAT problem. Section 7 concludes the paper, and discusses some interesting open problems for future work.

2   Parameter m

In this section we show how to use the seminal result of Lenstra regarding integer linear programming to devise a fixed-parameter algorithm with respect to m for the CSW problem. Our algorithm closely follows the ideas of Gramm *et al.* [13] who gave an FPT algorithm for CS with respect to m.

In the Integer Linear Programming Feasibility problem, we are given a set of variables some of which might be constrained to be integer, and a set of linear constraints, and the question is whether there exists an assignment to the variables which satisfies each of the constraints.

Theorem 1 (Lenstra [16]). *Integer Linear Programming Feasibility can be solved in $O(p^{9p/2}N)$ time, where N is the total input size, and p is the number of variables in the program.*

In what follows, let $[m] = \{1, \ldots, m\}$. Gramm *et al.* [13] first show that they can reduce the input set of strings to a set of strings over an alphabet of size m. Since this reduction works also for the case of strings with wildcards, we have the following lemma:

Lemma 1. *Let Σ be an arbitrary alphabet. There is a linear time algorithm that converts any set of strings $S = \{s_1, \ldots, s_m\} \subseteq (\Sigma \cup \{*\})^n$ to a set of strings $S' = \{s'_1, \ldots, s'_m\} \subseteq ([m] \cup \{*\})^n$ such that for any integer $d \geq 0$, any string $s \in \Sigma^n$ with $d(s, S) \leq d$ can be converted in linear time to a string $s' \in [m]^n$ with $d(s, S') \leq d$, and vice versa.*

The main idea of the algorithm by Gramm *et al.* is to utilize the fact that if we view our input set of strings as an $m \times n$ matrix, than each one of the n columns of this matrix (or more precisely, its transpose) is an element in $([m] \cup \{*\})^m$. Thus, there are at most $(m + 1)^m$ different *column types* in our input. We can write an equivalent instance of Integer Linear Program Feasibility (ILPF) from our CSW instance using this observation. Let T denote the set of all column types, and for each type $t \in T$, let n_t denote the number of columns in our input of type t. Then $n = \sum_{t \in T} n_t$. Also, let $T_i \subseteq T$ denote the set of all column types which have a wildcard at position i, for each $1 \leq i \leq m$, and let $\sigma_{t,i} \in \Sigma \cup \{*\}$ denote the letter at position i in a given column type t.

The instance of ILPF is defined over a set of $|T| \cdot |\Sigma| \leq (m+1)^{m+1}$ variables, one variable $x_{t,\sigma}$ for each $t \in T$ and $\sigma \in \Sigma$. The intended meaning of the variable $x_{t,\sigma}$ is to denote the number of columns of type t in the solution string which will be set to the letter σ. The ILPF instance then consists of the following constraints:

$$
\begin{array}{ll}
\sum_{t \in T \setminus T_i} \sum_{\sigma \in \Sigma \setminus \{\sigma_{t,i}\}} x_{t,\sigma} \leq d & \forall 1 \leq i \leq m \\
\sum_{\sigma \in \Sigma} x_{t,\sigma} = n_t & \forall 1 \leq t \leq T \qquad \text{(ILPF)} \\
x_{t,\sigma} \in \{0, 1, 2, \ldots\} & \forall 1 \leq t \leq T, \forall \sigma \in \Sigma
\end{array}
$$

Example 1. Consider an example set of three input strings $s_1 = 01*1$, $s_2 = *000$, and $s_3 = *101$. Then there are three column types $\alpha := [0, *, *]$, $\beta := [1, 0, 1]$, and $\gamma := [*, 0, 0]$. The three constraints of the first type in the corresponding ILPF instance are: $x_{\alpha,1} + x_{\beta,0} \leq d$, $x_{\beta,1} + x_{\gamma,1} \leq d$, and $x_{\beta,0} + x_{\gamma,1} \leq d$. The three constraints of the second type are: $x_{\alpha,0} + x_{\alpha,1} = 1$, $x_{\beta,0} + x_{\beta,1} = 2$, and $x_{\gamma,0} + x_{\gamma,1} = 1$.

Let s denote a string formed from a feasible solution to the above program. The i-th constraint of the first type insures that $d(s, s_i) \leq d$. The t-th constraint of the second type enforces that all columns of type t are set to some letter. Thus, it is not difficult to verify that the above program has a feasible solution iff our CSW instance has a solution. Plugging this into Theorem 1, gives us the following:

Theorem 2. *The* Closet String with Wildcards *problem can be solved in* $M^{O(M)}n$ *time, where* $M = (m+1)^{m+1}$.

3   Parameter n

In this section we consider the length n of our input strings as a parameter for the CSW problem. We observe that when the alphabet size $|\Sigma|$ is constant (or taken as a parameter) the problem can be trivially solved in FPT time by enumerating all $|\Sigma|^n$ possible solution strings. Thus, we will focus in the section on alphabets with arbitrary size.

For future purposes, we are interested in solving a slightly more general problem than CSW which we call the Generalized Closest String with Wildcard problem (GCSW). In this variant, each input string has its own distance bound, and the goal is to find a solution string which satisfies all bounds simultaneously. Thus, the input to the problem is a set of n pairs (s_i, d_i), $s_i \in \Sigma \cup \{*\}$ and $d_i \in \{0, 1, 2, \ldots\}$, and the goal is determine whether there exists a string s with $d(s, s_i) \le d_i$ for each i, $1 \le i \le n$.

Our algorithm is a rather simple recursive procedure. For a string s_i, and a set of positions $j_1, \ldots, j_\ell \in \{1, \ldots, n\}$ in s_i, we let $s_i \setminus \{j_1, \ldots, j_\ell\}$ denote the string obtained by s_i by removing all letters in positions $\{j_1, \ldots, j_\ell\}$.

- Let $s = \sigma^n$ for some arbitrary $\sigma \in \Sigma$.
- If $d_i < 0$ or some i, return NO.
- If $n \le d_i$ for all i, return s. Otherwise, assume *wlog* $d_1 < n$.
- Guess $j_1, \ldots, j_{n-d_i} \in \{1, \ldots, n\}$ positions in s_1.
  - Set $s[j_x] = s_1[j - x]$ for all $x \in \{1, \ldots, n - d_1\}$.
  - Set $d_i' = d_i - |\{s[j_x] \ne s_i[j_x] : 1 \le x \le n - d_1\}|$.
  - Recurse on $(s_1 \setminus \{j_1, \ldots, j_{n-d_i}\}, d_1'), \ldots, (s_m \setminus \{j_1, \ldots, j_{n-d_i}\}, d_m')$.

It is not difficult to see that the algorithm correctly computes a solution string s with $d(s, s_i) \le d_i$ for all i. Indeed, the correctness of the first steps before the branching are obvious. Furthermore, in the branch step we branch on all possibilities of $n - d_1$ positions in s_1 which have to match the corresponding positions in our solution string s. Once these are decided, we can safely recurse on all the remaining positions of the input strings with the updated distance bounds d_i'.

To bound the running time of our algorithm, we note that in each node in the recursion tree we branch to at most $2^{n/2}$ child nodes which have smaller length. Thus, the depth of the recursion is at most n, and the total number of recursive calls is at most $(2^{n/2})^n = 2^{n^2/2}$. Since the total amount of computation done in each recursive call (excluding the branching) is linear in mn, the time bound stated in the following theorem follows.

Theorem 3. *The* Closet String with Wildcards *problem can be solved in* $O(2^{n^2/2} \cdot mn)$ *time.*

4 Parameter $k + d$

In this section we consider the aggregated parameter $k + d$, where as usual k denotes the minimum number of wildcards in any input string and d denotes the required distance bound from the solution string. We begin by first recalling the $O(d^d mn)$ time algorithm by Gramm *et al.* [13] for the Closest String problem.

The algorithm in [13] uses any arbitrary input string s_i to create the set of all solution strings as follows: It first selects as a solution $s = s_i$, and searches for another input string with distance at least $d + 1$ to the current solution string s. If the no such string is found the algorithm has found a solution. Otherwise, the algorithm recursively branches on all possibilities of correcting one of the $d + 1$ "errors" in s. In each recursive call the algorithm decrease d by 1 to account for the distance from the current solution string s to the original input string s_i. Thus, the total size of the recursion tree of the algorithm is $(d + 1)^d$, and in time $O((d + 1)^d mn)$ the algorithm finds all possible solution strings s. This time bound can also be reduced to $O((d + 1)^d mn)$ with a slight modification. The important thing to notice here is that this strategy also works when there are wildcards in the input strings, so long as we start with a solution string s_i that has no wildcards. This fact is formalized in the following lemma:

Lemma 2. *One can compute all $O(d^d)$ solution strings to an input of the Closest String with Wildcards problem in $O(d^d mn)$ time, assuming the set of input strings contains a string with no wildcards.*

From the lemma above one immediately obtains an $O(|\Sigma|^k d^d mn)$ time algorithm for CSW by taking a string s_i which has k wildcards, and enumerating all possibilities for replacing these wildcards with letters from Σ. Thus, for constant size alphabets (or when Σ is taken as an additional parameter), we get an FPT algorithm with respect to $k + d$. We next describe an FPT algorithm for the same parameterization when Σ is arbitrarily large.

Our algorithm utilizes the fact that Lemma 2 above computes *all* solution strings in the case where one of the input strings contains no wildcards. Assume w.l.o.g. that s_1 is an input string with k wildcards, and that these wildcards all occur in the first k positions of s_1. Write a_i for the length k prefix of s_i, and b_i for the length $n - k$ suffix of s_i, and let $A := \{a_1, \ldots, a_m\}$ and $B := \{b_1, \ldots, b_m\}$. Note that b_1 has no wildcards by our assumptions. We thus can apply the algorithm of Lemma 2 to obtain the set $SOL(B)$ of all solution strings for the instance (B, d) of CSW. If $SOL(B) = \emptyset$, then we output that there is no solution for our original CSW instance. Otherwise, for each $s \in SOL(B)$, we create an instance $(a_1, d - d_1), \ldots, (a_m, d - d_m)$ of GCSW by computing the Hamming distance d_i ($\leq d$) of s to each input string s_i. We then use the algorithm of Theorem 3 to solve each GCSW instance in $O(2^{k^2/2} mn)$ time. Any solution found for one of these instances can be combined with the solution string

$s \in SOL(B)$ the instance originated from to obtain a solution for our original instance.

Correctness of this strategy is immediate from the correctness of Lemma 2 and Theorem 3. A simple calculation shows that its total time complexity is $O(2^{k^2/2}d^d mn)$. Thus, we obtain the following theorem:

Theorem 4. *The* Closet String with Wildcards *problem can be solved in* $O(|\Sigma|^k d^d mn)$ *time, and in* $O(2^{k^2/2}d^d mn)$ *time.*

5   Parameters k and d

In the previous section we gave an FPT algorithm for CSW using the aggregate parameterization of $k + d$. In this section, we show that (assuming P is not equal NP) no such result is possible when either of these parameters are used alone in a parameterization for CSW. For parameter k this is immediate. Since CS, *i.e.* the CSW problem when the input contains no wildcards, is known to be NP-hard, an FPT algorithm with respect to k for CSW will have to solve any CS instance in $f(0) \cdot n^{O(1)} = O(n^{O(1)})$ time, implying that P = NP. Below we show a similar result for parameter d by proving the following:

Theorem 5. *The* Closet String with Wildcards *problem is NP-hard for* $d \geq 2$, *even when the strings are over a binary alphabet.*

The proof of the theorem above is via a reduction from the classical 3-SAT problem which was shown to be NP-complete by Karp [14]. In the 3-SAT problem, we are given a set of clauses $c_1, \ldots, c_m$ defined over a set of Boolean variables $x_1, \ldots, x_n$, where each clause c_i is a disjunction of 3 distinct literals (variables or their negation). The goal is to find an assignment to the variables that satisfies all clauses. Given such an instance of 3-SAT, we construct a set S of m strings of length n, such that there is a string with Hamming distance at most 2 to S iff the 3SAT formula is satisfiable.

For each clause c_i, we create a string s_i of length n as follows. Each position in the string corresponds to one variable. If x_j appears in c_i, then the j'th position of s_i will be set to 1 if x_j appears in c_i in its positive form, otherwise we set this position to 0. If x_j does not appear in c_i at all, then the j'th position of s_i will contain a wildcard. The set S is taken as $S := \{s_i : 1 \leq i \leq m\}$.

Example 2. Assume that $n = 6$ and we have a clause $c_i = (x_1 \vee \overline{x_3} \vee \overline{x_6})$. The string s_i corresponding to this clause will be: "1*0**0".

Now suppose $\varphi : \{x_1, \ldots, x_n\} \to \{0, 1\}$ is an assignment satisfying all clauses $c_i, 1 \leq i \leq m$. We claim that the string $s := \phi(x_1)\phi(x_2)\cdots\phi(x_n)$ has Hamming distance at most $d = 2$ from each string $s_i, 1 \leq i \leq m$. To this, consider an arbitrary string s_i with its corresponding clause c_i. Then there is a variable x_j for which $\phi(x_j) = 1$ if x_j appears in positive form in c_i, and otherwise $\phi(x_j) = 0$. By construction, s_j has the letter $\phi(x_j)$ in position j, and so $d(s, s_i) \leq 2$. Conversely, one can show using similar arguments that a string $s \in \{0, 1\}^n$ with

Hamming distance at most 2 from some string s_i corresponds to an assignment $\varphi : \{x_1, \ldots, x_n\} \to \{0, 1\}$ satisfying the clause c_j. Thus, a solution to the CSW instance $(S, 2)$ corresponds to a solution to our 3-SAT instance.

This proves that the CSW problem with $d = 2$ is NP-complete. To prove that the problem is NP-complete for larger distances, we show a reduction from NP-hardness for some d implies hardness for $d + 1$. Assume $s_1, \ldots, s_m$ are the input strings for a CSW instance with some arbitrary distance bound d. To each string s_i we add a suffix $s'_i \in \{0, 1\}^m$ which has 1 in position j and 0's in all other positions. It is not difficult to verify that a string s has distance d to all strings $s_1, \ldots, s_m$ iff the string $s0^n$ has distance $d + 1$ to all strings $s_1 s'_1, \ldots, s_m s'_m$. This shows that if CSW is NP-hard for d then it is also NP-hard for $d + 1$. Combining this with our proof for $d = 2$ completes the proof of Theorem 5.

6   The Case of $d = 1$

In the previous section we proved that CSW is NP-complete when $d \geq 2$. In this section we complement this result by presenting a polynomial-time algorithm for the case when $d = 1$ and the strings are binary strings. This case complements particulary nicely when considering the proof of Theorem 5, since here CSW can be reduced to the 2-SAT problem. Aspvall *et al.* [2] and Even *et al.* [8] both presented algorithms that determine in linear-time (with respect to the number of variables and clauses) whether a 2-SAT formula is satisfiable.

We consider the set of input strings $s_1, \ldots, s_m \in \{0, 1\}^n$ as a matrix of size $m \times n$. Our 2SAT instance will have n variables, one for each column, and up to four clauses for each pair of columns. Thus, the total size of the formula will be $O(n^2)$. For each pair of columns $i, j \in \{1, \ldots, n\}$, we add clauses according to the possible combination of bit pairs that appear in the column pair:

- If 00 appears in the column pair, then we the clause $(\overline{x_i} \vee \overline{x_j})$.
- If 01 appears in the column pair, then we the clause $(\overline{x_i} \vee x_j)$.
- If 10 appears in the column pair, then we the clause $(x_i \vee \overline{x_j})$.
- If 11 appears in the column pair, then we the clause $(x_i \vee x_j)$.

The main idea behind our algorithm is that any assignment which does not satisfy a clause represents exactly two mismatches in one of the strings that caused the clause to be constructed. Conversely, if there is an assignment that satisfies all clauses, then assignment corresponds to a binary string which is at Hamming distance at most 1 from all the input strings. Thus, the resulting 2-SAT is satisfiable iff there is a string s with $d(s, S) \leq 1$. Since our construction can be carried out easily in $O(mn^2)$ time, and using one of the linear-time algorithms for 2-SAT [2,8], this yields the following:

Theorem 6. *The* Closet String with Wildcards *problem on binary strings can be solved in* $O(mn^2)$*-time when* $d = 1$.

7   Conclusions and Open Problems

In this paper we introduced a new variant of the CS problem, called CSW, where the input strings are allowed to contain wildcards that match every letter of the alphabet. We analyzed this problem within the framework of parameterized complexity, and showed for each possible combination of parameters m, n, d, and k, whether or not this parameterization yields a fixed-parameter algorithm for CSW. We also showed that the problem is polynomial-time solvable when $d = 1$ and the strings are binary. A natural open question in this context is whether this can be generalized to arbitrary alphabets.

All our results focused on the scenario where we are interested only in optimal exact solutions. Another interesting direction for future work is to design approximation algorithms (or alternatively, inapproxability results) for the optimization version of CSW. Here there is more than one natural option for what we wish to optimize: It could be, for instance, the minimum possible distance bound d such that the solution string has Hamming distance at most d from all input strings, or the maximum number $m' \leq m$ of input strings that are within distance at most some specified d from the solution string. Note that it is likely that most techniques for CS will not carry through to CSW, since the Hamming distance function between strings with wildcards is not a metric; it does not obey the triangle inequality. For example, we have $d(0^n, *^n) = 0$ and $d(*^n, 1^n) = 0$, but $d(0^n, 1^n) = n$.

References

1. Amir, A., Paryenty, H., Roditty, L.: Approximations and partial solutions for the consensus sequence problem. In: Grossi, R., Sebastiani, F., Silvestri, F. (eds.) SPIRE 2011. LNCS, vol. 7024, pp. 168–173. Springer, Heidelberg (2011)
2. Aspvall, B., Plass, M.F., Tarjan, R.E.: A linear-time algorithm for testing the truth of certain quantified Boolean formulas. Information Processing Letters 8(3), 121–123 (1979)
3. Chen, Z.Z., Wang, L.: Fast exact algorithms for the closest string and substring problems with application to the planted (l, d)-motif model. IEEE/ACM Transactions on Computational Biology and Bioinformatics 8(5), 1400–1410 (2011)
4. Chen, Z.Z., Ma, B., Wang, L.: A three-string approach to the closest string problem. Journal of Computer and System Sciences 78(1), 164–178 (2012)
5. Chimani, M., Woste, M., Böcker, S.: A closer look at the closest string and closest substring problem. In: ALENEX, pp. 13–24 (2011)
6. Dopazo, J., Rodríguez, A., Sáiz, J.C., Sobrino, F.: Design of primers for pcr ampification of highly variable genomes. Computer Applications in the Biosciences: CABIOS 9(2), 123–125 (1993)
7. Downey, R.G., Fellows, M.R.: Parameterized Complexity. Springer (1999)
8. Even, S., Itai, A., Shamir, A.: On the complexity of timetable and multicommodity flow problems. SIAM Journal of Computing 5(4), 691–703 (1976)
9. Fellows, M.R., Gramm, J., Niedermeier, R.: On the parameterized intractability of motif search problems*. Combinatorica 26(2), 141–167 (2006)
10. Flum, J., Grohe, M.: Parameterized Complexity Theory. Springer (2006)

11. Frances, M., Litman, A.: On covering problems of codes. Theory of Computing Systems 30(2), 113–119 (1997)
12. Gramm, J., Hüffner, F., Niedermeie, R.: Closest strings, primer design, and motif search. In: Currents in Computational Molecular Biology, Poster Abstracts of RECOMB, vol. 2002, pp. 74–75 (2002)
13. Gramm, J., Niedermeier, R., Rossmanith, P.: Fixed-parameter algorithms for closest string and related problems. Algorithmica 37(1), 25–42 (2003)
14. Karp, R.M.: Reducibility among combinatorial problems. Complexity of Computer Computations, 85–103 (1972)
15. Lanctot, J.K., Li, M., Ma, B., Wang, S., Zhang, L.: Distinguishing string selection problems. In: Proceedings of the Tenth Annual ACM-SIAM Symposium on Discrete Algorithms, pp. 633–642 (1999)
16. Lenstra, W.: Integer programming with a fixed number of variables. Mathematics of Operations Research, 538–548 (1983)
17. Li, M., Ma, B., Wang, L.: On the closest string and substring problems. Journal of the ACM (JACM) 49(2), 157–171 (2002)
18. Ma, B., Sun, X.: More efficient algorithms for closest string and substring problems. SIAM Journal on Computing 39(4), 1432–1443 (2009)
19. Marx, D.: Closest substring problems with small distances. SIAM Journal on Computing 38(4), 1382–1410 (2008)
20. Niedermeier, R.: Invitation to Fixed-Parameter Algorithms. Oxford University Press (2006)
21. Proutski, V., Holmes, E.C.: Primer master: A new program for the design and analysis of pcr primers. Computer Applications in the Biosciences: CABIOS 12, 253–255 (1996)
22. Stojanovic, N., Berman, P., Gumucio, D., Hardison, R., Miller, W.: A linear-time algorithm for the 1-mismatch problem. In: Rau-Chaplin, A., Dehne, F., Sack, J.-R., Tamassia, R. (eds.) WADS 1997. LNCS, vol. 1272, pp. 126–135. Springer, Heidelberg (1997)
23. Wang, L., Zhu, B.: Efficient algorithms for the closest string and distinguishing string selection problems. In: Deng, X., Hopcroft, J.E., Xue, J. (eds.) FAW 2009. LNCS, vol. 5598, pp. 261–270. Springer, Heidelberg (2009)
24. Zhao, R., Zhang, N.: A more efficient closest string problem. In: Bioinformatics and Computational Biology, pp. 210–215 (2010)

Computing Palindromic Factorizations
and Palindromic Covers On-line

Tomohiro I[1,2], Shiho Sugimoto[1], Shunsuke Inenaga[1],
Hideo Bannai[1], and Masayuki Takeda[1]

[1] Department of Informatics, Kyushu University, Japan
{tomohiro.i,shiho.sugimoto,inenaga,bannai,takeda}@inf.kyushu-u.ac.jp
[2] Japan Society for the Promotion of Science (JSPS)

Abstract. A *palindromic factorization* of a string w is a factorization of
w consisting only of palindromic substrings of w. In this paper, we present
an on-line $O(n \log n)$-time $O(n)$-space algorithm to compute smallest
palindromic factorizations of all prefixes of w, where n is the length of a
given string w. We then show how to extend this algorithm to compute
smallest maximal palindromic factorizations of all prefixes of w, consisting only of maximal palindromes (non-extensible palindromic substring)
of each prefix, in $O(n \log n)$ time and $O(n)$ space, in an on-line manner.
We also present an on-line $O(n)$-time $O(n)$-space algorithm to compute
a smallest palindromic cover of w.

1 Introduction

A *factorization* of a string w is a sequence $w_1, \ldots, w_k$ of non-empty strings such
that $w = w_1 \cdots w_k$, i.e., a decomposition of w into non-empty substrings of w.

Recently, Alatabbi et al. [2] introduced a new kind of factorizations, called
maximal palindromic factorizations. A factorization of string w is called a maximal palindromic factorization of w, if every string in the factorization is a maximal palindrome (i.e., a non-extensible palindromic substring) in w, and contains
fewest maximal palindromes as possible. They presented an off-line $O(n)$-time
algorithm to compute a maximal palindrome, where n is the length of a given
string.

In this paper, we introduce yet another kind of factorization, called *palindromic factorizations*. A factorization of string w is called a palindromic
factorization of w, if every string in the factorization is a palindrome (not necessarily maximal), and contains fewest palindromes in it. We present an on-line
$O(n \log n)$-time $O(n)$-space algorithm to compute a palindromic factorization of
a given string of length n. In addition, we show how to extend this algorithm
to obtain an on-line $O(n \log n)$-time $O(n)$-space algorithm to compute a maximal palindromic factorization. We achieve the $O(n \log n)$-time bound in our
solutions using combinatorial properties of palindromic suffixes of strings. We
remark that the algorithm of Alattabi et al. [2] is off-line, and a naïve extension
of their algorithm to the on-line scenario leads to an $O(n^2)$-time bound.

A.S. Kulikov, S.O. Kuznetsov, and P. Pevzner (Eds.): CPM 2014, LNCS 8486, pp. 150–161, 2014.

Also, we consider the problem of covering a given string with fewest palindromes. We show how to compute such covers in $O(n)$ time in an off-line fashion, and later describe how to extend it to an on-line $O(n)$-time algorithm. Both of the algorithms use $O(n)$ space. This solves an open problem of Alatabbi et al. [2].

Related Work

Various kinds of factorization of strings, and efficient algorithms to compute them, have been proposed. Examples are the following:

Ziv and Lempel [27] proposed the LZ77 factorization, which was originally intended for data compression. The LZ77 factorization plays a central role in the linear-time algorithm for finding maximal repetitions (a.k.a. runs) of strings [16], and in an approximation algorithm to the smallest grammar problem [22]. Due to its importance, efficient algorithms for computing the LZ77 factorization have been proposed, both in off-line manner [1,14,9,13,12] and in on-line manner [21,23,26]. Some variants of the LZ77 factorization also exist; the LZEnd factorization [17] allows faster random access in compressed strings, and the reversed LZ factorization [15,24] is useful for finding gapped palindromes in strings.

Ziv and Lempel [28] also proposed the LZ78 factorization, later Welch [25] proposed its variant called the LZW factorization. These factorizations are used in data compression and efficient computation of alignments of two strings [7]. The LZ78/LZW factorization of a string of length n over an alphabet of size σ can be computed in $O(n \log \sigma)$ time. A more efficient algorithm for computing the LZ78/LZW factorization was proposed by Jansson et al. [11].

The Lyndon factorization [5] of a string is a sequence of Lyndon words arranged in lexicographical order, where a string x is called a Lyndon word if x itself is the lexicographically smallest suffix of x. The Lyndon factorizations are used in a bijective variant of Burrows-Wheeler transform [18]. Recently, it was shown that the size of Lyndon factorization of a string is a lower bound of the smallest grammar that generates only the string [10]. The Lyndon factorization of a string can be computed in linear time [8].

2   Preliminaries

2.1   Notations on Strings

Let Σ be the alphabet. An element of Σ^* is called a string. For string w, if $w = xyz$, then x is called a prefix, y is called a substring, and z is called a suffix of w, respectively. The sets of substrings and suffixes of w are denoted by $Substr(w)$ and $Suffix(w)$, respectively. The length of string w is denoted by $|w|$. The empty string ε is a string of length 0, that is, $|\varepsilon| = 0$. For $1 \leq i \leq |w|$, $w[i]$ denotes the i-th character of w. For $1 \leq i \leq j \leq |w|$, $w[i..j]$ denotes the substring of w that begins at position i and ends at position j. An integer d $(0 < d \leq |w|)$ is a *period* of string w if $w[i] = w[i + d]$ for any $1 \leq i \leq |w| - d$. Let w^{rev} denote the reversed string of s, that is, $w^{\mathrm{rev}} = w[|w|] \cdots w[2]w[1]$. For any $1 \leq i \leq j \leq |w|$, note $w[i..j]^{\mathrm{rev}} = w[j]w[j-1] \cdots w[i]$.

A string x is called a palindrome if $x = x^{\text{rev}}$. The radius of palindrome x is $\frac{|x|}{2}$. Let $P \subset \Sigma^*$ be the set of palindromes. String x is said to be a substring palindrome (resp. a palindromic suffix) of string w, if $x \in Substr(w) \cap P$ (resp. $x \in Suffix(w) \cap P$). The center of substring palindrome $w[i..j]$ of string w is $\frac{i+j}{2}$. A substring palindrome $w[i..j]$ is called the *maximal palindrome* at the center $\frac{i+j}{2}$, if no other palindromes at the center $\frac{i+j}{2}$ have a larger radius than $w[i..j]$, i.e., if $w[i-1] \neq w[j+1]$, $i = 1$, or $j = |w|$. Since a string w of length $n \geq 1$ has $2n - 1$ centers $(1, 1.5, \ldots, n - 0.5, n)$, w has exactly $2n - 1$ maximal palindromes. Note that every palindromic suffix of w is a maximal palindrome of w.

2.2 Palindromic Factorization and Palindromic Cover of String

A *factorization* of string w is a sequence $f_1, \ldots, f_k$ of non-empty strings such that $w = f_1 \cdots f_k$. Each f_i is called a factor of the factorization. The size of the factorization is the number k of factors in it. A factorization $f_1, \ldots, f_k$ of w is called a *palindromic factorization* of w, if $f_i \in P$ for all $1 \leq i \leq k$. A palindromic factorization $f_1, \ldots, f_k$ is called a *maximal palindromic factorization* of w, if each f_i is the maximal palindrome at center $|f_1 \cdots f_{i-1}| + \frac{|f_i|+1}{2}$ of string w. Since any single character $a \in \Sigma$ is a palindrome, any string has a palindromic factorization. On the other hand, there is a sequence of strings that have no maximal palindromic factorizations, e.g., string $\mathsf{a(baca)}^k$ with $k \geq 1$ has no maximal palindromic factorization.

For a positive integer n, let $[1, n] = \{1, \ldots, n\}$. A set $\{[b_1, e_1], \ldots, [b_h, e_h]\}$ of subintervals of $[1, n]$ is called a *cover* of interval $[1, n]$, if $\bigcup_{i=1}^{h}[b_i, e_i] = [1, n]$. The size of the cover is the number h of subintervals in it. A cover $\{[b_1, e_1], \ldots, [b_h, e_h]\}$ of $[1, n]$ is said to be a *palindromic cover* of string w of length n, if $w[b_i..e_i] \in P$ for all $1 \leq i \leq h$. Note that any palindromic factorization of string w is a palindromic cover of w, and hence any string has a palindromic cover.

In this paper, we give efficient solutions to the following problems.

Problem 1. Given a string w of length n, compute smallest palindromic factorizations of all prefixes of w.

Problem 2. Given a string w of length n, compute smallest maximal palindromic factorizations of all prefixes of w.

Problem 3. Given a string w of length n, compute smallest maximal palindromic covers of all prefixes of w.

2.3 Tools

To solve the above problems efficiently, we make use of the following known results on palindromes, periods and advanced data structures.

For any string w, let $Spals(w)$ be the set of palindromic suffixes of w.

Lemma 1 ([4,20]). *For any string w of length n, the lengths of all strings in $Spals(w)$ can be represented by $O(\log n)$ arithmetic progressions.*

Lemma 2 ([19]). *Given a string w of length n, we can compute the maximal palindromes for all centers in w in $O(n)$ time.*

Lemma 3 (Periodicity Lemma [6]). *Let d and d' be two periods of a string w. If $d + d' - \gcd(d, d') \le |w|$, then $\gcd(d, d')$ is also a period of w.*

For any two nodes u and v in the same path of a weighted rooted tree, let $\min(u, v)$ be a query that returns a node in the path with minimum weight.

Lemma 4 ([3]). *Under a RAM model, a dynamic tree can be maintained in linear space so that a $\min$ query and an operation of adding a leaf to the tree are both supported in the worst-case $O(1)$ time.*

3   Combinatorial Properties of Palindromic Suffixes

Here we introduce some combinatorial properties of palindromic suffixes as well as some notations which will be used for designing online algorithms in Sections 4.1 and 4.2. Some of the properties were stated in [4,20] in a different form.

The next lemma shows a mutual relation between palindromic suffixes and periods.

Lemma 5. *For any palindrome x, $z \in Spals(x)$ iff $|x| - |z|$ is a period of x.*

Proof. $\Rightarrow$: Since x is a palindrome and z is a suffix of x, z^{rev} is a prefix of x. Since z is a palindrome, $z^{rev} = z$ is a prefix and also a suffix of x, which means $|x| - |z|$ is a period of x.

$\Leftarrow$: Since $|x| - |z|$ is a period of x, z is a prefix of x. It follows from $x = x^{rev}$ that z (prefix of length $|z|$ of x) and z^{rev} (prefix of length $|z|$ of x^{rev}) are equivalent, which means $z \in Spals(x)$. $\square$

Lemma 5 leads to the following lemmas.

Lemma 6. *For any palindrome x, let z be the largest palindromic suffix of x with $|z| < |x|$. Then $|x| - |z|$ is the smallest period of x.*

Lemma 7. *For any string w and $y, z \in Spals(w)$ with $2|z| > |y| > |z|$, the suffix of length $2|z| - |y|$ is also in $Spals(w)$.*

For any string w of length n, let $LSpals(w)$ denote the lengths of the palindromic suffixes of w, i.e., $LSpals(w) = \{|z| \mid z \in Spals(w)\}$. Thanks to Lemma 1, $LSpals(w)$ can be represented by $O(\log n)$ arithmetic progressions. Let $LSpals(w) = \{q_1, q_2, \ldots, q_{|LSpals(w)|}\}$ with $q_1 < q_2 < \cdots < q_{|LSpals(w)|}$. As a consequence of Lemma 7, it is known that the differences between two adjacent elements in $LSpals(w)$ are monotonically non-decreasing.

We consider partitioning $LSpals(w)$ into $O(\log n)$ groups as follows: For any $1 \le j \le |LSpals(w)|$, q_j is contained in an *AP-group* of w with common difference d iff $q_j - q_{j-1} = q_{j+1} - q_j$ or $q_{j+1} - q_j = q_{j+2} - q_{j+1}$, where either of the equalities is ignored when $q_{j'}$ is used for $j' < 1$ or $j' > |LSpals(w)|$, respectively. Namely, if there exist more than two consecutive elements with common difference d of $LSpals(w)$, they, except the largest one, belong to the same group. An element of $LSpals(w)$ that does not belong to any AP-group makes a *single-group* that consists only of itself. Each group can be represented by $\langle q, d, q' \rangle$, where q (resp. q') is the smallest (resp. largest) element of the group and d is its common difference, i.e., $\langle q, d, q' \rangle = \{q, q + d, \dots, q'\}$, where $d = 0$ for single-groups. Let $X(w)$ denote the set of groups of w, that is $LSpals(w) = \bigcup_{\langle q,d,q' \rangle \in X(w)} \{q, q + d, \dots, q'\}$.

Let $P(w) = \{|w| - q \mid q \in LSpals(w)\}$, which represents the set of positions p s.t. $p + 1$ is the beginning position of a palindromic suffix of w. For any $\langle q, d, q' \rangle \in X$, let $P(w, \langle q, d, q' \rangle)$ denote a subset of P that restricts palindromic suffixes to ones corresponding to $\langle q, d, q' \rangle$, i.e., $P(w, \langle q, d, q' \rangle) = \{|w| - q, |w| - (q+d), \dots, |w| - q'\}$. The next lemma shows that the characters attached to the left of palindromic suffixes corresponding to $\langle q, d, q' \rangle$ are identical.

Lemma 8. *For any string w, $\langle q, d, q' \rangle \in X(w)$ and $p, p' \in P(w, \langle q, d, q' \rangle)$, $w[p] = w[p']$.*

Proof. Since it is clear when $|\langle q, d, q' \rangle| = 1$, consider the case where $|\langle q, d, q' \rangle| > 1$. It follows from $\langle q, d, q' \rangle \in X(w)$ and the definition of $X(w)$, $q' + d, q' \in LSpals(w)$. By Lemma 5, d is a period of the suffix z of length $q' + d$ of w. Since positions p, p' are in z and $|p - p'|$ is dividable by d, the lemma holds. $\square$

Lemma 9. *Let w be a string of length n and $a \in \Sigma$. Given a sorted list of $X(w)$, we can compute a sorted list of $X(wa)$ in $O(\log n)$ time.*

Proof. A simple but important observation is that for any $w[i..n+1] \in Spals(wa)$ with $i < n$, $w[i + 1..n] \in Spals(w)$. Then except for $w[n + 1]$ and $w[n..n + 1]$ we can compute $Spals(wa)$ by expanding palindromic suffixes of w.

After adding $w[n + 1]$ and $w[n..n + 1]$ (if $w[n] = a$) to a tentative list X of $X(wa)$, we process $\langle q, d, q' \rangle \in X(w)$ in increasing order of their lengths. Thanks to Lemma 8, we can process each $\langle q, d, q' \rangle$ in $O(1)$ time, that is, we add $\langle q + 2, d, q' + 2 \rangle$ to X iff $w[n-q] = a$. After processing all groups in $X(w)$ we check the consecutive groups in X and merge them into one AP-group if needed. Therefore we can get a sorted list of $X(wa)$ in $O(|X(w)|) = O(\log n)$ time. $\square$

4 Algorithms

4.1 Computing Smallest Palindromic Factorizations On-line

In this subsection, we present an $O(n \log n)$-time online algorithm that solves Problem 1 of computing smallest palindromic factorization of all prefixes of a string w of length n. More precisely, we compute an array F s.t. for each position $1 \le i \le n$, $i - F[i]$ gives the length of the last factor of a smallest palindromic factorization of $w[1..i]$ (when more than one factorization exists, choose arbitrary one). Notice that using F, given any position $1 \le i \le n$ one can compute a smallest palindromic factorization of $w[1..i]$ by computing the lengths of the factors from right to left in $O(k_i)$ time, where k_i is the number of factors. Our algorithm will also compute k_i's online.

We use the following abbreviation. For any $1 \leq i \leq n$, let $L_i = LSpals(w[1..i])$, $X_i = X(w[1..i])$ and $P_i = P(w[1..i])$. Also, for any $\langle q, d, q' \rangle \in X_i$ let $P_i \langle q, d, q' \rangle = P(w[1..i], \langle q, d, q' \rangle)$.

Suppose that we have processed positions $1, \ldots, i - 1$ and now processing i. Note that $F[i] = \arg\min_{p \in P_i}\{k_p\}$ and $k_i = \min_{p \in P_i}\{k_p + 1\}$, where $k_0 = 0$ for convenience. However checking all elements in P_i to compute the minimum value will take $O(i)$ time since $|P_i| = O(i)$ in the worst case.

In order to achieve our aim, we utilize X_i representation. Now we focus on the following subproblem: given $\langle q, d, q' \rangle \in X_i$, compute $\arg\min_{p \in P_i \langle q, d, q' \rangle}\{k_p\}$. We show how to solve this in constant time. When $|\langle q, d, q' \rangle|$ is small enough as we can treat it as a constant, we can solve this in constant time naïvely. However $|\langle q, d, q' \rangle| = O(i)$ in the worst case. In what follows, we consider how to process efficiently the case where $|\langle q, d, q' \rangle|$ is large. The next lemma gives a key observation.

Lemma 10. *For any $\langle q, d, q' \rangle \in X_i$ with $|\langle q, d, q' \rangle| \geq 3$, $\langle q, d, q' - d \rangle \in X_{i-d}$.*

Proof. First we show $q, q + d, \ldots, q'$ is a subsequence of L_{i-d}. By definition of X_i, the suffix z of length $q' + d$ of $w[1..i]$ is a palindrome. It follows from Lemma 5 that d is a period of z, which means that palindromic structures of $w[i - |z| + d + 1..i]$ are identical to those of $w[i - |z| + 1..i - d]$. In other words, $L_i \cap [1, q'] = L_{i-d} \cap [1, q']$. Hence $q, q + d, \ldots, q'$ is a subsequence of L_{i-d}. Another consequence of the equality $L_i \cap [1, q'] = L_{i-d} \cap [1, q']$ is that the largest element of L_{i-d} which is smaller than q is not $q - d$.

Next we show that $q' + d \notin L_{i-d}$. Assume on the contrary that $q' + d \in L_{i-d}$, i.e., the suffix y of length $q' + d$ of $w[1..i - d]$ is a palindrome. It follows from Lemma 5 that d is a period of y. Let x be the suffix of length q' of y. Note that x is also a prefix of z. Let $y = y'x$ and $z = xz'$. Since x is a palindrome and has y' as a prefix and z' as a suffix, $y' = (z')^{\mathrm{rev}}$. Therefore $y'xz'$, which is the suffix of length $q' + 2d$ of $w[1..i]$, is a palindrome. This implies that $q, q + d, \ldots, q' + d, q' + 2d$ is a subsequence of L_i, which contradicts that $\langle q, d, q' \rangle \in X_i$.

Putting all together, $q, q + d, \ldots, q'$ with $|\{q, q + d, \ldots, q'\}| \geq 3$ is a subsequence of L_{i-d} that is maximal with common difference d, which leads to the argument. $\square$

Since $P_{i-d}\langle q, d, q' - d \rangle = \{i - d - q, i - 2d - q, \ldots, i - q'\} = P_i \langle q, d, q' \rangle \setminus \{i - q\}$, Lemma 10 implies that at position $i - d$ we actually computed $p' = \arg\min_{p \in P_i \langle q, d, q' \rangle \setminus \{i-q\}}\{k_p\}$. Then if we keep this information, it suffices for us to compare $k_{p'}$ and k_{i-q}, and take the smaller one, which requires constant time.

More precisely, we maintain the following data structures dynamically: If there exists $\langle q, d, q' \rangle \in X_i$ with $|\langle q, d, q' \rangle| \geq 5$, our algorithm maintains an array A_d of length d s.t. for $j \equiv i \pmod{d}$,

just Before Processing Position i: $A_d[j]$ is undefined if $|\langle q, d, q' \rangle| = 5$, and
$A_d[j] = \arg\min_{p \in P_{i-d}\langle q, d, q' - d \rangle}\{k_p\}$ if $|\langle q, d, q' \rangle| \geq 6$.

just After Processing Position i: $A_d[j]$ is updated to be $\arg\min_{p \in P_i \langle q,d,q' \rangle}\{k_p\}$.

This array A_d is used for AP-groups $\langle q, d, q' \rangle \in X$ with $|\langle q, d, q' \rangle| \geq 5$ of adjacent positions. It is allocated when necessary and discarded as soon as an AP-group with common difference d disappears from X while processing positions.

The next lemma ensures that when A_d is allocated there exists AP-groups with common difference d for the d immediately preceding positions of i, which will be used for analyzing the cost for allocation/deallocation of A_d arrays.

Lemma 11. *Let* $\langle q, d, q' \rangle \in X_i$ *with* $|\langle q, d, q' \rangle| \geq 5$. *For any* $1 \leq h \leq d$, *there exists an AP-group of* $w[1..i - h]$ *with common difference* d.

Proof. We show that there exist at least three consecutive palindromic suffixes of $w[1..i-h]$ with common difference d. From the assumption, there exist at least six palindromic suffixes of $w[1..i]$ with common difference d. Let us take a look at the largest three of them of lengths $q' + d, q'$ and $q' - d$, i.e., $w[i - (q' + d) + 1..i]$, $w[i - q' + 1..i]$ and $w[i - (q' - d) + 1..i]$. Since $q' - d > 3d$, we can obtain three palindromes that ends at $i - h$, $x = w[i - (q' + d) + 1 + h..i - h]$, $y = w[i - q' + 1 + h..i - h]$ and $z = w[i - (q' - d) + 1 + h..i - h]$. Notice that $|x| - |y| = |y| - |z| = d$ and $|z| > d$.

Now we will show that there is no palindromic suffix u of $w[1..i - h]$ s.t. $|x| > |u| > |y|$ or $|y| > |u| > |z|$. Assume on the contrary that such u exists. Let $d' = |u| - |y|$ if $|u| > |y|$ and $d' = |u| - |z|$ if $|y| > |u|$. Let v be the prefix of length d of z. It follows from Lemma 5 that d' and $d - d'$ are periods of v. From the periodicity lemma (Lemma 3), $\gcd(d', d - d')$ is also a period of v. However, Lemma 6 implies that the suffix of length $q' + d$ of $w[1..i]$ has the smallest period d, and hence its substring of length d cannot have a period $d'' (< d)$ which can divide d, a contradiction. □

Theorem 1. *Problem 1 can be solved in* $O(n \log n)$ *time and* $O(n)$ *space in an online manner.*

Proof. Suppose that we have processed positions $1, \ldots, i - 1$, i.e., we now have X_{i-1} and A_d arrays are properly maintained. At the beginning of processing i, we compute X_i from X_{i-1} as well as allocate/deallocate A_d arrays if necessary. Next, for each $\langle q, d, q' \rangle \in X_i$ we compute $\arg\min_{p \in P_i \langle q,d,q' \rangle}\{k_p\}$ by naïvely checking if $|\langle q, d, q' \rangle| \leq 5$, and using A_d arrays if $|\langle q, d, q' \rangle| \geq 6$. If $|\langle q, d, q' \rangle| \geq 5$, we also update its corresponding value of A_d. Then $F[i]$ and k_i can be obtained from $\arg_p \min_{\langle q,d,q' \rangle \in X_i, p \in P_i \langle q,d,q' \rangle}\{k_p\}$.

By Lemma 9, computing X_i for all positions $1 \leq i \leq n$ takes a total of $O(n \log n)$ time. For each $\langle q, d, q' \rangle \in X_i$, $\arg\min_{p \in P_i \langle q,d,q' \rangle}\{k_p\}$ can be computed in constant time by using A_d arrays, and hence it takes $O(n \log n)$ time in total. It follows from Lemma 11 that the cost for allocation/deallocation of each position of A_d arrays can be attributed to distinct $\langle q, d, q' \rangle \in X_i$ for some position i. Therefore the algorithm runs in $O(n \log n)$ time.

During the computation we maintain A_d arrays dynamically so that for any position i its space usage is bounded by $\sum\{d \mid \langle q, d, q' \rangle \in X_i\} = O(i) = O(n)$. □

4.2   Computing Smallest Maximal Palindromic Factorizations On-line

In this subsection, we present an $O(n \log n)$-time online algorithm that solves
Problem 2 of computing smallest maximal palindromic factorization of all prefixes of a string w of length n. Our high-level strategy is similar to that of
previous subsection, i.e., checking palindromic suffixes of $w[1..i]$ for each position $1 \leq i \leq n$. However, in this problem we have to be careful not to use a
non-maximal palindrome as a factor.

For any position $1 \leq i \leq n$ let h_i be the number of factors of a maximal
palindromic factorization of $w[1..i]$ of the smallest size if such exists, and ∞
otherwise. For any position $1 \leq i < n$, let h'_i be the number of factors in a smallest
palindromic factorization of $w[1..i]$ which consists only of maximal palindromes
of $w[1..i+1]$ if such exists, and ∞ otherwise. Note that the values of h_i and h'_i
may differ, since we can use any palindromic suffix of $w[1..i]$ for h_i while for h'_i
we cannot use it if it is not maximal in $w[1..i+1]$.

For any position $1 \leq i < n$ let P'_i denote the set of positions p s.t. $w[p+1..i]$
is a maximal palindrome of w, that is $P'_i = \{p \in P_i \mid w[p] \neq w[i+1]\}$. It holds
that $h_i = \min_{p \in P_i} \{h'_p + 1\}$ and $h'_i = \min_{p \in P'_i} \{h'_p + 1\}$, where $\infty + 1 = \infty$. Similar
to array F of Section 4.1, in this problem we compute two arrays G and G' s.t.
for each position i, $G[i] = \arg\min_{p \in P_i} \{h'_p\}$ and $G'[i] = \arg\min_{p \in P'_i} \{h'_p\}$, where
$h'_0 = 0$ for convenience. Using these arrays, given any position $1 \leq i \leq n$, one
can compute a maximal palindromic factorization of $w[1..i]$ of the smallest size
(if $h_i \neq \infty$) in $O(h_i)$ time as follows; the h_ith factor is $w[G[i]+1..i]$, and for any
$1 < j \leq h_i$ the $(j-1)$th factor is $w[G'[i_j]+1..i_j]$, where $i_j + 1$ is the beginning
position of the jth factor.

In light of Lemma 8, for any $\langle q, d, q' \rangle \in X_i$ we can know in constant time if
this group corresponds to P'_i or not by just investigating $w[p]$ and $w[i+1]$ for
arbitrary position $p \in P_i \langle q, d, q' \rangle$. Hence during the computation of arrays G
and G' we can process each $\langle q, d, q' \rangle \in X_i$ in constant time.

Everything else can be managed in the same way as the algorithm proposed
in Section 4.1. As to the maintenance of A_d arrays, a minor remark is that we
use h'_p instead of k_p, i.e., the value of A_d is updated to be $\arg\min_{p \in P_i \langle q, d, q' \rangle} \{h'_p\}$.
Therefore we get the following theorem.

Theorem 2. *Problem 2 can be solved in $O(n \log n)$ time and $O(n)$ space in an
online manner.*

4.3   Computing Smallest Palindromic Covers On-line

In this subsection, we present an $O(n)$-time on-line algorithm that solves Problem 3 of computing smallest palindromic covers of all prefixes of a string w
of length n. We begin this subsection with a simpler problem of computing a
smallest palindromic cover of w, and present an $O(n)$-time offline algorithm that
solves the problem. The following observation is a key to our solution.

Observation 1. *If $\{[b_1, e_1], \ldots, [b_k, e_k]\}$ is a palindromic cover of string w, then
there is a palindromic cover $\{[b'_1, e'_1], \ldots, [b'_k, e'_k]\}$ of w such that $b'_i = b_i - d$ and*

$e_i' = e_i + d$ *for some $d \geq 0$ and $w[b_i'..e_i']$ is the maximal palindrome at center* $\frac{b_i'+e_i'}{2} = \frac{b_i+e_i}{2}$ *for all $1 \leq i \leq k$.*

The above observation says that for any palindromic cover of string w, there always exists a palindromic cover of w of the same size which consist only of maximal palindromes. Hence, to compute a palindromic cover of w of the smallest size, it suffices for us to consider covers which are composed only of maximal palindromes.

Theorem 3. *Given a string w of length n, we can compute a smallest palindromic cover of w in $O(n)$ time.*

Proof. We use an array R of length n such that $R[i]$ stores the beginning position of the maximal palindrome that contains position i with the least (leftmost) beginning position. We compute a palindromic cover of w from array R in a greedy manner, from right to left; The longest palindromic suffix of w is the rightmost palindrome of a smallest palindromic cover of w. Given a smallest palindromic cover of suffix $w[b..n]$, we add into the cover the maximal palindrome stored in $R[b-1]$, the maximal palindrome that contains position $b-1$ with the leftmost beginning position. The procedure terminates when we obtain from R a palindromic prefix of w, i.e., a maximal palindrome that begins at position 1 in w.

Let C be the maximal palindromic cover the above algorithm computes. Let C_k denote the subset of C that contains the last k intervals (the k consecutive intervals from right) in C. Let b_k be the beginning position of the leftmost interval of C_k. We show that C is a smallest maximal palindromic cover of w by induction on k. If $k = 1$, then C_1 contains the longest palindromic suffix of w. Then, clearly $C_1 = \{[b_1, n]\}$ is the smallest palindromic cover of $w[b_1..n]$. Assume that C_k is the smallest palindromic cover of $w[b_k..n]$ for $k \geq 1$, computed greedily as above. Then, $C_{k+1} = [R[b_k - 1], e] \cup C_k$, where e denotes the ending position of the corresponding maximal palindrome starting at $R[b_k - 1]$. By definition of $R[b_k - 1]$, C_{k+1} is a smallest palindromic cover of $w[b_{k+1}..n]$. Since $b_{|C|} = 1$, $C_{|C|} = C$ is a smallest palindromic cover of $w[b_{|C|}..n] = w$.

Let us analyze the time complexity of the algorithm. The maximal palindromes can be computed in $O(n)$ time by Lemma 2. To compute array R, we sort the maximal palindromes in increasing order of their beginning positions, and consider only the longest one for each beginning position. This can be done in $O(n)$ time by using bucket sort. Then R can be obtained in $O(n)$ time from the maximal palindromes above. Obviously it takes $O(n)$ time to greedily choose the maximal palindromes from R. This completes the proof.   □

Now, we extend the algorithm of Theorem 3 to compute smallest palindromic covers of all prefixes of a given string in an online manner (Problem 3). A basic idea is to compute the longest palindromic suffix of each prefix $w[1..i]$ of a given string w, which is the last (rightmost) palindrome of a smallest palindromic cover of $w[1..i]$.

Theorem 4. *Given a string w of length n, we can compute an $O(n)$-size representation of smallest palindromic covers of all prefixes of w in $O(n)$ time. Given a position j $(1 \leq j \leq n)$ in w, the representation returns the size s_j of a smallest palindromic cover of $w[1..j]$ in $O(1)$ time, and allows us to compute a smallest palindromic cover in $O(s_j)$ time.*

Proof. For any $1 \leq i \leq n$, let ℓ_i denote the length of the longest palindromic suffix of $w[1..i]$, and let s_i be the size of a smallest palindromic cover of $w[1..i]$. If $i = 1$, then clearly $[1, 1]$ is the smallest palindromic cover of $w[1..1] = w[1]$. Hence $s_1 = 1$ and $\ell_1 = 1$. Consider the case where $i \geq 2$. There are two cases to consider.

- When $|\ell_i| = |\ell_{i-1}| + 2$, namely, the longest palindromic suffix of $w[1..i]$ is an extension of that of $w[1..i]$. In this case, $s_i = \min\{s_{i-1}, s_{i-\ell_i} + 1\}$. (See also Fig. 1)
- When $|\ell_i| < |\ell_{i-1}| + 2$, namely, the longest palindromic suffix of $w[1..i]$ is not an extension of that of $w[1..i]$. In this case, $s_i = \min\{s_k \mid i - \ell_i \leq k < i\} + 1$.

For all $1 \leq i \leq n$, we can compute ℓ_i in $O(n)$ time in an on-line manner in increasing order of i [24]. We also compute s_i in increasing order of i. When $|\ell_i| = |\ell_{i-1}| + 2$, we can easily compute s_i in $O(1)$ time. To efficiently compute s_i also in the case where $|\ell_i| < |\ell_{i-1}| + 2$, we maintain a list such that its ith element is s_i. Now, given ℓ_i and the list that stores s_k for all $1 \leq k < i$, which is augmented with the data structure of Lemma 4, each s_i can be computed in $O(1)$ time. We append s_i to the list and the data structure of Lemma 4 in $O(1)$ time. Hence, the total time complexity is $O(n)$.

To answer a smallest palindromic cover of a prefix $w[1..j]$ for a given position $1 \leq j \leq n$ in $O(s_j)$ time, we do the following. For each $1 \leq i \leq n$ let $b_i = i - \ell_i + 1$, i.e., b_i is the beginning position of the longest palindromic suffix of $w[1..i]$. We maintain a list that represents b_i for all $1 \leq i \leq n$ in an on-line manner, which is also augmented with the data structure of Lemma 4. After computing ℓ_i and b_i for position i, we compute $r = \min\{b_k \mid b_i - 1 \leq k < i\}$ in $O(1)$ time by Lemma 4. Then, we have $r = R[b_i]$, where R is the array used in Theorem 3. Therefore, given a position j, we can compute a smallest palindromic cover of $w[1..j]$ from right to left, in $O(s_j)$ time. □

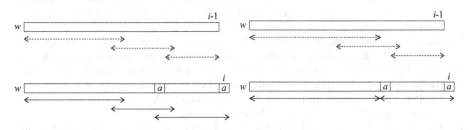

Fig. 1. To the left is the case where $s_i = s_{i-1}$, and to the right is the case where $s_i = s_{i-\ell_i} + 1$

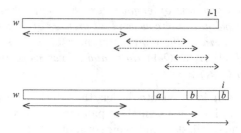

Fig. 2. The case where $s_i = \min\{s_k \mid i - \ell_i \leq k < i\} + 1$

References

1. Al-Hafeedh, A., Crochemore, M., Ilie, L., Kopylov, J., Smyth, W., Tischler, G., Yusufu, M.: A comparison of index-based Lempel-Ziv LZ77 factorization algorithms. ACM Computing Surveys 45(1), Article 5 (2012)
2. Alatabbi, A., Iliopoulos, C.S., Rahman, M.S.: Maximal palindromic factorization. In: Proc. PSC 2013, pp. 70–77 (2013)
3. Alstrup, S., Holm, J.: Improved algorithms for finding level ancestors in dynamic trees. In: Welzl, E., Montanari, U., Rolim, J.D.P. (eds.) ICALP 2000. LNCS, vol. 1853, pp. 73–84. Springer, Heidelberg (2000)
4. Apostolico, A., Breslauer, D., Galil, Z.: Parallel detection of all palindromes in a string. Theoretical Computer Science 141(1&2), 163–173 (1995)
5. Chen, K.T., Fox, R.H., Lyndon, R.C.: Free differential calculus. iv. the quotient groups of the lower central series. Annals of Mathematics 68(1), 81–95 (1958)
6. Crochemore, M., Rytter, W.: Text Algorithms. Oxford University Press, New York (1994)
7. Crochemore, M., Landau, G.M., Ziv-Ukelson, M.: A subquadratic sequence alignment algorithm for unrestricted scoring matrices. SIAM J. Comput. 32(6), 1654–1673 (2003)
8. Duval, J.P.: Factorizing words over an ordered alphabet. J. Algorithms 4(4), 363–381 (1983)
9. Goto, K., Bannai, H.: Simpler and faster Lempel Ziv factorization. In: Proc. DCC 2013, pp. 133–142 (2013)
10. I, T., Nakashima, Y., Inenaga, S., Bannai, H., Takeda, M.: Faster Lyndon factorization algorithms for SLP and LZ78 compressed text. In: Kurland, O., Lewenstein, M., Porat, E. (eds.) SPIRE 2013. LNCS, vol. 8214, pp. 174–185. Springer, Heidelberg (2013)
11. Jansson, J., Sadakane, K., Sung, W.-K.: Compressed dynamic tries with applications to LZ-compression in sublinear time and space. In: Arvind, V., Prasad, S. (eds.) FSTTCS 2007. LNCS, vol. 4855, pp. 424–435. Springer, Heidelberg (2007)
12. Kärkkäinen, J., Kempa, D., Puglisi, S.J.: Lightweight Lempel-Ziv parsing. In: Bonifaci, V., Demetrescu, C., Marchetti-Spaccamela, A. (eds.) SEA 2013. LNCS, vol. 7933, pp. 139–150. Springer, Heidelberg (2013)
13. Kärkkäinen, J., Kempa, D., Puglisi, S.J.: Linear time Lempel-Ziv factorization: Simple, fast, small. In: Fischer, J., Sanders, P. (eds.) CPM 2013. LNCS, vol. 7922, pp. 189–200. Springer, Heidelberg (2013)
14. Kempa, D., Puglisi, S.J.: Lempel-Ziv factorization: Simple, fast, practical. In: Proc. ALENEX 2013, pp. 103–112 (2013)

15. Kolpakov, R., Kucherov, G.: Searching for gapped palindromes. Theoretical Computer Science 410(51), 5365–5373 (2009)
16. Kolpakov, R., Kucherov, G.: Finding maximal repetitions in a word in linear time. In: Proc. FOCS 1999, pp. 596–604 (1999)
17. Kreft, S., Navarro, G.: LZ77-like compression with fast random access. In: Proc. DCC 2010, pp. 239–248 (2010)
18. Kufleitner, M.: On bijective variants of the Burrows-Wheeler transform. In: Proc. PSC 2009, pp. 65–79 (2009)
19. Manacher, G.K.: A new linear-time "on-line" algorithm for finding the smallest initial palindrome of a string. J. ACM 22(3), 346–351 (1975)
20. Matsubara, W., Inenaga, S., Ishino, A., Shinohara, A., Nakamura, T., Hashimoto, K.: Efficient algorithms to compute compressed longest common substrings and compressed palindromes. Theoretical Computer Science 410(8-10), 900–913 (2009)
21. Okanohara, D., Sadakane, K.: An online algorithm for finding the longest previous factors. In: Halperin, D., Mehlhorn, K. (eds.) ESA 2008. LNCS, vol. 5193, pp. 696–707. Springer, Heidelberg (2008)
22. Rytter, W.: Application of Lempel-Ziv factorization to the approximation of grammar-based compression. Theoretical Computer Science 302(1-3), 211–222 (2003)
23. Starikovskaya, T.: Computing Lempel-Ziv factorization online. In: Rovan, B., Sassone, V., Widmayer, P. (eds.) MFCS 2012. LNCS, vol. 7464, pp. 789–799. Springer, Heidelberg (2012)
24. Sugimoto, S., I, T., Inenaga, S., Bannai, H., Takeda, M.: Computing reversed Lempel-Ziv factorization online. In: Proc. PSC 2013. pp. 107–118 (2013)
25. Welch, T.A.: A technique for high-performance data compression. IEEE Computer 17(6), 8–19 (1984)
26. Yamamoto, J., I, T., Bannai, H., Inenaga, S., Takeda, M.: Faster compact on-line Lempel-Ziv factorization. To appear in Proc. STACS 2014 (2013), preprint is availabe at http://arxiv.org/abs/1305.6095
27. Ziv, J., Lempel, A.: A universal algorithm for sequential data compression. IEEE Transactions on Information Theory IT-23(3), 337–349 (1977)
28. Ziv, J., Lempel, A.: Compression of individual sequences via variable-length coding. IEEE Transactions on Information Theory 24(5), 530–536 (1978)

Compactness-Preserving Mapping on Trees[*]

Jan Baumbach[1], Jiong Guo[2], and Rashid Ibragimov[3]

[1] University of Southern Denmark, Campusvej 5, 5230 Odense M, Denmark
jan.baumbach@imada.sdu.dk
[2] Universität des Saarlandes, Campus E 1.7, Saarbrücken 66123, Germany
jguo@mmci.uni-saarland.de
[3] Max Planck Institute für Informatik, Saarbrücken 66123, Germany
ribragim@mpi-inf.mpg.de

Abstract. We introduce a generalization of the graph isomorphism problem. Given two graphs $G_1 = (V_1, E_1)$ and $G_2 = (V_2, E_2)$ and two integers l and d, we seek for a one-to-one mapping $f : V_1 \to V_2$, such that for every $v \in V_1$, it holds that $L'_v - L_v \leq d$, where $L_v := \sum_{u \in N^l_{G_1}(v)} \mathrm{dist}_{G_1}(v, u)$, $L'_v := \sum_{u \in N^l_{G_1}(v)} \mathrm{dist}_{G_2}(f(v), f(u))$, and $N^i_G(v)$ denotes the set of vertices which have distance at most i to v in a graph G. We call this problem COMPACTNESS-PRESERVING MAPPING (CPM). In the paper we study CPM with input graphs being trees and present a dichotomy of classical complexity with respect to different values of l and d. CPM on trees can be solved in polynomial time only if $l \leq 2$ and $d \leq 1$.

1 Introduction

Applications associated with graph isomorphism and related problems arise in many fields such as computer vision, pattern recognition, bioinformatics, chemistry [1,2,3]. Here one usually seeks for a one-to-one mapping between vertices of two graphs that preserves edges and/or satisfies some other constraints, such as vertex labeling.

In this paper we introduce and study a new generalization of the graph isomorphism problem, called Compactness-Preserving Mapping (CPM). Given two graphs $G_1 = (V_1, E_1)$ and $G_2 = (V_2, E_2)$, CPM asks if there is a one-to-one mapping $f : V_1 \to V_2$, such that for every vertex v from G_1 the vertices that are "close" to v in G_1 are mapped to the vertices in G_2 that are close in total to $f(v)$. Formally, the problem is defined as follows. Here $N^i_G(v)$ denotes the i-neighborhood of v, i.e., the set of vertices that have distance at most i to v in the graph G. L_v and L'_v are defined as follows. Given a graph G and an integer l, the proper distance L_v of vertex $v \in V(G)$ is defined as $L_v := \sum_{u \in N^l_G(v)} \mathrm{dist}_G(v, u)$. Given an integer l, two graphs G_1 and G_2 and a mapping $f : V(G_1) \to V(G_2)$, the image distance L'_v of vertex $v \in V(G_1)$ is defined as $L'_v := \sum_{u \in N^l_{G_1}(v)} \mathrm{dist}_{G_2}(f(v), f(u))$. The formal definition of CPM is as follows:

[*] Supported by the DFG Excellence Cluster MMCI and the International Max Planck Research School.

A.S. Kulikov, S.O. Kuznetsov, and P. Pevzner (Eds.): CPM 2014, LNCS 8486, pp. 162–171, 2014.

Input: Graphs $G_1 = (V_1, E_1)$ and $G_2 = (V_2, E_2)$ with $|V_1| = |V_2|$, non-negative integers l and d.

Question: Is there a one-to-one mapping $f : V_1 \to V_2$ such that $\forall v \in V_1$, $L'_v - L_v \leq d$?

We call a solution mapping f of CPM *compactness-preserving*.

The study of CPM is mainly motivated by the Protein-Protein Interaction (PPI) Network Alignment problem. The input of the problem consists of two graphs, i.e., PPI networks of two species, where vertices represent proteins and edges correspond to interactions between pairs of vertices. The output is a mapping between input graphs, which ideally maps proteins (i.e. vertices) of the same biological function and/or of the same evolutionary origin. We use CPM to model the network alignment problem for the following reasons. Since throughout the evolution many interactions (i.e. edges of the graphs) remain conserved, that, as consequence, results in mostly conserved neighborhoods of the vertices, a biologically meaningful mapping between two PPI networks should also map larger part of the corresponding interactions. However, the requirement that all edges have to be mapped exactly is too strict, since not all interactions are conserved and the data is noisy and incomplete. The parameter l of CPM determines the radius of the conserved neighborhoods of the vertices, while the parameter d can be used to reflect the degree of noise and incompleteness of data. Thus, given two PPI networks as input a mapping of CPM with large value of l and small value of d could represent a biologically meaningful alignment. The similar can be applied to the backbone subtrees extracted from PPI networks. Based on this, we study the classical complexity of CPM with input graphs being trees. In particular, we give a complexity dichotomy of CPM on trees. CPM on trees is polynomial-time solvable only if $l \leq 2$ and $d \leq 1$; all other cases turn out to be NP-complete.

A number of matching problems have been studied on trees. For instance, one considered two labeled rooted trees with ordered or unordered children as the input. The aim is to compute the tree edit distance, i.e. a mapping that respects the hierarchy of vertices in the trees and has minimal editing cost over all possible mappings. For the ordered version an algorithm running in $O(n^3)$ time was developed in [4] and with some improvements in [5], while the unordered version was shown to be NP-hard [6]. A closed variant of CPM, called NEIGHBORHOOD-PRESERVING MAPPING (NPM) has been recently studied for trees [7]. NPM on trees takes two unrooted, unlabeled trees T_1, T_2 and three integers l, d, k as input, and seeks for a mapping f from $V(T_1)$ to $V(T_2)$ such that for every vertex $v \in V(T_1)$ and its l-neighbors, with the exception of some deleted vertices $D \in V(T_1)$, the images of the neighbors are not too far from $f(v)$, i.e., within distance d, and $|D| \leq k$. NPM on trees is NP-hard, if $k > 0$. Note that CPM can easily be generalized to an approximate matching version by introducing a parameter as k. As for NPM, the approximate version of CPM on trees can be shown to be NP-complete for all values of l and d by a reduction from the subforest isomorphism problem [8].

Due to lack of space, we defer some proofs and details to the full version.

Preliminaries. Throughout this paper, we consider only simple, undirected graphs without self-loops. Given a graph G, we use $V(G)$ and $E(G)$ to denote the vertex and edge sets of G, respectively. The number of vertices in a graph G is denoted by $|G|$ or $|V(G)|$. The direct neighborhood of a vertex v in a graph G, denoted by $N_G(v)$, is the set of all vertices adjacent to v. The *degree* of v is $|N_G(v)|$. The length of a path P is the number of edges in P. The distance $\text{dist}_G(u,v)$ between two vertices $u, v \in V(G)$ is the length of the shortest path between u and v in G. The *exact i-neighborhood* $\hat{N}^i_G(v)$ of vertex $v \in G$ is a set of vertices that have distance exactly i to v in G. The degree-one vertices of a tree are called leaves. For a vertex u of a rooted tree T, we use $T(u)$ to denote the subtree rooted at u. For vertices u and v of an unrooted tree T, the subtree $T(u)$ with T rooted at v is denoted by $T(u,v)$. The size of a star S is the number of edges in S. The graph P_2 is a path of length two. A P_2-*packing* of size q in a graph G is a collection of q vertex-disjoint P_2's in G.

2   NP-Hardness Results

We first present the proofs of the NP-hard cases. Note that, given a graph G, for $v \in V(G)$ we have $L_v := \sum_{u \in N^l_{G_1}(v)} \text{dist}_G(v,u) = \sum_{i=1}^l i \cdot |\hat{N}^i_G(v)|$. The correctness of the following lemma is not difficult to prove:

Lemma 1. *If there is a compactness-preserving mapping f from T_1 to T_2 with $l = 1$, then for any $v \in V_1$ it holds $|N_{T_2}(f(v))| + d \geq |N_{T_1}(v)|$.*

Theorem 1. *CPM on trees with $l = 1$ and $d \geq 2$ is NP-complete.*

Proof. Clearly, given a mapping from T_1 to T_2, it can be tested in polynomial time whether it is compactness-preserving.

In order to show the NP-hardness we provide a reduction from 3-PARTITION, where given a set of $3n$ non-negative integers $A = \{a_1, \ldots, a_{3n}\}$ and an integer B, we ask whether there exists a *3-partition* of A? That is, can A be partitioned into n disjoint sets $A_1, A_2, \ldots, A_n$ such that for every $1 \leq j \leq n$ we have $|A_j| = 3$ and $\sum_{a \in A_j} a = B$? Note that 3-PARTITION remains strongly NP-hard even if $B/4 < a_i < B/2$ for every $1 \leq i \leq 3n$ [8]. In the following, given an instance (A, B) of 3-PARTITION with all elements of A being even, we construct the corresponding instance $(T_1, T_2, l = 1, d \geq 2)$ of CPM on trees.

We first introduce two graph gadgets. A *donor tree* $\rho := \text{DT}_1(a,b)$ consists of one vertex, called the *center vertex* of ρ, and a stars, each having b leaves. The centers of the stars are made adjacent to the center vertex of ρ. A *recipient tree* $\tilde{\rho} := \text{RT}_1(a,b,c)$ consists of a path with $c + 1$ vertices, $a - c$ stars, each having b leaves, and c stars, each having $b - 1$ leaves. The centers of all stars are connected to an end-vertex v of the path. We call v the *center vertex* of $\tilde{\rho}$, and the other end-vertex of the path the *tail* of $\tilde{\rho}$. Also note that the number of vertices in a recipient tree $\text{RT}_1(a,b,c)$ is independent of c and equal to $|\text{DT}_1(a,b)| = a \cdot (b+1) + 1$.

The tree T_1 consists of the following components:
- one star S_0 with the center c and $3n$ leaves,
- one "big" donor tree $D_b := DT_1(N_2, N_1)$,
- a set $\mathcal{D}_m$ of $3n$ "middle" donor trees $DT_1(N_4, N_3)$,
- a set $\mathcal{D}_s$ of nB "small" donor trees $DT_1(N_6, N_5)$, and
- a set $\mathcal{P}$ of $3n$ paths; the i-th path for $i = 1, \ldots, 3n$ corresponds to one particular element a_i of A and has a_i vertices.

Here $N_i := Bdn^{8-i}$, for $i = 1, \ldots, 6$.

The components are connected to T_1 as follows. Every vertex of a path from $\mathcal{P}$ is connected to the center vertex of one particular donor tree from $\mathcal{D}_s$. Then, for every leaf v of S_0, we connect to v a middle donor tree $\rho \in \mathcal{D}_m$ by adding an edge between v and the center of ρ. We also add en edge between v and one end-vertex of a path in $\mathcal{P}$. Finally, the big donor tree is connected to S_0 by adding an edge between its center vertex and c.

The tree T_2 consists of the following components:
- one star $\tilde{S}_0$ with the center $\tilde{c}$ and $3n$ leaves,
- one "big" recipient tree $\tilde{R}_b := RT_1(N_2, N_1, d+1)$,
- a set $\tilde{\mathcal{R}}_{m_1}$ of n "middle" recipient trees $RT_1(N_4, N_3, d+1)$,
- a set $\tilde{\mathcal{R}}_{m_2}$ of $2n$ "middle" recipient trees $RT_1(N_4, N_3, d-1)$,
- a set $\tilde{\mathcal{R}}_s$ of nB "small" recipient trees $RT_1(N_6, N_5, d-1)$, and
- a set $\tilde{\mathcal{P}} := \bigcup_{j=1}^n \tilde{\mathcal{P}}_j$, where $\tilde{\mathcal{P}}_j$ is a set of length-1 paths with $|\tilde{\mathcal{P}}_j| = B/2$.

The components are combined to T_2 as follows. Every vertex $v \in V(\tilde{\mathcal{P}})$ is connected to one particular recipient tree $\tilde{\rho}$ from $\tilde{\mathcal{R}}_s$ by adding an edge between v and the tail of $\tilde{\rho}$. Note that the distance between v and the center vertex of $\tilde{\rho}$ is equal to $d-1$. The vertex $\tilde{c}$ is made adjacent to the tail of $\tilde{R}_b$; every leaf of $\tilde{S}_0$ is made adjacent to the tail of one particular tree from $\tilde{\mathcal{R}}_{m_1} \cup \tilde{\mathcal{R}}_{m_2}$. By $\mathcal{L}_{m_1}$ and $\mathcal{L}_{m_2}$ we denote the sets of the leaves of $\tilde{S}_0$ that are adjacent to the recipient trees of $\tilde{\mathcal{R}}_{m_1}$ and $\tilde{\mathcal{R}}_{m_2}$, respectively. Note that the distance between the center vertex of a recipient tree $\tilde{\rho} \in \tilde{\mathcal{R}}_{m_1}$ and the corresponding leaf in $\mathcal{L}_{m_1}$ is $d+1$, which is equal to the distance between the tail of $\tilde{\rho}$ and the centers of the stars of $\tilde{\rho}$. Similarly, the distance between the center vertex of a recipient tree $\tilde{\rho} \in \tilde{\mathcal{R}}_{m_2}$ and the corresponding leaf in $\mathcal{L}_{m_2}$ is equal to $d-1$. Finally, we add an edge between the j-th vertex in $\mathcal{L}_{m_1}$ and one end-vertex of every path from $\tilde{\mathcal{P}}_j$, $j = 1, \ldots, n$.

Obviously, the time needed to construct T_1 and T_2 is polynomial in n, B, and d. $\qquad\square$

Theorem 2. *CPM on trees with $l = 2$ and $d \geq 2$ is NP-hard.*

Theorem 3. *CPM on trees with $l = 3$ and $d \geq 0$ is NP-hard.*

3   Polynomial-Time Solvable Cases

Case $l = 1$, $d = 0$. In the case of CPM on trees with $l = 1$ and $d = 0$ the problem is equivalent to TREE ISOMORPHISM, which can be solved in polynomial time [9].

Theorem 4. *CPM on trees with $l = 1$, $d = 0$ can be solved in polynomial time.*

Case $l = 2, d = 0$.

Lemma 2. *If there is a compactness-preserving mapping f with $l = 2$ and $d = 0$ from T_1 to T_2, then for every $v \in V_1$, $|N_{T_1}(v)| = |N_{T_2}(f(v))|$.*

Proof. Assume that there is a vertex $v \in V_1$ with $|N_{T_1}(v)| > |N_{T_2}(f(v))|$. Then, with $L_v = |N_{T_1}(v)| + 2|\hat{N}_{T_1}^2(v)|$ and $L'_v \geq |N_{T_2}(f(v))| + 2(|N_{T_1}^2(v)| - |N_{T_2}(f(v))|) = 2|N_{T_1}^2(v)| - |N_{T_2}(f(v))|$, we have $L'_v - L_v \geq 2|N_{T_1}^2(v)| - 2|\hat{N}_{T_1}^2(v)| - |N_{T_1}(v)| - |N_{T_2}(f(v))| = |N_{T_1}(v)| - |N_{T_2}(f(v))| > 0 = d$, which is a contradiction to the compactness-preserving property of $f(v)$. Thus, $|N_{T_1}(v)| \leq |N_{T_2}(f(v))|$.

Now assume $|N_{T_1}(v)| < |N_{T_2}(f(v))|$. Since $\sum_{y \in V_1} |N_{T_1}(y)| = \sum_{y' \in V_2} |N_{T_2}(y')| = 2|E_1|$, there exists $x \in V_1$ with $|N_{T_1}(x)| > |\hat{N}_{T_2}(f(x))|$, which contradicts to the previous conclusion. Thus, for each $v \in V_1$, $|N_{T_1}(v)| = |N_{T_2}(f(v))|$. □

Lemma 3. *If there is a compactness-preserving mapping f with $l = 2$ and $d = 0$ from T_1 to T_2, then for every $v \in V_1$ and $u \in N_{T_1}^2(v)$, $f(u) \in N_{T_2}^2(f(v))$.*

Proof. Assume that there exists $u \in N_{T_1}^2(v)$ for a vertex $v \in V_1$, such that $f(u) \notin N_{T_2}^2(f(v))$. Thus, $\Delta = \text{dist}_{T_2}(f(v), f(u)) \geq 3$. For the vertex v, we have $L_v = |N_{T_1}(v)| + 2|\hat{N}_{T_1}^2(v)|$ and, assuming that $|N_{T_1}(v)|$ many vertices in $N_{T_1}^2(v)$ are mapped to $N_{T_2}(f(v))$ and the remaining vertices, except the vertex u, are mapped to $\hat{N}_{T_2}^2(f(v))$, $L'_v \geq |N_{T_2}(f(v))| + 2(|\hat{N}_{T_1}^2(v)| - 1) + \Delta$. Then, it holds $L'_v - L_v \geq |N_{T_2}(f(v))| + 2(|\hat{N}_{T_1}^2(v)| - 1) + \Delta - (|N_{T_1}(v)| + 2|\hat{N}_{T_1}^2(v)|) = |N_{T_2}(f(v))| - |N_{T_1}(v)| + \Delta - 2 = \Delta - 2 > 0 = d$, which is a contradiction to the compactness-preserving property of $f(v)$, since $\Delta \geq 3$. □

The following two lemmas follow directly from Lemmas 2-3.

Lemma 4. *If there is a compactness-preserving mapping f with $l = 2$ and $d = 0$ from T_1 to T_2, then for every $v \in V_1$, $|\hat{N}_{T_1}^2(v)| = |\hat{N}_{T_2}^2(f(v))|$.*

Lemma 5. *If there is a compactness-preserving mapping f with $l = 2$ and $d = 0$ from T_1 to T_2, then for every $v \in V_1$, $u \in N_{T_1}(v)$, $\text{dist}_{T_2}(f(v), f(u)) \leq 2$.*

Lemma 6. *Let T_1 be a tree with diameter greater than 3. If there is a compactness-preserving mapping f with $l = 2$ and $d = 0$ from T_1 to T_2, then for every $(v, u) \in E_1$, $(f(v), f(u)) \in E_2$.*

Proof. Assume the claim is not true. Let f be the compactness-preserving mapping. According to Lemma 2, the leaves from T_1 can only be mapped to the leaves in T_2. Let $x \in \text{leaves}(T_1)$, $x' \in \text{leaves}(T_2)$ with $x' = f(x)$. Suppose T_1 and T_2 are rooted such that x' has the maximal depth in T_2. Let u and u' be the parent vertices of x and x', respectively; v and v' be the parent vertices of u and u', respectively.

According to the above lemmas, the leaves from $\hat{N}_{T_1}^2(x)$ have to be mapped to the leaves within distance two from x'. The only non-leaf vertices in $N_{T_2}^2(x')$ are u' and v'. If u is mapped to v', then $|N_{T_1}(u)| = |N_{T_2}(v')|$ and v can only be mapped to u'. Then, all vertices in $N_{T_1}^2(v) \setminus N_{T_1}(u)$ have to be mapped either to the leaf children of u' or to the vertices in $N_{T_2}(v')$. Since the diameter of T_1 is greater than 3, the former one is not possible. For the latter one, since all leaf children of u' can only be mapped to the children of u and $|N_{T_1}(v)| = |N_{T_2}(u')|$, the children of u must be all mapped to the children of u'. This means $|N_{T_1}(u)| = |N_{T_2}(u')|$ and $|N_{T_1}(v)| = |N_{T_2}(v')|$, which excludes the second possibility. Therefore, $f(u) = u'$. Thus, the vertex u can only be mapped to u', i.e. $u' = f(u)$. If the vertex v is not mapped to v', then $\text{dist}_{T_2}(x', f(v)) > 2$, contradicting to Lemma 5. Hence, it holds $v' = f(v)$.

Thus, the resulting so far mapping f keeps u and its neighbors adjacent in T_2, i.e. for every $t \in N_{T_1}(u)$, it holds $(f(t), f(u)) \in E_2$. Further, every $q \in \hat{N}^2_{T_1}(u)$ has to be mapped to $N_{T_2}(v') \setminus \{u'\}$; also note that $|\hat{N}^2_{T_1}(u)| = |N_{T_2}(v') \setminus \{u'\}|$. Thus, for $(q, v) \in E_1$, we have $(f(q), f(v)) \in E_2$. The similar has to be satisfied for $\hat{N}^2_{T_1}(v)$, which is mapped to $\hat{N}^2_{T_1}(v')$, and, continuing, for the remaining vertices in T_1 and T_2. In the conclusion, for every edge $(t_1, t_2) \in E_1$, we have $(f(t_1), f(t_2)) \in E_2$. $\quad\square$

Theorem 5. *CPM on trees with $l = 2$, $d = 0$ can be solved in polynomial time.*

Proof. By Lemma 6, if the diameter of T_1 is greater than 3, then the problem is equivalent to TREE ISOMORPHISM. We consider now the cases of diameter being 2 and 3. If T_1 is a star, then, by Lemma 2, T_2 must also be a star. If the diameter of T_1 is equal to 3, then there is a path in T_1 with 4 vertices x, u, v, t and all other vertices in T_1 are leaves and adjacent to u or v. Assume that the leaf x is mapped to a leaf $x' \in T_2$. Then, there are two non-leaf vertices $u', v' \in N^2_{T_2}(x')$ with $(x', u') \in E_2$. Obviously, the leaves adjacent to u have to be mapped to the leaves adjacent to u'. Consider first, u is mapped to u', then v has to be mapped to v' and the leaves adjacent to v must be mapped to the leaves adjacent to v'. Thus, the diameter of T_2 is also equal to 3 and $|N_{T_1}(u)| = |N_{T_2}(u')|$, $|N_{T_1}(v)| = |N_{T_2}(v')|$. Consider now, u is mapped to v', then v has to be mapped to u'. By Lemma 2, $|N_{T_2}(v')| = |N_{T_1}(u)|$ and $|N_{T_2}(u')| = |N_{T_1}(v)|$. By Lemma 4, we have $|\hat{N}^2_{T_2}(x')| = |\hat{N}^2_{T_1}(x)|$, which implies that $|N_{T_2}(u')| = |N_{T_1}(u)|$ and that the leaves adjacent to v can only be mapped to the leaves adjacent to v'. Thus, in this case the diameter of T_2 is also equal to 3 and non-leaf vertices of both trees have the same degrees. This is clearly checkable in polynomial time. $\quad\square$

Case $l = 1$, $d = 1$. We need the following observations.

Lemma 7. *Let $v, u \in V_1$ with $(v, u) \in E_1$. If there is a compactness-preserving mapping f from T_1 to T_2, then $\text{dist}_{T_2}(f(v), f(u)) \leq 2$.*

Proof. Assume that there exist $v \in V_1$ and $u \in N_{T_1}(v)$ such that $\Delta := \text{dist}_{T_2}(f(v), f(u)) \geq 3$. We have $L'_v \geq (|N_{T_1}(v)| - 1) + \Delta$, and $L'_v - L_v \geq (|N_{T_1}(v)| - 1) + \Delta - |N_{T_1}(v)| = \Delta - 1 > 1$. $\quad\square$

Lemma 8. *Let $v \in V_1$ and $p := |\{u \in N_{T_1}(v) : \text{dist}_{T_2}(f(v), f(u)) = 2\}|$. If there is a compactness-preserving mapping f from T_1 to T_2, then $p \leq 1$.*

Proof. Assume $p \geq 2$. We have $L'_v \geq (|N_{T_1}(v)| - p) + 2p$, and $L'_v - L_v \geq (|N_{T_1}(v)| - p) + 2p - |N_{T_1}(v)| = p > 1$.   □

Lemma 9. *Let $v, u, x \in V_1$ with $(v, u) \in E_1$, $(u, x) \in E_1$, and $(f(v), f(u)) \in E_2$. If there is a compactness-preserving mapping f from T_1 to T_2, then either for every $y \in V(T_1(x, u))$, $f(y) \in V(T_2(f(u), f(v)))$ or for every $y \in V(T_1(x, u))$, $f(y) \in V(T_2(f(v), f(u)))$.*

Lemma 10. *If there is a compactness-preserving mapping f from T_1 to T_2, then for every $v \in V_1$, $|N_{T_2}(f(v))| + 1 \geq |N_{T_1}(v)|$.*

In the following we assume that T_2 is rooted at an arbitrary vertex r' and tree T_1 remains unrooted.

Our algorithm firstly works from the leaves of T_2 to its root. At each vertex $v' \in V_2$, we iterate over all vertices $v \in V_1$ and decide whether v can be mapped to v'. To make this decision, we distinguish several cases, based on whether there is a neighbor u of v, such that one vertex $z \in T(u, v)$ could be mapped to the parent t' of v'. Let $u_2 \in N_{T_1}(v) \setminus \{u\}$, u'_1, u'_2, u'_3 be three children of v', and $s' \in N_{T_2}(t') \setminus \{v'\}$. We know by Lemmas 8-10 $z \in N^3_{T_1}(v)$. In the first case, we assume that there is a $u \in N_{T_1}(v)$ mapped to t'. Then, at most one neighbor x of u can be mapped to a child u'_1 of v'. If vertex x does not exist, we say that x and u'_1 are null elements, denoted by $\emptyset$. In the second case, we assume that there is a vertex $z \in T(u, v)$ mapped to t' and $z \in \hat{N}^2_{T_1}(v)$, i.e. $z \in N_{T_1}(u) \setminus \{v\}$. Then u has to be mapped to either s' or u'_1. If u is mapped to s', then there is at most one neighbor y of z that can be mapped to a child u'_1 of v'. In the third case, we assume $z \in T(u, v)$ and $z \in \hat{N}^3_{T_1}(v)$, i.e. z is a neighbor of $x \in N_{T_1}(u) \cap V(T(u, v))$ Then, either u is mapped to a child x'_1 of u'_1 and x is mapped to u'_1 or u is mapped to s' and x is mapped to $t'' \in N_{T_2}(s')$. For the first and second cases, where $z \in N^2_{T_1}(v)$ and u mapped to $N_{T_2}(v')$, there exists at most one neighbor u_2 of v that can be mapped to s' or to a child x'_2 of u'_2. If there is a neighbor u_2 of v mapped to x'_2, then, there exist at most one neighbor x_2 of u_2 and at most one neighbor y_2 of x_2 that can be mapped to two children u'_2 and u'_3 of v', respectively. Note if v' is the root r' of T_2, then z is a null element.

Given v and v', for every possible z, we guess all possible combinations of the vertices u, x, y, u_2, x_2, y_2 and $u'_1, u'_2, u'_3, x'_1, s', t''$ corresponding to the above cases to decide whether there can exist a compactness-preserving mapping f with $v' = f(v)$.

In the algorithm we encode the above combinations with configurations. A *configuration* $C = (c_1, c_2, c_3, c_4, c_5, c_6)$ means $c_1, c_3, c_5 \in V_1$ can be mapped to $c_2, c_4, c_6 \in V_2$, respectively, where $c_3 \in N^3_{T_1}(c_1)$, $c_5 \in N_{T_1}(c_1)$, c_2 is a child of c_4 and $\text{dist}_{T_2}(c_2, c_6) = 2$. Given a configuration C, the algorithm computes two sets $\bar{N}_1$ and $\bar{N}_2$ that contain neighbors of v and v', respectively, that can be mapped to the configuration C. That is, by the definition of the cases, we have $\bar{N}_1 := \{u, u_2\}$ and $\bar{N}_2 := \{t', u'_1, u'_2, u'_3\}$. Notice, that depending on the configuration under consideration some of the vertices in $\bar{N}_1$, $\bar{N}_2$ can be null elements.

For v, v' and configuration C, let $\mathcal{N}_1 := N_{T_1}(v) \setminus \bar{N}_1$ and $\mathcal{N}_2 := N_{T_2}(v') \setminus \bar{N}_2$. We say that there exists a matching $F(C)$ for a configuration C, if there are a set $B \subseteq \mathcal{N}_1 \cup N_{T_1}(\mathcal{N}_1) \setminus \{v\}$ and a bijection g between B and $\mathcal{N}_2$, such that the subtree $T_1(x)$ for each $x \in B$ can be mapped to the subtree $T_2(g(x))$. For a vertex $x \in B$, if $N_{T_1}(x) \cap B = \emptyset$, then $T_1(x) := T_1(x, v)$. Otherwise, let y be the only element in $N_{T_1}(x) \cap B$; in the case of $x \in N_{T_1}(v)$, $T_1(x) := T_1(x, v) \setminus T_1(y, x)$; in the case of $y \in N_{T_1}(v)$, $T_1(x) := T_1(x, y)$. A configuration C is *realizable* if there exists such a matching $F(C)$. A *confirming configuration* of a configuration C, denoted by $\hat{C}$, is a configuration $(\hat{c}_1, \hat{c}_2, \hat{c}_3, \hat{c}_4, \hat{c}_5, \hat{c}_6)$ with $\hat{c}_3 = c_3$, $\hat{c}_4 = c_4$, $\hat{c}_1 = c_5$, $\hat{c}_2 = c_6$, $\hat{c}_5 = c_1$, $\hat{c}_6 = c_2$.

The Algorithm: The algorithm works in two phases. In the first phase we process bottom-up and, for every vertex $v' \in V_2$, we compute the set of all realizable configurations by combining the sets of realizable configurations of the children of v'. We check for a configuration C^* whether there exists a matching F^*. If yes, then it saves C^* together with F^* for vertex v'; otherwise, it discards C^*.

In the second phase, given that there is at least one realizable configuration at the root of T_2, we construct the actual compactness-preserving mapping f in a top-down manner. Let $\mathcal{C}$ be the set of realizable configurations of all vertices. Note that, if the set of realizable configurations at a vertex $v' \in V_2$ is empty, then no compactness-preserving mapping exists and the algorithm returns "no".

Phase 1: We first compute realizable configurations for the leaves of T_2. Let v' be a leaf of T_2. According to Lemma 10, v' can only be mapped to a vertex $v \in V_1$ with degree at most 2. Consider first v is a leaf. Given the only neighbor u of v, we only need to test configurations $(v, v', u, t', \emptyset, \emptyset)$ and (v, v', x, t', u, s'), where $x \in N_{T_1}(u) \setminus \{v\}$ and $s' \in N_{T_2}(t') \setminus \{v'\}$. Now let $v \in V_1$ be of degree two and u_1, u_2 are the only neighbors of v. We test configurations $(v, v', u_i, t', u_{3-i}, s')$, where $i \in \{1, 2\}$ and $s' \in N_{T_2}(t') \setminus \{v'\}$.

Consider now a non-leaf vertex $v' \in V_2$ and a vertex $v \in V_1$. Let $\mathcal{C}_{v,v'} := \{C \in \mathcal{C} : c_4 = v', v = c_3\}$. Given v is mapped to v', we combine sets of realizable configurations of the children of v' to compute realizable configurations for v'. For v and v' we generate every possible configuration C^* according to the described cases and test if C^* is realizable.

Next, we describe how to check whether a configuration C^* is realizable. First, we compute $\bar{N}_1, \bar{N}_2$ and a set M, which contains all configurations for the vertices in $N_{T_1}^2(v) \setminus \bar{N}_1$ specifying how they should be mapped to $\mathcal{N}_2$ given that C^* is realizable. The set M is defined as $M := \{C \in \mathcal{C}_{v,v'} : c_1 \in \mathcal{N}_1, c_6 \notin \bar{N}_2$, and if $c_6 \in N_{T_2}(v')$ then, there exist a confirming configuration $\hat{C}$ for $C \in \mathcal{C}_{v,v'}\}$. Note that, a configuration $C \in \mathcal{C}_{v,v'}$ satisfying $c_6 \in N_{T_2}(v')$ is generated by assuming that there is a vertex $c_5 \in N_{T_1}(c_1)$ such that the subtree $T_1(c_5, c_1)$ is completely mapped to the subtree $T_2(c_6)$, where c_6 is a child of $c_4 = v'$. Now, while moving from c_2 to its parent v', we have to verify whether this assumption is sound to create realizable configurations for v'. This can be done by checking the existence of the confirming configurations for C.

The assumptions of all cases require that the vertices from $\mathcal{N}_1$ have to be mapped to the vertices in $\mathcal{N}_2$. Moreover, depending on the configurations in M, for every $u \in \mathcal{N}_1$, there could be at most one vertex in $N_{T_1}(u) \setminus \{v\}$, which is mapped to $\mathcal{N}_2$ as well.

With the help of M, we can compute a matching for C^*. Let $u_1, \ldots, u_m$ be the neighbors of v for which there is a configuration $C \in M$ with $c_1 = u_i$ for $1 \leq i \leq m$ and let $M_i := \{C \in M : c_1 = u_i\}$ for $i = 1, \ldots, m$. Given the definitions of configuration C^* and the corresponding set M, C^* is realizable if there is a "matching" set M^* of configurations $C^1, \ldots, C^m$ with $C^i \in M_i$ such that (1) $m = |\mathcal{N}_1|$, (2) $\{c_2, c_6 : C^i \in M^*\} \cap \{c_2, c_6 : C^j \in M^*\} = \emptyset$ for $i \neq j$, and (3) $\{c_2, c_6 : C \in M^*\} = \mathcal{N}_2$.

In the following we construct a weighted bipartite graph $B = B_1 \cup B_2$ based on the sets $M_1, \ldots, M_m$. For every configuration $C \in M_i$, we add a corresponding vertex $\hat{q}_i^j$, $j = 1, \ldots, |M_i|$ to B_1. For every vertex $u_i' \in \mathcal{N}_2$ we add to B_2 the corresponding vertex q_i. Additionally we add $2 \cdot (|M_i| - 1)$ vertices p_i^j, $j = 1, \ldots, 2 \cdot (|M_i| - 1)$ to B_2. Let $w_0 := 1$ and $w_i := 1 + 2n \cdot \sum_{\ell=0}^{i-1} w_\ell$. Further, we add an edge with weight w_i between every $\hat{q}_i^{j_1}$ and every $p_i^{j_2}$, $j_1 = 1, \ldots, |M_i|$, $j_2 = 1, \ldots, 2 \cdot (|M_i| - 1)$. For every $\hat{q}_i^j$ that corresponds to configuration $C \in M_i$, we add an edge with weight 1 between $\hat{q}_i^j$ and the vertex $p' \in B_2$ that corresponds to c_2 of C. Further, if $c_6 = \emptyset$, then we add an additional "dummy" vertex and connect it to $\hat{q}_i^j$ by adding an edge with weight 1; otherwise we add an edge with weight 1 between $\hat{q}_i^j$ and the vertex $p'' \in B_2$ that corresponds to c_6. The construction of B is completed with applying the above procedure to every $i = 1, \ldots, m$.

Now, we compute the matching set M^* by solving so called CONSTRAINED WEIGHTED P_2-PACKING ON BIPARTITE GRAPHS (CWPB) on B, where given a weighted bipartite graph $B = (B_1 \cup B_2, E_{12})$ in which each edge is associated with a positive weight, and an integer q, we seek for a P_2-packing $\mathcal{P}$ of size q with end-vertices in B_2 such that the weight of $\mathcal{P}$ is maximum over all such P_2-packings. CWPB can be solved in $O(q \cdot (|E_{12}| + |V(B)| \cdot \log |V(B)|))$ time [10]. The correctness of the following lemma follows from the construction of the graph B and the weights of its edges.

Lemma 11. *The configuration C^* is realizable if and only if there is a P_2-packing for the instance $(B, |M|)$ of CWPB.*

Given an output $\mathcal{P}$ for $(B, |M|)$, we derive the corresponding matching set M^* with the dummy vertices in B representing empty sets. Consequently, we save the matching $F(C^*)$ corresponding to M^* together with C^*.

Phase 2: Given a realizable configuration C at root r' of T_2, we first construct a compactness-preserving mapping f for r' and its children. We assign $r' = f(c_1)$, $c_5 = f(c_6)$ and use assignments from the matching $F(C)$. Then, we proceed top-down and construct f for children of every assigned vertex v'. Consider a realizable configuration $b = (f^{-1}(v'), v', f^{-1}(t'), t', b_3, b_3')$. If $b_3' \in T_2 \setminus (T_2(v') \cup \{t'\})$, then we use $F(b)$ to construct f for the children of v'; otherwise, we build f for the children of v' with $b_3 = f(b_3')$. If a child u' of v' is already assigned with $u = f^{-1}(u')$, then for the children of v' we retrieve a matching $F(b)$ with an additional requirement that u is matched to u'.

Theorem 6. *CPM on trees with $l = 1$, $d = 1$ can be solved in polynomial time.*

Proof. In both phases we iterate over vertices $v' \in V_2$. The vertex v' can have at most $|V_1|$ vertices that can possibly be mapped to v'. Then, for every pair v' and v, we generate all possible configurations that requires at most $O(n^6)$ combinations. In order to check if a configuration is realizable, we construct graph B for CWPB, whose size is clearly polynomial in n. Thus, the overall running time is polynomial. □

For the case of $l = 2$ and $d = 1$, we can observe similar properties as Lemmas 7-10. Thus, the basic ideas of the above algorithm could also apply, but with more complicated case analysis. Due to lack of space, we defer the detail to the full version.

Corollary 1. *CPM on trees with $l = 2$, $d = 1$ is polynomial-time solvable.*

4   Outlook

Given NP-hardness of the most cases of CPM, the next step would be to study the approximability of the optimization version of CPM. Hereby, the optimization function could be minimizing $\max\{L'_v - L_v\}$ over all $v \in V_1$. Concerning the solvability of CPM in practice, efficient heuristics are required, even for the cases we proved to be polynomial-time solvable; the polynomial factor here is very high. Finally, it would be interesting to study the variant of CPM, where the goal is to minimize the total difference of the proper and image distances over all vertices.

References

1. Heath, A.P., Kavraki, L.E.: Computational challenges in systems biology. Computer Science Review 3(1), 1–17 (2009)
2. Bunke, H., Riesen, K.: Recent advances in graph-based pattern recognition with applications in document analysis. Pattern Recognition 44(5), 1057–1067 (2011)
3. Conte, D., Foggia, P., Sansone, C., Vento, M.: Thirty years of graph matching in pattern recognition. IJPRAI 18(3), 265–298 (2004)
4. Demaine, E.D., Mozes, S., Rossman, B., Weimann, O.: An optimal decomposition algorithm for tree edit distance. In: Arge, L., Cachin, C., Jurdziński, T., Tarlecki, A. (eds.) ICALP 2007. LNCS, vol. 4596, pp. 146–157. Springer, Heidelberg (2007)
5. Pawlik, M., Augsten, N.: RTED: A Robust Algorithm for the Tree Edit Distance. PVLDB 5(4), 334–345 (2011)
6. Zhang, K., Statman, R., Shasha, D.: On the editing distance between unordered labeled trees. Information Processing Letters 42(3), 133–139 (1992)
7. Baumbach, J., Guo, J., Ibragimov, R.: Neighborhood-preserving mapping between trees. In: Dehne, F., Solis-Oba, R., Sack, J.-R. (eds.) WADS 2013. LNCS, vol. 8037, pp. 427–438. Springer, Heidelberg (2013)
8. Garey, M.R., Johnson, D.S.: Computers and Intractability, A Guide to the Theory of NP-Completeness. W. H. Freeman and Company, San Francisco (1979)
9. Aho, A.V., Hopcroft, J.E., Ullman, J.D.: The Design and Analysis of Computer Algorithms. Addison-Wesley Longman Publishing Co., Inc. (1974)
10. Feng, Q., Wang, J., Chen, J.: Matching and P_2-Packing: Weighted Versions. In: Fu, B., Du, D.-Z. (eds.) COCOON 2011. LNCS, vol. 6842, pp. 343–353. Springer, Heidelberg (2011)

Shortest Unique Substring Query Revisited*

Atalay Mert İleri[1], M. Oğuzhan Külekci[2], and Bojian Xu[3],**

[1] Department of Computer Engineering, Bilkent University, Turkey
[2] TÜBİTAK National Research Institute of Electronics and Cryptology, Turkey
[3] Department of Computer Science, Eastern Washington University, WA 99004, USA
aileri@bilkent.edu.tr, oguzhan.kulekci@tubitak.gov.tr, bojianxu@ewu.edu

Abstract. We revisit the problem of finding shortest unique substring (SUS) proposed recently by Pei *et al.* (ICDE'13). We propose an optimal $O(n)$ time and space algorithm that can find an SUS for every location of a string of size n and thus significantly improve their $O(n^2)$ time complexity. Our method also supports finding all the SUSes covering every location, whereas theirs can find only one SUS for every location. Further, our solution is simpler and easier to implement and can also be more space efficient in practice, since we only use the inverse suffix array and the longest common prefix array of the string, while their algorithm uses the suffix tree of the string and other auxiliary data structures. Our theoretical results are validated by an empirical study that shows our method is much faster and more space-saving.

Keywords: shortest unique substring, repetitiveness, regularity.

1 Introduction

Repetitive structure and regularity finding [1] has received much attention in stringology due to its comprehensive applications in different fields, including natural language processing, computational biology and bioinformatics, security, and data compression. However, finding the shortest unique substring (SUS) covering a given string location was not studied, until recently it was proposed by Pei *et al.* [5]. As pointed out in [5], SUS finding has its own important usage in search engines and bioinformatics. We refer readers to [5] for its detailed discussion on the applications of SUS finding. Pei *et al.* proposed a solution that costs $O(n^2)$ time and $O(n)$ space to find a SUS for every location of a string of size n. In this paper, we propose an optimal $O(n)$ time and space algorithm for SUS finding. Our method uses simpler data structures that include the suffix array, the inverse suffix array, and the longest common prefix array of the given string, whereas the method in [5] is built upon the suffix tree data structure. Our

---

* Part of the work was done while the first and the third authors were with TÜBİTAK-BİLGEM-UEKAE Mathematical and Computational Sciences Labs in 2013 summer. All missed proofs and pseudocode can be found in the full version of this paper [2].
** Corresponding author. Supported in part by EWU's Faculty Grants for Research and Creative Works.

A.S. Kulikov, S.O. Kuznetsov, and P. Pevzner (Eds.): CPM 2014, LNCS 8486, pp. 172–181, 2014.
© Springer International Publishing Switzerland 2014

algorithm also provides the functionality of finding all the SUSes covering every location, whereas the method of [5] searches for only one SUS for every location. Our method not only improves their results theoretically, the empirical study also shows that our method gains space saving by a factor of 20 and a speedup by a factor of four. The speedup gained by our method can become even more significant when the string becomes longer due to the quadratic time cost of [5]. Due to the very high memory consumption of [5], we were not able to run their method with massive data on our machine.

2   Preliminary

We consider a **string** $S[1 \ldots n]$, where each character $S[i]$ is drawn from an alphabet $\Sigma = \{1, 2, \ldots, \sigma\}$. A **substring** $S[i \ldots j]$ of S represents $S[i]S[i + 1] \ldots S[j]$ if $1 \leq i \leq j \leq n$, and is an empty string if $i > j$. String $S[i' \ldots j']$ is a **proper substring** of another string $S[i \ldots j]$ if $i \leq i' \leq j' \leq j$ and $j' - i' < j - i$. The **length** of a non-empty substring $S[i \ldots j]$, denoted as $|S[i \ldots j]|$, is $j - i + 1$. We define the length of an empty string is zero. A **prefix** of S is a substring $S[1 \ldots i]$ for some i, $1 \leq i \leq n$. A **proper prefix** $S[1 \ldots i]$ is a prefix of S where $i < n$. A **suffix** of S is a substring $S[i \ldots n]$ for some i, $1 \leq i \leq n$. A **proper suffix** $S[i \ldots n]$ is a suffix of S where $i > 1$. We say the character $S[i]$ occupies the string **location** i. We say the substring $S[i \ldots j]$ **covers** the kth location of S, if $i \leq k \leq j$. For two strings A and B, we write $\mathbf{A = B}$ (and say A is **equal** to B), if $|A| = |B|$ and $A[i] = B[i]$ for $i = 1, 2, \ldots, |A|$. We say A is lexicographically smaller than B, denoted as $\mathbf{A < B}$, if (1) A is a proper prefix of B, or (2) $A[1] < B[1]$, or (3) there exists an integer $k > 1$ such that $A[i] = B[i]$ for all $1 \leq i \leq k - 1$ but $A[k] < B[k]$. A substring $S[i \ldots j]$ of S is **unique**, if there does not exist another substring $S[i' \ldots j']$ of S, such that $S[i \ldots j] = S[i' \ldots j']$ but $i \neq i'$. A substring is a **repeat** if it is not unique.

Definition 1. *For a particular string location $k \in \{1, 2, \ldots, n\}$, the **shortest unique substring (SUS) covering location k**, denoted as $\mathbf{SUS_k}$, is a unique substring $S[i \ldots j]$, such that (1) $i \leq k \leq j$, and (2) there is no other unique substring $S[i' \ldots j']$ of S, such that $i' \leq k \leq j'$ and $j' - i' < j - i$.*

For any string location k, SUS_k must exist, because the string S itself can be SUS_k if none of the proper substrings of S is SUS_k. Also there might be multiple candidates for SUS_k. For example, if $S = \mathsf{abcbb}$, then SUS_2 can be either $S[1, 2] = \mathsf{ab}$ or $S[2, 3] = \mathsf{bc}$.

For a particular string location $k \in \{1, 2, \ldots, n\}$, the **left-bounded shortest unique substring (LSUS) starting at location** k, denoted as $\mathbf{LSUS_k}$, is a unique substring $S[k \ldots j]$, such that either $k = j$ or any proper prefix of $S[k \ldots j]$ is not unique. Note that $LSUS_1 = SUS_1$, which always exists. However, if S is not suffixed by an artificial terminator character $\$ \notin \Sigma$, then for an arbitrary location $k \geq 2$, $LSUS_k$ may not exist. For example, if $S = \mathsf{abcabc}$, then none of $\{LSUS_4, LSUS_5, LSUS_6\}$ exists. An **up-to-j extension of LSUS$_k$** is the substring $S[k \ldots j]$, where $k + |LSUS_k| \leq j \leq n$.

The **suffix array** $SA[1\ldots n]$ of the string S is a permutation of $\{1, 2, \ldots, n\}$, such that for any i and j, $1 \le i < j \le n$, we have $S[SA[i]\ldots n] < S[SA[j]\ldots n]$. That is, $SA[i]$ is the starting location of the ith suffix in the sorted order of all the suffixes of S. The **rank array** $Rank[1\ldots n]$ is the inverse of the suffix array. That is, $Rank[i] = j$ iff $SA[j] = i$. The **longest common prefix (lcp) array** $LCP[1\ldots n+1]$ is an array of $n+1$ integers, such that for $i = 2, 3, \ldots, n$, $LCP[i]$ is the length of the lcp of the two suffixes $S[SA[i-1]\ldots n]$ and $S[SA[i]\ldots n]$. We set $LCP[1] = LCP[n+1] = 0$. In the literature, the lcp array is often defined as an array of n integers. We include an extra zero at $LCP[n+1]$ is just to simplify the description of our upcoming algorithms. The next Lemma 1 shows that, by using the rank array and the lcp array of the string S, it is easy to calculate any $LSUS_i$ if it exists or to detect that it does not exist.

Lemma 1. *For $i = 1, 2, \ldots, n$:*

$$LSUS_i = \begin{cases} S[i\ldots i + L_i], & \text{if } i + L_i \le n \\ \text{not existing}, & \text{otherwise} \end{cases}$$

where $L_i = \max\{LCP[Rank[i]], LCP[Rank[i] + 1]\}$.

3   SUS Finding for One Location

In this section, we want to find the SUS covering a given location k using $O(n)$ time and space. We start with finding the leftmost one if k has multiple SUSes. In the end, we will show an extension to find all the SUSes covering location k with the same time and space complexities, if k has multiple SUSes.

Lemma 2. *Every SUS is either an LSUS or an extension of an LSUS.*

Example 1: $S = \text{abcbca}$, then $SUS_2 = S[1, 2] = \text{ab}$, which is $LSUS_1$. Example 2: $S = \text{abcbc}$, then $SUS_2 = S[1, 2] = \text{ab}$, which is an extension of $LSUS_1 = S[1]$ to location 2.

By Lemma 2, we know SUS_k is either an LSUS or an extension of an LSUS, and the starting location of that LSUS must be on or before location k. Then the algorithm for finding SUS_k for any given string location k is simply to calculate $LSUS_1, \ldots, LSUS_k$ if existing, using Lemma 1. During this calculation, if any LSUS does not cover the location k, we simply extend that LSUS up to location k. We will pick the shortest one among all the LSUS or their up-to-k extensions as SUS_k. We resolve the tie by picking the leftmost candidate. It is possible this procedure can early stop if it finds an LSUS does not exist, because that indicates all the other remaining LSUSes do not exist either. Due to the page limit, we give the pseudocode of this procedure in the full version of this paper [2].

Lemma 3. *Given a string location k and the rank and the lcp array of the string S, we can find SUS_k using $O(k)$ time. If multiple SUS_k exist, the leftmost one is returned.*

Adding the linear time cost for the construction of the suffixe array, the rank array, and the lcp array, we have the following theorem.

Theorem 1. *For any location k in the string S, we can find SUS_k using $O(n)$ time and space. If multiple SUS_k exist, the leftmost one is returned.*

It is trivial to extend the one-SUS finding algorithm to find all the SUSes covering a particular location k as follows. We will first find the leftmost SUS_k. Then we start over again to recheck $LSUS_1 \ldots LSUS_k$ or their up-to-k extensions, and return those whose length is equal to the length of SUS_k. The pseudocode of this new procedure is given in [2]. This procedure clearly costs an extra $O(k)$ time. Combining the claim in Theorem 1, we get the following theorem.

Theorem 2. *We can find all the SUSes covering any given location k using $O(n)$ time and space.*

4   SUS Finding for Every Location

In this section, we want to find SUS_k for every location $k = 1, 2, \ldots, n$. If k has multiple SUSes, the leftmost one will be returned. In the end, we will show an extension to return all SUSes for every location.

A natural solution is to iteratively use the algorithm for finding the SUS covering a particular location as a subroutine to find every SUS_k, for $k = 1, 2, \ldots, n$. However, the total time cost of this solution will be $O(n) + \sum_{k=1}^{n} O(k) = O(n^2)$, where $O(n)$ captures the time cost for the construction of the suffix array, the rank array, and the lcp array, and $\sum_{k=1}^{n} O(k)$ is the total time cost for the n instances of the one-SUS finding algorithm. We want to have a solution that costs a total of $O(n)$ time and space, which implies that the amortized cost for finding each SUS is $O(1)$.

By Lemma 2, we know that every SUS must be an LSUS or an extension of an LSUS. The next Lemma 4 further says if SUS_k is an extension of an LSUS, it has special properties and can be quickly obtained from SUS_{k-1}.

Lemma 4. *For any $k \in \{2, 3, \ldots, n\}$, if SUS_k is an extension of an LSUS, then (1) SUS_{k-1} must be a substring whose right boundary is the character $S[k-1]$, and (2) SUS_k is the substring SUS_{k-1} appended by the character $S[k]$.*

4.1   The Overall Strategy

We are ready to present the overall strategy for finding SUS of every location, by using Lemma 2 and 4. We will calculate all the SUS in the order of $SUS_1, SUS_2, \ldots, SUS_n$. That means when we want to calculate SUS_k, $k \geq 2$, we have had SUS_{k-1} calculated already. Note that $SUS_1 = LSUS_1$, which is easy to calculate using Lemma 1. Now let's look at the calculation of a particular SUS_k, $k \geq 2$. By Lemma 2, we know SUS_k is either an LSUS or an extension of an LSUS. By Lemma 4, we also know if SUS_k is an extension of an LSUS, then the right boundary of SUS_{k-1} must be $S[k-1]$ and SUS_k is just SUS_{k-1} appended by the character $S[k]$. Suppose when we want to calculate SUS_k, we have already calculated the shortest LSUS covering location k or have known the fact that no LSUS covers location k. Then, by using SUS_{k-1}, which has been calculated by then, and the shortest LSUS covering location k, we will be able to calculate SUS_k as follows:

Case 1: If the right boundary of SUS_{k-1} is not $S[k-1]$, then SUS_k cannot be an extension of an LSUS (the contrapositive of Lemma 4). Thus, SUS_k is just the shortest LSUS covering location k, which must be existing in this case.

Case 2: If the right boundary of SUS_{k-1} is $S[k-1]$, then SUS_k may or may not be an extension of an LSUS. We will consider two possibilities: (1) If the shortest LSUS covering location k exists, we will compare its length with $|SUS_{k-1}|+1$, and pick the shorter one as SUS_k. If both have the same length, we resolve the tie by picking the one whose starting location index is smaller. (2) If no LSUS covers location k, SUS_k will just be SUS_{k-1} appended by $S[k]$.

Therefore, the real challenge, by the time we want to calculate SUS_k, $k \geq 2$, is to ensure that we would already have calculated the shortest LSUS covering location k or we would already have known that no LSUS covers location k.

4.2   Preparation

We now focus on the calculation of the shortest LSUS covering every string location k, denoted by $\mathbf{SLS_k}$. Let $\mathbf{Candidate_i^k}$ denote the shortest one among those of $\{LSUS_1, \ldots, LSUS_k\}$ that exist and cover location i. The leftmost one will be picked if multiple choices exist for both SLS_k and $Candidate_i^k$. For an arbitrary k, $1 \leq k \leq n$, SLS_k may not exist, because the location k may not be covered by any LSUS. However, if SLS_k exists, by the definition of SLS and $Candidate$, we have:

Fact 1. *If* SLS_k *exists:* $SLS_k = Candidate_k^k = Candidate_k^{k+1} = \cdots = Candidate_k^n$

Our goal is to ensure SLS_k will have been known when we want to calculate SUS_k, so we calculate every SLS_k following the same order $k = 1, 2, \ldots, n$, at which we calculate all SUSes. Because we need to know every $LSUS_i$, $i \leq k$ in order to calculate SLS_k (Fact 1), we will walk through the string locations $k = 1, 2, \ldots, n$: at each walk step k, we calculate $LSUS_k$ and maintain $Candidate_i^k$ for every string location i that has been covered by at least one of $\{LSUS_1, LSUS_2, \ldots, LSUS_k\}$. Note that $Candidate_i^k = SLS_i$ for every $i \leq k$ (Fact 1). Those $Candidate_i^k$ with $i \leq k$ would have been used as SLS_i in the calculation of SUS_i. So, after each walk step k, we will only need to maintain the candidates for locations after k.

Lemma 5. *(1)* $LSUS_1$ *always exists. (2) If* $LSUS_k$ *exists, then* $\{LSUS_1, LSUS_2, \ldots, LSUS_k\}$ *all exist. (3) If* $LSUS_k$ *does not exist, then none of* $\{LSUS_k, LSUS_{k+1}, \ldots, LSUS_n\}$ *exists.*

We know after k walk steps, we have calculated $LSUS_1, LSUS_2, \ldots, LSUS_k$. By Lemma 5, we know that there exists some ℓ_k, $1 \leq \ell_k \leq k$, such that

$LSUS_1, \ldots, LSUS_{\ell_k}$ all exist, but $LSUS_{\ell_k+1} \ldots LSUS_k$ do not exist. If $\ell_k = k$, that means $LSUS_1, \ldots, LSUS_k$ all exist. Let γ_k denote the right boundary of $LSUS_{\ell_k}$, i.e., $LSUS_{\ell_k} = S[\ell_k \ldots \gamma_k]$. We know every location $j = 1, \ldots, \gamma_k$ has its candidate $Candidate_j^k$ calculated already, because every such location j has been covered by at least one of the LSUSes among $LSUS_1, \ldots, LSUS_{\ell_k}$. We also know if $\gamma_k < n$, every location $j = \gamma_k + 1, \ldots, n$ still does not have its candidate calculated, because every such location j has not been covered by any LSUS from $LSUS_1, \ldots, LSUS_{\ell_k}$ that we have calculated at the end of the kth walk step.

Lemma 6. *At the end of the kth walk step, if $\gamma_k > k$, then for any i and j, $k \leq i < j \leq \gamma_k$, $Candidate_j^k$ also covers location i.*

Lemma 7. *At the end of the kth walk step, if $\gamma_k > k$, then $\mid Candidate_k^k \mid \leq \mid Candidate_{k+1}^k \mid \leq \ldots \leq \mid Candidate_{\gamma_k}^k \mid$.*

The next lemma shows that the right boundary of $LSUS_i$ will be on or after the right boundary of $LSUS_{i-1}$, if $LSUS_i$ exists.

Lemma 8. *For each $i = 2, 3, \ldots, n$: $\mid LSUS_i \mid \geq \mid LSUS_{i-1} \mid - 1$*

4.3 Finding *SLS* for Every Location

Invariant. We calculate SLS_k for $k = 1, 2, \ldots, n$ by maintaining the following invariant at the end of every walk step k: (A) If $\gamma_k > k$, locations $\{k + 1, k + 2, \ldots, \gamma_k\}$ will be cut into chunks, such that: (A.1) All locations in one chunk have the same candidate. (A.2) Each chunk will be represented by a linked list node of four fields: {ChunkStart, ChunkEnd, start, length}, respectively representing the start and end location of the chunk and the start and length of the candidate shared by all locations of the chunk. (A.3) All nodes representing different chunks will be connected into a linked list, which has a head and a tail, referring to the two nodes that represent the lowest positioned chunk and the highest positioned chunk. (B) If $\gamma_k \leq k$, the linked list is empty.

Maintenance of the Invariant. We describe in an inductive manner the procedure that maintains the invariant. Algorithm 1 shows the pseudocode. We start with an empty linked list.

Base Step: $k = 1$. We are walking the first step. We first calculate $LSUS_1$ using Lemma 1. We know $LSUS_1$ must exist. Let's say $LSUS_1 = S[1 \ldots \gamma_1]$ for some $\gamma_1 \leq n$. Then, $Candidate_i^1 = LSUS_1$ for every $i = 1, 2, \ldots, \gamma_1$. We record all these candidates by using a single node $(1, \gamma_1, 1, \gamma_1)$. This is the only node in the linked list and is pointed by both head and tail. We know $SLS_1 = Candidate_1^1$ (Fact 1), so we return SLS_1 by returning (head.start, head.length) = $(1, \gamma_1)$. We then update head.ChunkStart from 1 to be 2. If it turns out head.ChunkEnd = $\gamma_1 < 2$, meaning $LSUS_1$ really covers location 1 only, we delete the head node from the linked list, which will then become empty.

Algorithm 1. Function calls $FindSLS(1)$, ..., $FindSLS(n)$ return SLS_1, ..., SLS_n, if the corresponding SLS exists; otherwise, `null` is returned

---

1 Construct $Rank[1\ldots n]$ and $LCP[1\ldots n]$ of the string S;
2 Initialize an empty $List$;   // Each node has four fields: {ChunkStart, ChunkEnd, start, length}.
3 $head \leftarrow 0$; $tail \leftarrow 0$;   // Reference to the head and tail node of the $List$

4 $FindSLS(k)$
    /* Process $LSUS_k$, if it exists.   */
5      $L \leftarrow \max\{LCP[Rank[k]], LCP[Rank[k]+1]\}$;
6   **if** $k + L \leq n$ **then**   // $LSUS_k$ exists.
       // Add a new list element at the tail, if necessary.
7   **if** $head = 0$ **then** $List[1] \leftarrow (k, k+L, k, L+1)$; $head \leftarrow 1$; $tail \leftarrow 1$; // $List$ was empty.
8   **else if** $k + L > List[tail].ChunkEnd$ **then**
9           $tail + +$; $List[tail] \leftarrow (List[tail-1].ChunkEnd+1, k+L, k, L+1)$;

       /* Update candidates and merge the nodes whose candidates can be shorter. Resolve the tie by picking the leftmost one.   */
10        $j \leftarrow tail$;
11   **while** $j \geq head$ and $List[j].length > L+1$ **do** $j --$;
12   ;
13        $List[j+1] \leftarrow (List[j+1].ChunkStart, List[tail].ChunkEnd, k, L+1)$; $tail \leftarrow j+1$;

14   **if** $head \neq 0$ **then** $SLS_k \leftarrow (head.start, head.length)$;   // The list is not empty.
15   **else** $SLS_k \leftarrow (\text{null}, \text{null})$;   // SLS_k does not exist.
16   ;

     /* Discard the information about location k from the $List$.   */
17   **if** $head > 0$ **then**   // $List$ is not empty
18   **if** $List[head].ChunkEnd \leq k$ **then**
19          $head + +$;   // Delete the current head node
20   **if** $head > tail$ **then** $head \leftarrow 0$; $tail \leftarrow 0$; ;   // $List$ becomes empty
21   **else** $List[head].ChunkStart \leftarrow k+1$;
22   ;
23   **return** SLS_k

Inductive Step: $k \geq 2$. We are walking the kth step. We first calculate $LSUS_k$. Case 1: $LSUS_k$ does not exist. (1) If **head** does not exist. It means that location k is covered neither by any of $LSUS_1, \ldots, LSUS_{k-1}$ nor by $LSUS_k$, so SLS_k simply does not exist, and we will simply return $(\text{null}, \text{null})$ to indicate that SLS_k does not exist. (2) If **head** exists, $(\text{head.start}, \text{head.length})$ will be returned as SLS_k, because $Candidate_k^k = SLS_k$ (Fact 1). Then we will remove the information about location k from the head by setting $head.ChunkStart = k+1$. After that, we will remove the **head** node if **head.ChunkEnd** < **head.ChunkStart**.

Case 2: $LSUS_k$ exists and $LSUS_k = S[k\ldots\gamma_k]$, $\gamma_k \leq n$. By Lemma 5, we know $LSUS_1, \ldots, LSUS_{k-1}$ all exist. Let γ_{k-1} denote the right boundary of $LSUS_1, \ldots, LSUS_{k-1}$. By Lemma 8, we know $\gamma_k \geq \gamma_{k-1}$ and γ_{k-1} is also the right boundary of $LSUS_{k-1}$, i.e., $LSUS_{k-1} = S[k-1\ldots\gamma_{k-1}]$. Note that both $\gamma_{k-1} < k$ and $\gamma_{k-1} \geq k$ are possible. (1) If **head** does not exist, it means $\gamma_{k-1} < k$ and none of locations $\{k\ldots\gamma_k\}$ is covered by any of $LSUS_1, \ldots, LSUS_{k-1}$. We will insert a new node $(k, \gamma_k, k, \gamma_k - k + 1)$, which will be the only node in the linked list. (2) If head exists, it means $\gamma_{k-1} \geq k$. If $\gamma_k > \text{tail.ChunkEnd} = \gamma_{k-1}$, we will first insert a new node $(\text{tail.ChunkEnd} + 1, \gamma_k, k, \gamma_k - k + 1)$ at the tail side of the linked list to record the candidate information for locations in the chunk after γ_{k-1} through γ_k. After the work in either (1) or (2) is finished, we then travel through the nodes in the linked list from the tail side toward the head. We stop when we meet a node whose candidate is shorter than or equal to $LSUS_k$ or when we reach the head end of the linked list. This travel is valid because of Lemma 7. We will merge all the nodes whose candidates are longer than $LSUS_k$ into one node. The chunk covered by the new node is the union of the chunks covered by the merged nodes, and the candidate of the new node obtained from merging is $LSUS_k$. This merge process ensures every location maintains its best (shortest) candidate by the end of each walk step, and also resolves ties of multiple candidates by picking the leftmost one. We will return (head.start, head.length) as SLS_k, because $Candidate_k^k = SLS_k$ (Fact 1). Finally, we will remove the information about location k from the head by setting $head.ChunkStart = k + 1$. We will remove the head node if it turns out that head.ChunkEnd > head.ChunkStart.

Lemma 9. *Given the lcp array and the rank array of S, the amortized time cost of FindSLS() is $O(1)$ over the sequence of function calls FindSLS(1), FindSLS(2), ..., FindSLS(n).*

4.4 Finding SUS for Every Location

Once we are able to sequentially calculate every SLS_k or detect it does not exist, we are ready to calculate every SUS_k at the order of $k = 1, 2, \ldots, n$, by using the strategy described in Section 4.1. Due to the page limit, the pseudocode describing this procedure is given in [2].

Theorem 3. *We can find $SUS_1, SUS_2, \ldots, SUS_n$ of string S using a total of $O(n)$ time and space.*

4.5 Extension: Finding all the SUSes for every Location

It is possible that a particular location can have multiple SUSes. For example, if $S = \text{abcbb}$, then SUS_2 can be either $S[1, 2] = \text{ab}$ or $S[2, 3] = \text{bc}$. The algorithm of Theorem 3 only returns one of them. However, we can easily modify the algorithm to return all the SUSes of every location, without changing Algorithm 1.

Suppose a particular location k has multiple SUSes. We know, at the end of the kth walk step but before the linked list update, SLS_k returned by Algorithm 1 is recorded by the head node and is the leftmost one among all the SUSes that are LSUS and cover location k. Because every string location maintains its shortest candidate and due to Lemma 7, all the other SUSes that are LSUS and cover location k are being recorded by other linked list nodes that are immediately following the head node. This is because if those other SUSes are not being recorded, that means the location right after the head node's chunk

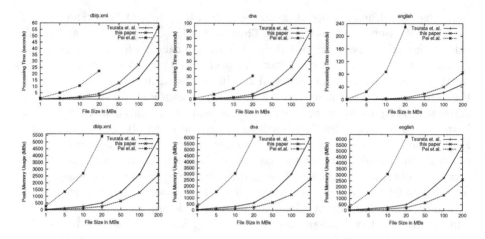

Fig. 1. Processing speed and peak memory consumption of RSUS, OSUS, and ours

has a candidate longer than SUS_k or does not have a candidate calculated yet, but that location is indeed covered by a SUS_k at the end of the kth walk step. It's a contradiction. Same argument can be made to the other next neighboring locations that are covered by SUS_k.

Therefore, finding all the SUSes covering location k becomes easy—simply go through the linked list nodes from the **head** node toward the **tail** node and report all the LSUSes, whose lengths are equal to the length of SUS_k, which we have found. If the rightmost character of SUS_{k-1} is $S[k-1]$ and the substring SUS_{k-1} appended by $S[k]$ has the same length, that substring will be reported too. Due to the page limit, the pseudocode describing this procedure is given in [2]. The overall time cost of maintaining the linked list data structure (the sequence of function calls $FindSLS(1), FindSLS(2), \ldots, FindSLS(n)$) is still $O(n)$. The time cost of reporting the SUSes covering a particular location becomes $O(occ)$, where occ is the number of SUSes that cover that location.

5   Experiments

We have implemented our proposal in C++ without best engineering effort, using the libdivsufsort[1] library for the suffix array construction and Kasai *et al.*'s method [3] to compute the LCP array. We have compared our work against Pei *et al.*'s RSUS [5] and Tsurata *et al.*'s [6] OSUS implementations, a recent independent work obtained via personal communication after we posted our work at arXiv. Notice that OSUS also computes the suffix array with the same libdivsufsort package.

RSUS was prepared with an R interface. We stripped off that R interface and built a standalone C++ executable for the sake of fair benchmarking. OSUS was

---

[1] Available at: https://code.google.com/p/libdivsufsort

developed in C++. We run it with the -l option to compute a single leftmost SUS for a given position rather than its default configuration of reporting all SUSs. We also commented the sections that print the results to the screen on all three programs so as to measure the algorithmic performance better.

We run the tests on a machine that has Intel(R) Core(TM) i7-3770 CPU @ 3.40GHz processor with 8192 KB cache size and 16GB memory. The operating system was Linux Mint 14. We used the Pizza&Chili corpus in the experiments by taking the first 1, 5, 10, 20, 50, 100, and 200 MBs of the largest *dblp.xml*, *dna*, and *English* files. The results are shown in Figure 1.

It was not possible to run the RSUS on large files, since RSUS requires more memory than that our machine has, and thus, only up to 20MB files were included in the RSUS benchmark. Compared to RSUS, we have observed that our proposal is more than 4 times faster and uses 20 times less memory. The experimental results revealed that OSUS is on the average 1.6 times faster than our work, but in contrast, uses 2.6 times more memory.

The asymptotic time and space complexities of both ours and OSUS are same as being linear (note that the x axis in both figures uses log scale). The peak memory usage of OSUS and ours are different although they both use suffix array, rank array (inverse suffix array), and the LCP array, and computing these arrays are done with the same library (libdivsufsort). The difference stems from different ways these studies follow to compute the SUS. OSUS computes the SUS by using an additional array, which is named as the meaningful minimal unique substring array in the corresponding study. Thus, the space used for that additional data structure makes OSUS require more memory.

Both OSUS and our scheme present stable running times on all dblp, dna, and english texts and scale well on increasing sizes of the target data conforming to their linear time complexity. On the other hand RSUS exhibits its $O(n^2)$ time complexity on all texts, and especially its running time on english text takes much longer when compared to other text types.

References

1. Crochemore, M., Rytter, W.: Jewels of stringology. World Scientific (2003)
2. İleri, A.M., Külekci, M.O., Xu, B.: Shortest unique substring query revisited, http://arxiv.org/abs/1312.2738
3. Kasai, T., Lee, G.H., Arimura, H., Arikawa, S., Park, K.: Linear-time longest-common-prefix computation in suffix arrays and its applications. In: Amir, A., Landau, G.M. (eds.) CPM 2001. LNCS, vol. 2089, pp. 181–192. Springer, Heidelberg (2001)
4. Ko, P., Aluru, S.: Space efficient linear time construction of suffix arrays. Journal of Discrete Algorithms 3(2-4), 143–156 (2005)
5. Pei, J., Wu, W.C.H., Yeh, M.Y.: On shortest unique substring queries. In: Proceedings of the 2013 IEEE International Conference on Data Engineering (ICDE), pp. 937–948 (2013)
6. Tsuruta, K., Inenaga, S., Bannai, H., Takeda, M.: Shortest unique substrings queries in optimal time. In: Geffert, V., Preneel, B., Rovan, B., Štuller, J., Tjoa, A.M. (eds.) SOFSEM 2014. LNCS, vol. 8327, pp. 503–513. Springer, Heidelberg (2014)

A *really* Simple Approximation of Smallest Grammar

Artur Jeż[1,2,⋆]

[1] Max Planck Institute für Informatik, Saarbrücken, Germany
[2] Institute of Computer Science, University of Wrocław, Poland
aje@cs.uni.wroc.pl

Abstract. We present a really simple linear-time algorithm construct-ing a context-free grammar of size $\mathcal{O}(g\log(N/g))$ for the input string, where N is the size of the input string and g the size of the optimal grammar generating this string. The algorithm works for arbitrary size alphabets, but the running time is linear when the alphabet Σ of the input string can be identified with numbers from $\{1,\ldots,N\}$. Algorithms with such an approximation guarantee and running time are known, however all of them were non-trivial and their analyses involved. The here presented algorithm computes the LZ77 factorisation (of size ℓ) and transforms it in phases to a grammar. In each phase it maintains an LZ77-like factorisation of the word with at most ℓ factors as well as ad-ditional $\mathcal{O}(\ell)$ letters. In one phase in a greedy way (by a left-to-right sweep) we choose a set of pairs of consecutive letters to be replaced with new symbols, i.e. nonterminals of the constructed grammar. We choose at least 2/3 of the letters in the word and there are $\mathcal{O}(\ell)$ many differ-ent pairs among them. Hence there are $\mathcal{O}(\log N)$ phases, each introduces $\mathcal{O}(\ell)$ nonterminals. A more precise analysis yields a bound $\mathcal{O}(\ell\log(N/\ell))$. As $\ell \leq g$, this yields $\mathcal{O}(g\log(N/g))$.

Keywords: Grammar-based compression, Construction of the smallest grammar, SLP, compression.

1   Introduction

Grammar based compression. In the grammar-based compression text is repre-sented by a context-free grammar (CFG) generating exactly one string. Such an approach was first considered by Rubin [7], though he did not mention CFGs explicitly. In general, the idea behind this approach is that a CFG can compactly represent the structure of the text, even if this structure is hidden. Furthermore, the natural hierarchical definition of the CFGs makes such a representation suit-able for algorithms, without the need of explicit decompression. A recent survey by Lohrey [6] gives a comprehensive description of several areas of in which grammar-based compression was successfully applied.

⋆ Supported by the Humboldt Foundation Postdoctoral grant.

A.S. Kulikov, S.O. Kuznetsov, and P. Pevzner (Eds.): CPM 2014, LNCS 8486, pp. 182–191, 2014.

The main drawback of the grammar-based compression is that producing the smallest CFG for a text is *intractable*: the size of the smallest grammar for the input string cannot be approximated within some small constant factor [1].

Previous approximation algorithms. The first two algorithms with an approximation ratio $\mathcal{O}(\log(N/g))$ were developed simultaneously by Rytter [8] and Charikar et al. [1]. Rytter's algorithm [8] creates an LZ77 representation of the input string and then transforms it to an $\mathcal{O}(\ell \log(N/\ell))$ size grammar, where ℓ is the size of the LZ77 representation. It is easy to show that $\ell \leq g$ and as $f(x) = x \log(N/x)$ is increasing, the bound $\mathcal{O}(g \log(N/g))$ on the size of the grammar follows. The crucial part is the requirement that the derivation tree of the intermediate constructed grammar satisfies the AVL condition, which turns out to be nontrivial.

Charikar et al. [1] followed more or less the same path, with a different condition imposed on the grammar: it is required that its derivation tree is length-balanced, i.e. for a rule $X \to YZ$ the lengths of words generated by Y and Z are within a certain multiplicative constant factor from each other. For such trees efficient implementation of merging, splitting etc. operations were defined by the authors and so the same running time as in the case of the AVL grammars was obtained. Since all the operations were defined from scratch, the obtained algorithm is also quite involved and the analysis is even more non-trivial.

Sakamoto [9] proposed a different algorithm, based on RePair [5], which is one of the practical algorithm for grammar-based compression. His algorithm iteratively replaces pairs of different letters and maximal repetitions of letters (a^ℓ is a *maximal repetition* if that cannot be extended by a to either side). A special pairing of the letters was devised, so that it is 'synchronising': if u has 2 disjoint occurrences in w, then those two occurrences can be represented as $u_1 u' u_2$, where $u_1, u_2 = \mathcal{O}(1)$, such that both occurrences of u' in w are paired and compressed in the same way. The analysis was based on considering the LZ77 representation of the text and proving that due to 'synchronisation' the factors of LZ77 are compressed very similarly as the text to which they refer. Constructing such a pairing is involved and the analysis non-trivial.

Recently, the author proposed another algorithm [2]. Similarly to the Sakamoto's algorithm it iteratively replaces pairs of different letters and maximal repetition of letters. Though the choice of pairs to be replaced was simpler, still the construction was involved. The feature of the algorithm was its analysis based on the re-compression technique, which allowed avoiding the connection of SLPs and LZ77 compression. This made it possible to generalise this approach also to grammars generating trees [3]. On the downside, the analysis is quite complex.

Contribution of this paper. We present a very simple algorithm together with a straightforward and natural analysis. It chooses the pairs to be replaced in the word during a left-to-right sweep using the information given by an LZ77 factorisation. We require that any pair that is chosen to be replaced is either inside a factor of length at least 2 or consists of two factors of length 1 and that the factor is paired in the same way as its definition. To this end we modify the LZ77 factorisation during the sweep. After the choice, the pairs are replaced and the new word inherits the factorisation from the original word. This is repeated until a trivial word is obtained.

2   The Algorithm

Notions

LZ77 factorisation. An LZ77 factorisation (called simply factorisation in the rest of the paper) of a word w is a representation $w = f_1 f_2 \cdots f_\ell$, where each f_i is either a single letter (called *free letter* in the following) or $f_i = w[j \mathinner{.\,.} j + |f_i| - 1]$ for some $j \le |f_1 \cdots f_{i-1}|$, in such a case f is a *factor* and $w[j \mathinner{.\,.} j + |f_i| - 1]$ is called the *definition* of this factor. A factor can have exactly one letter though when we find such a factor we demote it to a free letter. The *size* of the LZ77 factorisation $f_1 f_2 \cdots f_\ell$ is ℓ. There are several efficient linear-time algorithms for computing the smallest LZ77 factorisation of a word, see [4] and references therin.

SLP. Straight Line Programme (SLP) is a CFG with nonterminals $X_1, \ldots, X_g$ and with each rule either of the form $X_i \to a$ or $X_i \to X_j X_k$, where $j, k < i$. The *size* of the SLP is the number of its nonterminals (here: g).

The size of the smallest grammar for the input word cannot be approximated within some small constant factor [1]. On the other hand, several algorithms with approximation ratio $\mathcal{O}(\log(|w|/g))$, where g is the size of the smallest grammar generating w, are known [1,8,9,2]. Most of those constructions use the inequality $\ell \le g$, where ℓ is the size of the smallest LZ77 factorisation for w [8].

Intuition

Pairing. Let us identify each nonterminal generating a single letter with this letter. Suppose that we already have an SLP for w. Consider the derivation tree for w and the nodes that have only leaves as children (they correspond to nonterminals that have only letters on the right-hand side). Such nodes define a *pairing* on w, in which each letter can be paired with one of the neighbouring letters (such pairing is of course a symmetric relation). Construction of the grammar can be naturally identified with iterative pairing: for a word w_i we find a pairing, replace pairs of letters with 'fresh' letters (different occurrences of a pair ab can be replaced with the same letter though this is not essential), obtaining w_{i+1} and continue the process until a word $w_{i'}$ has only one letter. The fresh letters from all pairings are the nonterminals of the constructed SLP and its size is twice the number of different introduced letters.

Creating a pairing. Suppose that we are given a word w and its factorisation. We try the naive pairing: the first letter is paired with second, third with fourth and so on. If we now replace all pairs with new letters, we get a word that is 2 times shorter so $\log N$ such iterations give an SLP for w. However, in the worst case there are $N/2$ different pairs already in the first pairing and so we cannot give any better bound on the grammar size than $\mathcal{O}(N)$.

A better estimation uses the LZ77 factorisation. Let $w = f_1 f_2 \cdots f_\ell$ and consider a factor f_i. It is equal to $w[j \mathinner{\ldotp\ldotp} j + |f_i| - 1]$ and so all pairs occurring in f_i already occur in $w[j \mathinner{\ldotp\ldotp} j + |f_i| - 1]$ unless the parity is different, i.e. j and $|f_1 \cdots f_{i-1}| + 1$ are of different parity. We want to fix this: it seems a bad idea to change the pairing in $w[j \mathinner{\ldotp\ldotp} j + |f_i| - 1]$, so we change it in f_i: it is enough to shift the pairing by one letter, i.e. leave the first letter of f_i unpaired and pair the rest as in $w[j + 1 \mathinner{\ldotp\ldotp} j + |f_i| - 1]$. Note that the last letter in the factor definition may be paired with the letter to the right, which may be impossible inside f_i. As a last observation note that since we alter each f_i, instead of creating a pairing at the beginning and modifying it we can create the pairing while scanning the word from left to right.

There is one issue: after the pairing we want to replace the pairs with fresh letters. This may make some of the the factor definitions improper: when f_i is defined as $w[j \mathinner{\ldotp\ldotp} j + |f_i| - 1]$ it might be that $w[j]$ is paired with letter to the left. To avoid this situation, we replace the factor f_i with $w[j]f_i'$ and change its definition accordingly. Similar operation may be needed at the end of the factor. This increases the size of the LZ77 factorisation, but the number of factors stays the same (i.e. only the number of free letters increases). Additionally, we pair two neighbouring free letters, whenever this is possible.

Using a pairing. When the appropriate pairing is created, we replace each pair with a new letter. If the pair is within a factor, we replace it with the same symbol as the pair in the definition of the factor. In this way only pairs that are formed from free letters may contribute a fresh letter. As a result we obtain a new word together with a factorisation in which there are ℓ factors.

The Algorithm

Stored data. The word is stored in a table. The table *start* stores the information about the beginnings of factors: $start[i] = j$ means that i is the first letter of a factor and j is the first letter of its definition; otherwise $start[i] = \mathbf{false}$. Similarly $end[i]$ is a bit flag that tells whether $w[i]$ is the last letter of a factor.

When we replace the pairs with new letters, we reuse the same tables. Additionally, a table *newpos* stores the corresponding positions: $newpos[i] = j$ means that letter on position i was either unpaired and the position of the corresponding letter in the new word is on position j or that letter on position i was paired with a letter to the right and the corresponding letter in the new word is on position j; $newpos[i]$ is undefined when position i was the second element in the pair.

Technical assumption. Our algorithm makes a technical assumptions: a factor f_i starting at position j cannot have $start[j] = j - 1$. This is verified and repaired while sweeping through w: if $w[j \mathinner{\ldotp\ldotp} j + |f_i| - 1] = w[j - 1 \mathinner{\ldotp\ldotp} j + |f_i| - 2]$ then $f_i = a^{|f_i|}$. We split f_i: we make $w[j]$ a free letter and set $start[j + 1] = j - 1$.

Pairing. We are going to devise a pairing which has the following properties:

(P1) there are no two consecutive letters that are both unpaired;

(P2) the first (last) two letters of any factor f are paired with each other;

(P3) if $f = w[i \mathinner{.\,.} i + |f| - 1]$ has a definition $w[start[i] \mathinner{.\,.} start[i] + |f| - 1]$ then letters in f and in $w[start[i] \mathinner{.\,.} start[i] + |f| - 1]$ are paired in the same way.

The pairing is found incrementally by a left-to-right scan through w: we read w and when we are at letter i we make sure that the word $w[1 \mathinner{.\,.} i]$ satisfies (P1)–(P3). To this end we not only devise the pairing but also modify the factorisation a bit (by replacing a factor f with af or by fb, where a is the first and b the last letter of f). If during the sweep some f is shortened so that $|f| = 1$ then we demote it to a free letter.

The pairing is recorded in table: $pair[i]$ can be set to *first*, *second* or *none*, meaning that $w[i]$ is the first, second in the pair or it is unpaired, respectively.

Creation of pairing. If we read $w[i]$ that is a free letter then we check, whether the previous letter is not paired. If so, then we pair them. Otherwise we continue to the next position.

If i is a first letter of a factor, we check whether the length of this factor is one; if so, we change $w[i]$ into a free letter. If the factor has definition only one position to the left (i.e. at $i - 1$) then we split the factor: we make $w[i]$ a free letter and set $w[i + 1]$ as a first letter of a factor with a definition starting at $i - 1$. Otherwise we check whether $w[start[i]]$ is indeed the first letter of a pair. If not (i.e. it is a second letter of a pair or an unpaired letter) then we split the factor: we make $w[i]$ a free letter and $w[i + 1]$ the beginning of a factor with a definition beginning at $start[i] + 1$; we view the factor beginning at $w[i + 1]$ as a modified factor that used to begin at $w[i]$. If for any reason we turned $w[i]$ into a free letter, we re-read this letter, treating it accordingly. If $w[start[i]]$ is a first letter of a pair, we copy the pairing from the whole factor's definition to the factor starting at i.

Afterwards we ensure that the factor ends with a pair: if the last letter, say $w[i']$, of a factor is not the second in the pair, we again split the factor: we make $w[i']$ a free letter, we clear i''s pairing, decrease i' by 1 and make for $w[i']$ the end of the new factor. We iterate it until the $w[i']$ is a second letter of a factor.

When the pairing is done, we read w again and replace the pairs by letters. We keep two indices: i, which points at the first unread letter in the current word and i', which points at the first free position in the new word. Additionally, when reading i we store (in $newpos[i]$) the position of the corresponding letter in the new word, which is always i'.

Algorithm 3. TtoG

1: compute LZ77 factorisation of w

2: **while** $|w| > 1$ **do**

3:    compute a pairing of w using Pairing

4:    replace the pairs using PairReplacement

5: output the constructed grammar

Algorithm 1. Pairing

```
 1: pair[1] ← none, i ← 2
 2: while i ≤ |w| do
 3:     if start[i] then                                ▷ w[i] is the first element of a factor
 4:         if end[i] then                              ▷ This is one-letter factor
 5:             start[i] ← end[i] ← false               ▷ Turn it into a free letter
 6:         else if start[i] = i − 1 then
                                        ▷ The factor is aᵏ, its definition begins one position to the left
 7:             start[i + 1] ← i − 1, start[i] ← false
                                        ▷ Move the definition of the factor, make w[i] a free letter
 8:         else if pair[start[i]] ≠ first then
                                                ▷ The pairing of the definition of factor is bad
 9:             start[i + 1] ← start[i] + 1             ▷ Shorten the factor definition
10:             start[i] ← false                        ▷ Make w[i] a free letter
11:         else                                        ▷ Good factor
12:             j ← start[i]                            ▷ Factor's definition begins at j
13:             repeat                                  ▷ Copy the pairing from the factor definition
14:                 pair[i] ← pair[j]
15:                 i ← i + 1, j ← j + 1
16:             until end[i − 1]
17:             while pair[i − 1] ≠ second do    ▷ Looking for a new end of the factor
18:                 i ← i − 1
19:             end[i − 1] = true, end[i] = false, pair[i] = none     ▷ Shorten the
        factor and clear the pairing
20:     if not start[i] then                            ▷ w[i] a free letter
21:         if pair[i − 1] = none then                  ▷ If previous letter is not paired
22:             pair[i − 1] ← first, pair[i] ← second   ▷ Pair them
23:         else pair[i] ← none
24:         i ← i + 1
```

If $w[i]$ is a first letter in a pair and this pair consists of two free letters, in the new word we add a fresh letter and move two letters to the right in w (as well as one position in the new word). If $w[i]$ is unpaired and a free letter then we copy it to the new word, increasing both i and i' by 1. If $w[i]$ is first letter of a factor (and so also a first letter of a pair by (P2)), we copy the corresponding fragment of the new word (the first position is given by $newpos[start[i]]$), moving i and i' in parallel: i' is always incremented by 1, while i is moved by 2 when it reads a first letter of a pair and by 1 when it reads an unpaired letter. Also, we store the new beginning and end of the factor in the new word: for a factor beginning at i and ending at i' we set $start[newpos[i]] = newpos[start[i]]$ and $end[newpos[i' − 1]] = $ **true** (note that $i' − 1$ and i' are paired).

Analysis

Lemma 1. Pairing *runs in linear time. It creates a proper factorisation and returns a pairing that satisfies (P1)–(P3) (for this new factorisation). If there were m factors in w then* Pairing *creates at most 6m new free letters and the returned pairing has at most m factors.*

Algorithm 2. PairReplacement

1: $i \leftarrow i' \leftarrow 1$ ▷ i' is the position corresponding to i in the new word
2: **while** $i \leq |w|$ **do**
3:     **if** $start[i]$ **then**   ▷ $w[i]$ is the first element of a factor
4:        $start[i'] \leftarrow j' \leftarrow newpos[start[i]]$ ▷ Factor in new word begins at the
    position corresponding to the beginning of the current factor
5:        $start[i] \leftarrow$ **false**   ▷ Clearing obsolete information, note that $i' < i$
6:        **repeat**
7:           $newpos[i] \leftarrow i'$ ▷ Position corresponding to i
8:           $w[i'] \leftarrow w[j']$ ▷ Copy the letter according to new factorisation
9:           $i' \leftarrow i' + 1, \ j' \leftarrow j' + 1$
10:          **if** $pair[i] = first$ **then** $i \leftarrow i + 2$ ▷ We move left by the whole pair
11:          **else** $i \leftarrow i + 1$ ▷ We move left by the unpaired letter
12:        **until** $end[i - 1]$
13:        $end[i' - 1] \leftarrow$ **true**   ▷ End in the new word
14:        $end[i - 1] \leftarrow$ **false**   ▷ Clearing obsolete information
15:     **if not** $start[i]$ **then**   ▷ $w[i]$ a free letter
16:        $newpos[i] \leftarrow i'$
17:        **if** $pair[i] = none$ **then**
18:           $w[i'] \leftarrow w[i]$ ▷ We copy the unpaired letter
19:           $i \leftarrow i + 1, \ i' \leftarrow i' + 1$ ▷ We move by this letter to the right
20:        **else**
21:           $w[i'] \leftarrow$ fresh letter   ▷ Paired free letters are replaced by a fresh letter
22:           $i \leftarrow i + 2, \ i' \leftarrow i' + 1$ ▷ We move to the right by the whole pair

---

Proof. For the running time analysis note that a single letter can be considered at most twice: once as a part of a factor and once as a free letter.

We show the second claim of the lemma by induction: at all time the stored factorisation is proper, furthermore, when we processed $w[1 . . i]$ (i.e. we are at position $i + 1$, note that we can go back in which case position gets unprocessed) then we have a partial pairing, which differs from the pairing only in the fact that the position i may be assigned as first in the pair and $i + 1$ is not yet paired. This partial pairing satisfies (P1)–(P3) restricted to $w[1 . . i]$.

We first show that after considering i the modified factorisation is proper.

If in line 4 we have $start[i] = end[i]$ then $w[i]$ is a one-letter factor and so after replacing it with a free letter the factorisation stays proper. The verification in line 4 ensures that in each other case considered in lines 6–19 we deal with factors of length at least 2.

The modifications of the factorisation in line 7 results in a proper factorisation: the change is applied only when $start[i] = i - 1$, in which case $w[i . . i + |f| - 1] = w[i - 1 . . i + |f| - 2]$, which implies that $f = a^{|f|}$, where $a = w[i]$. Since $|f| \geq 2$, in such a case $w[i + 1 . . i + |f| - 1] = w[i - 1 . . i + |f| - 3]$ so we can split the factor $f = w[i . . i + |f| - 1]$ to $w[i]$ and a factor $w[i + 1 . . i + |f| - 1]$ defined as $w[i - 1 . . i + |f| - 3]$.

In line 9 we shorten the factor by one letter (and create a free letter), so the factorisation remains proper.

Concerning the symmetric shortening in line 18, it leaves a proper factorisation (as in case of line 9), as long as we do not move i before the beginning of the factor. However, observe that when we reach line 14 then the factor beginning at i has length at least 2, $start[i] < i-1$ and $pair[start[i]] = first$. Thus $start[i]+1 < i$ and so by induction assumption we already made a pairing for it. Since $start[i]$ is assigned $first$, $start[i] + 1$ is assigned $second$. So $i + 1$ is assigned $second$ as well. Since the end of the factor is at position $i' \geq i+1$, in our search for element marked with $second$ at positions i', $i' - 1$, ... we cannot move to the left more than to $i + 1$. Thus the factor remains (and has at least 2 letters).

We show that indeed we have a partial pairing. Firstly, if i is decreased, then as a result we get a partial pairing: the only nontrivial case is when $i - 1$ and i were paired then $i - 1$ is assigned as the first element in the pair but it has no corresponding element, which is allowed in the partial pairing. If i is increased then we need to make sure if $i - 1$ is assigned as a first element in a pair then i will be assigned as the second one (or the pairing is cleared). Note that $i - 1$ can be assigned in this way only when it is part of the factor, i.e. it gets the same status as some j. If i is also part of the same factor, then it is assigned the status of $j + 1$, which by inductive assumption is paired with j, so is the second element in the pair. In the remaining case, if $i - 1$ was the last element of the factor then in loop in line 17 we decrease i and so unprocess $i - 1$ (in particular, we cleared its pairing).

For (P2) observe that for the first two letters it is explicitly verified in line 8. Similarly, for the second part of (P2): we shorten the last factor in line 17 (ending at i) until $pair[i] = second$. We already shown that pairing is defined for $w[1 \ldots i]$ and when i is assigned $second$ then $i - 1$ is assigned $first$, as claimed.

Condition (P3) is explicitly enforced in loop in line 13.

Suppose that (P1) does not hold for $i - 1, i$, i.e. they are both unpaired after processing i. It cannot be that they are both within the same factor, as then the corresponding $w[j - 1]$ and $w[j]$ in the definition of the factor are also unpaired, which contradicts the induction assumption. Similarly, it cannot be that one of them is in a factor and the other outside this factor, as by (P2) (which holds for $w[1 .. i]$) a factor begins and ends with two paired letters. So they are both free letters. But then we needed to pass line 20 for i and both $w[i - 1]$ and $w[i]$ were free and unpaired at that time, which means that they should have been paired at that point, contradiction.

To see the third claim of the lemma, fix a factor f that begins at position i. When it is modified, we identify the obtained factor with f (which in particular shows that the number of factors does not increase). We show that it creates at most 6 new free letters in this phase.

If at any point the factor has only one letter then it is replaced with a free letter and afterwards cannot introduce any free letters (as f is no longer there). Hence at most one free letter is introduced by f due to condition in line 4.

If $start[i] = i - 1$ then it creates one free letter inside condition in line 6. It cannot introduce another free letter in this way (in this phase), as afterwards $start[i + 1] = i - 1$ and there is no way to decrease this distance (in this phase).

We show that condition in line 8 holds at most twice for a fixed factor f in a phase. Since we set $j = start[i]$ and increase both i and j by 1 until $pair[j] = first$, this can be viewed as searching for the smallest position $j' \geq j$ that is first in a pair and we claim that $j' \leq j + 2$. On the high-level, this should hold because (P1) holds for $w[1 .. i - 1]$, and so among three consecutive elements there is at least one that is the first element in the pair. This is more complicated, as some pairing may change during the search. The proof follows by case inspection of possible actions of the algorithm and is omitted due to space constraints. Similar analysis can be applied to the last letter of a factor.

It is left to show that after processing the whole w we have a proper factorisation and a pairing satisfying (P1)-(P3). From the inductive proof it follows that the kept factorisation is proper and the partial pairing satisfies (P1)-(P3) for the whole word. So it is enough to show that the last letter of w is not assigned as a first element of a pair. Consider, whether it is in a factor or a free letter. If it is in a factor then by (P2) it is the second element in a pair. If it is a free letter observe that we only pair free letters in line 22, which means that it is paired with the letter on the next position, contradiction.   □

Now, we show that when we have a pairing satisfying (P1)–(P3) then PairReplacement creates a word w' out of w together with a factorisation.

Lemma 2. *When a pairing satisfies (P1)–(P3) then* PairReplacement *runs in* $\mathcal{O}(|w|)$ *time and returns a word w' together with a factorisation; $|w'| \leq (2|w| + 1)/3$ and the returned factorisation of w' has the same number of factors as the factorisation of w. If p new letters were introduced then w' has p less free letters than w.*

Proof. The running time is obvious as we make one scan through w. Concerning the size of the produced word, by (P1) each unpaired letter (perhaps except the last letter of w) is followed by a pair. Thus, at least $\frac{1}{3}(|w| - 1)$ letters are removed from w.

Concerning the factorisation of w', observe that by an easy induction for each i the $newpos[i]$ is undefined, when i is second in a pair, or is the position of the corresponding letter in w'. Now, consider any factor f in w with a definition $w[j \ldots j + |f| - 1]$. By (P2) both the first and the last two letters of f are paired and by (P3) pairing of f is the same as the pairing of its definition. So it is enough to copy the letters in w' corresponding to $w[j \ldots j + |f| - 1]$, i.e. beginning with $newpos[j]$. When we consider a free letter, if it is unpaired, it should be copied (as it is not replaced), and when it is paired, the pair can be replaced with a fresh letter; in both cases the corresponding letter in the new word should be free.

Concerning the number of fresh letters introduced, suppose that ab is replaced with c. If ab is within some factor f then we use for the replacement the same letter as we use in the factor definition and so this letter is not fresh. If both this a and b are free letters then such a pair contributes one fresh letter, but one of those free letters is removed. The last possibility is that one letter from ab

comes from a factor and the other from outside this factor, but this contradicts (P2). □

Theorem 1. *Its approximation ratio is $\mathcal{O}(\log(N/g))$, where g is the size of the optimal grammar. TtoG runs in linear time and returns an SLP of size $\mathcal{O}(\ell + \ell \log(N/\ell))$.*

Proof. By Lemma 2 each introduction of a fresh letter reduces the number of free letters by 1, so we estimate the number of created free letters. In the initial LZ77 factorisation there are at most ℓ of them. For the free letters created during the Pairing let us fix a factor f of the original factorisation and estimate how many free letters it created. By (P1) the $|f|$ drops by a constant fraction in each phase and so it takes part in $\mathcal{O}(\log |f|)$ phases. In each phase it introduces at most 6 free letters, by Lemma 1. So we introduce $\mathcal{O}(\sum_{i=1}^{\ell} \log |f_i|)$ free letters during all phases. As $\sum_{i=1}^{\ell} |f_i| \leq N$, by standard tools of mathematical analysis $\sum_{i=1}^{\ell} \log |f_i| \leq \ell \log(N/\ell)$, which is the bound on the number of nonterminals introduced in this way to the grammar. Adding the ℓ for the free letters in the origianl LZ77 factorisations yields the claim.

For the running time, the creation of the LZ77 factorisation takes linear time, see [4]. In each phase the pairing and replacement of pairs takes linear time in the length of the current word. By (P1) the length of such a word is reduced by a constant fraction in each phase, hence the total running time is linear.   □

References

1. Charikar, M., Lehman, E., Liu, D., Panigrahy, R., Prabhakaran, M., Sahai, A., Shelat, A.: The smallest grammar problem. IEEE Transactions on Information Theory 51(7), 2554–2576 (2005)
2. Jeż, A.: Approximation of grammar-based compression via recompression. In: Fischer, J., Sanders, P. (eds.) CPM 2013. LNCS, vol. 7922, pp. 165–176. Springer, Heidelberg (2013)
3. Jeż, A., Lohrey, M.: Approximation of smallest linear tree grammar. In: Mayr, E., Portier, N. (eds.) STACS. LIPIcs, vol. 24, pp. 445–457. Schloss Dagstuhl — Leibniz-Zentrum fuer Informatik (2014)
4. Kärkkäinen, J., Kempa, D., Puglisi, S.J.: Linear time lempel-ziv factorization: Simple, fast, small. In: Fischer, J., Sanders, P. (eds.) CPM 2013. LNCS, vol. 7922, pp. 189–200. Springer, Heidelberg (2013)
5. Larsson, N.J., Moffat, A.: Offline dictionary-based compression. In: Data Compression Conference, pp. 296–305. IEEE Computer Society (1999)
6. Lohrey, M.: Algorithmics on SLP-compressed strings: A survey. Groups Complexity Cryptology 4(2), 241–299 (2012)
7. Rubin, F.: Experiments in text file compression. Commun. ACM 19(11), 617–623 (1976)
8. Rytter, W.: Application of Lempel-Ziv factorization to the approximation of grammar-based compression. Theor. Comput. Sci. 302(1-3), 211–222 (2003)
9. Sakamoto, H.: A fully linear-time approximation algorithm for grammar-based compression. J. Discrete Algorithms 3(2-4), 416–430 (2005)

Efficient Algorithms
for Shortest Partial Seeds in Words

Tomasz Kociumaka[1,*], Solon P. Pissis[2], Jakub Radoszewski[1],
Wojciech Rytter[1,3], and Tomasz Waleń[1]

[1] Faculty of Mathematics, Informatics and Mechanics,
University of Warsaw, Warsaw, Poland
{kociumaka,jrad,rytter,walen}@mimuw.edu.pl
[2] Department of Informatics, King's College London,
London WC2R 2LS, UK
solon.pissis@kcl.ac.uk
[3] Faculty of Mathematics and Computer Science,
Copernicus University, Toruń, Poland

Abstract. A factor u of a word w is a *cover* of w if every position
in w lies within some occurrence of u in w. A factor u is a *seed* of
w if it is a cover of a superstring of w. Covers and seeds extend the
classical notions of periodicity. We introduce a new notion of α-*partial
seed*, that is, a factor covering as a seed at least α positions in a given
word. We use the Cover Suffix Tree, introduced recently in the context
of α-*partial covers* (Kociumaka et al, CPM 2013); an $\mathcal{O}(n \log n)$-time
algorithm constructing such a tree is known. However it appears that
partial seeds are more complicated than partial covers—our algorithms
require algebraic manipulations of special functions related to edges of
the modified Cover Suffix Tree and the border array. We present an
algorithm for computing shortest α-partial seeds that works in $\mathcal{O}(n)$
time if the Cover Suffix Tree is already given.

1 Introduction

Periodicity in words is a fundamental topic in combinatorics on words and string
algorithms (see [5]). The concept of quasiperiodicity is a generalization of the
notion of periodicity [1]. Quasiperiodicity enables detecting repetitive structure
of words when it cannot be found using the classical characterizations of periods.
Several types of quasiperiods have already been introduced. It depends on the
type of quasiperiod what kinds of repetitive structure it allows to detect.

The best-known type of quasiperiodicity is the *cover* of word. A factor u of
a word w is said to be a cover of w if every position in w lies within some
occurrence of u in w, we also say that w is covered by u. An extension of the
notion of cover is the notion of *seed*, in this case the positions covered by a seed

---

* Supported by Polish budget funds for science in 2013-2017 as a research project
under the 'Diamond Grant' program.

A.S. Kulikov, S.O. Kuznetsov, and P. Pevzner (Eds.): CPM 2014, LNCS 8486, pp. 192–201, 2014.

u are also positions within overhanging occurrences of u. More formally, u is a seed of w if w is a factor of a word y covered by u.

Several algorithms are known for computation of covers and seeds. A linear-time algorithm for computing the shortest cover of a word was proposed by Apostolico et al. [2], and a linear-time algorithm for computing all the covers was proposed by Moore & Smyth [12]. Linear-time algorithms providing yet more complete characterizations of covers by so-called cover arrays were given in [3,11]. Seeds were first introduced by Iliopoulos, Moore, and Park [7] who presented an $\mathcal{O}(n \log n)$-time algorithm computing seeds. This result was improved recently by Kociumaka et al. [8] who gave a complex linear-time algorithm.

It remains unlikely that an arbitrary word has a cover or a seed shorter than the word itself. Due to this reason, relaxed variants of quasiperiodicity have been introduced. One of the ideas are *approximate covers* [13] and *approximate seeds* [4] that require each position to lie within an approximate occurrence of the corresponding quasiperiod. Another idea, introduced recently in [9], was the notion of *partial cover* that is required to cover a certain number of positions of a word. We extend the ideas of [9] and introduce the notion of *partial seed*.

Let $\mathcal{C}(u, w)$ denote the number of positions in w covered by (full) occurrences of the word u in w. The word u is called an α-*partial cover* of w if $\mathcal{C}(u, w) \geq \alpha$. We call a non-empty prefix of w that is also a suffix of u a *left-overhanging* occurrence of u in w. Symmetrically, a non-empty suffix of w which is a prefix of u is called a *right-overhanging* occurrence. Let $\mathcal{S}(u, w)$ denote the number of positions in w covered by full, left-overhanging, or right-overhanging occurrences of u in w. We call u an α-*partial seed* of w if $\mathcal{S}(u, w) \geq \alpha$. If the word w is clear from the context, we use the simpler notations of $\mathcal{C}(u)$ and $\mathcal{S}(u)$.

Example 1. If $w = $ aaaabaabaaaaaba, see also Fig. 1, then

$$\mathcal{S}(\text{abaa}) = 12, \ \mathcal{S}(\text{aba}) = 10, \ \mathcal{S}(\text{ab}) = 7, \ \mathcal{S}(\text{a}) = 12.$$

```
a b a a             a b a a           a b a a
  a b a a   a b a a                     a b a a
      a a a a b a a b a a a a a b a
```

Fig. 1. The positions covered by **abaa** as a partial seed are underlined. The word **abaa** is a 12-partial seed of w, it has four overhanging occurrences and two full occurrences. Note that **a** is the shortest 12-partial seed of w.

We study the following two related problems.

PARTIALSEEDS
Input: a word w of length n and a positive integer $\alpha \leq n$
Output: all shortest factors u of w such that $\mathcal{S}(u, w) \geq \alpha$

LIMITEDLENGTHPARTIALSEEDS
Input: a word w of length n and an interval $[\ell, r]$, $0 < \ell \leq r \leq n$
Output: a factor u of w, $|u| \in [\ell, r]$, which maximizes $\mathcal{S}(u, w)$

In [9] a data structure called the *Cover Suffix Tree* and denoted by $CST(w)$ was introduced. For a word w of length n the size of $CST(w)$ is $\mathcal{O}(n)$ and the construction time is $\mathcal{O}(n \log n)$. In this article, we obtain the following results.

Theorem 2. *Given* $CST(w)$, *the* LIMITEDLENGTHPARTIALSEEDS *problem can be solved in* $\mathcal{O}(n)$ *time.*

By applying binary search, Theorem 2 implies an $\mathcal{O}(n \log n)$-time solution to the PARTIALSEEDS problem. However, this solution can be improved to an $\mathcal{O}(n)$-time algorithm, provided that $CST(w)$ is known.

Theorem 3. [Main result] *Given* $CST(w)$, *the* PARTIALSEEDS *problem can be solved in* $\mathcal{O}(n)$ *time.*

Structure of the Paper. In Section 2 we introduce basic notation related to words and suffix trees and recall the Cover Suffix Tree. Next in Section 3 we extend CST to obtain its counterpart suitable for computation of partial seeds, which we call the Seed Suffix Tree (SST). In Section 4 we introduce two abstract problems formulated in terms of simple functions which encapsulate the intrinsic difficulty of the two types of PARTIALSEEDS problems. We present the solutions of the abstract problems in Section 5; this section is essentially the most involved part of our contribution. We summarize our results in the Conclusions section.

2   Preliminaries

Let us fix a word w of length n over a totally ordered alphabet Σ. For a factor v of w, by $Occ(v)$ we denote the set of positions where occurrences of v in w start. By $first(v)$ and $last(v)$ we denote $\min Occ(v)$ and $\max Occ(v)$, respectively.

By $w[i..j]$ we denote the factor starting at the position i and ending at the position j. Factors $w[1..i]$ are called *prefixes* of w, and factors $w[i..n]$ are called *suffixes* of w. Words shorter than w that are both prefixes and suffixes of w are called *borders* of w. By $\beta(w)$ we denote the length of the longest border of w. The border array $\beta[1..n]$ and reverse border array $\beta^R[1..n]$ of w are defined as follows: $\beta[i] = \beta(w[1..i])$ and $\beta^R[i] = \beta(w[i..n])$. The arrays β, β^R can be constructed in $\mathcal{O}(n)$ time [6].

The *suffix tree* of w, denoted by $ST(w)$, is the compacted suffix trie of w in which only branching nodes and suffixes are explicit. We identify the nodes of $ST(w)$ with the factors of w that they represent. An *augmented* suffix tree may contain some additional explicit nodes, called *extra* nodes. For an explicit node $v \neq \varepsilon$, we set $path(v) = (v_0, v_1, \ldots, v_k)$ where $v_0 = v$ and $v_1, \ldots, v_k$ are the implicit nodes on the path going upwards from v to its nearest explicit ancestor. E.g., in the right tree in Fig. 2 we have $path(v) = (v, v_1, v_2, v_3, v_4, v_5)$. We define the *locus* of a factor v' of w as a pair (v, j) such that $v' = v_j$ where $v_j \in path(v)$.

The Cover Suffix Tree is an augmented version of a suffix tree introduced in [9] that allows to efficiently compute $\mathcal{C}(v)$ for any explicit or implicit node, as shown in the following theorem.

Theorem 4 ([9]). *Let w be a word of length n. There exists an augmented suffix tree of size $\mathcal{O}(n)$, such that for each edge $path(v)$ we have $\mathcal{C}(v_j) = c(v) - j\Delta(v)$ for some positive integers $c(v), \Delta(v)$. Such a tree together with values $c(v), \Delta(v)$, denoted as $CST(w)$, can be constructed in $\mathcal{O}(n \log n)$ time and $\mathcal{O}(n)$ space.*

Actually [9] provides explicit formulas for $c(v), \Delta(v)$ in terms of $Occ(v)$. Their form is not important here; the only property which we use is that $1 \leq \Delta(v) \leq |Occ(v)|$.

3   Seed Suffix Tree

CST introduces some extra nodes to ST thanks to which the cover index $\mathcal{C}(v_j)$ on each edge becomes a linear function: $\mathcal{C}(v_j) = c(v) - j\Delta(v)$. With seed index $\mathcal{S}(v_j)$, the situation is more complex. However, if we make some more nodes explicit, then $\mathcal{S}(v_j)$ becomes a relatively simple function. We call the resulting tree the *Seed Suffix Tree*, denoted by $SST(w)$.

Lemma 5. *Let w be a word of length n. We can construct an augmented suffix tree, denoted by $SST(w)$, of size $\mathcal{O}(n)$ such that for each node v there exists a function $\phi_v(x) = a_v x + b_v + \min(c_v, \beta[x])$ and a range $R_v = (\ell_v, r_v]$ such that for all $v_j \in path(v)$ we have $\mathcal{S}(v_j) = \phi_v(r_v - j)$. Additionally, $0 \leq a_v \leq |Occ(v)|$. The tree $SST(w)$, together with the border array β and tuples $(a_v, b_v, c_v, \ell_v, r_v)$ representing ϕ_v, can be constructed in $\mathcal{O}(n)$ time given $CST(w)$.*

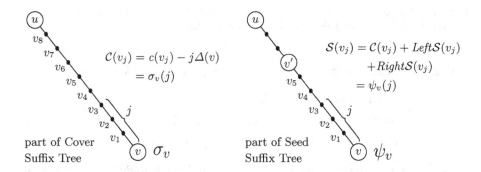

Fig. 2. In $CST(w)$ there is a constant-space description of a linear function σ_v associated with each explicit node v, which gives the values of $\mathcal{C}(v_j)$ for implicit nodes on the edge from v upwards. In $SST(w)$ there is a corresponding function ψ_v which is a combination of the linear function σ_v and two functions depending on border arrays. After suitable linear transformation of variable j, the function $\psi_v(j)$ is converted to a more convenient function $\phi_v(x)$. When transforming $CST(w)$ to $SST(w)$, some implicit nodes are made explicit to guarantee that ϕ_v has a simple form (v' on the figure).

Proof. For any factor v of w we define:

$$LeftS(v) = \min(\beta[first(v) + |v| - 1], first(v) - 1)$$
$$RightS(v) = \min(\beta^R[last(v)], n - |v| + 1 - last(v)).$$

The following observation relates these values to $S(v)$, see also Fig. 3.

Claim. $S(v) = C(v) + LeftS(v) + RightS(v)$.

Proof (of the claim). $C(v)$ counts all positions covered by full occurrences of v. We claim that the remaining positions covered by left-overhanging occurrences are counted by $LeftS(v)$. Let $p = first(v) + |v| - 1$. Note that $w[1..p]$ has v as a suffix, which means that $\beta[p]$ is the length of the longest left-overhanging occurrence of v. It covers $\beta[p]$ positions, but, among them, positions greater than or equal to $first(v)$ are already covered by a full occurrence of v. $RightS(v)$ has a symmetric interpretation.   □

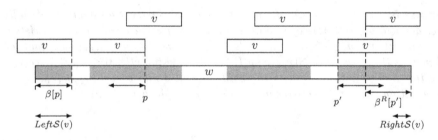

Fig. 3. The positions contained in $S(v)$ are marked gray. In this case $LeftS(v) = \beta[first(v) + |v| - 1]$, and $RightS(v) = n - |v| + 1 - last(v)$.

Consider an edge $path(v) = (v_0, \ldots, v_k)$. For any $0 \le j \le k$ we have:

$$C(v_j) = c(v) - j\Delta(v)$$
$$LeftS(v_j) = \min(\beta[first(v) + |v| - j - 1], first(v) - 1)$$
$$RightS(v_j) = \min(\beta^R[last(v)], n - |v| + j + 1 - last(v)).$$

We consider the function $S(v_j) = C(v_j) + LeftS(v_j) + RightS(v_j)$. Note that only $C(v_j)$ is a linear function of j. Also, $RightS(v_j)$ is relatively simple. It either already is a linear function in the whole $\{0, \ldots, k\}$, or it becomes one in both parts of $\{0, \ldots, k\}$ if we split it at $j \in \{0, \ldots, k\}$ such that $\beta^R[last(v)] = n - |v| + j + 1 - last(v)$. We subdivide each $path(v)$ at v_j if such j exists. Note that we can easily update values c and Δ for newly created edges. Also, we make explicit at most $\mathcal{O}(n)$ nodes (at most one per edge of $CST(w)$), so the resulting tree $SST(w)$ has $\mathcal{O}(n)$ explicit nodes in total.

It remains to show that after these transformations $S(v_j) = \phi_v(r_v - j)$ for $r_v = first(v) + |v| - 1$ and that the coefficients of ϕ_v can be efficiently computed.

We omit the explicit formulae in this version, they can be obtained with just a few simple algebraic transformations. The additional inequality $0 \leq a_v \leq |Occ(v)|$ follows from the property that $1 \leq \Delta(v) \leq |Occ(v)|$. □

The following observation is a direct consequence of Lemma 5.

Observation 6. *Given $SST(w)$ and a locus of u, a factor of w, one can compute $\mathcal{S}(u)$ in constant time.*

4   Reduction to Two Abstract Problems

We say that an integer array $A[1 \mathinner{.\,.} k]$ is a *linear-oscillation* array if

$$\sum_{i=1}^{k-1} |A[i] - A[i+1]| = \mathcal{O}(k).$$

Observation 7. *Any border array is a linear-oscillation array.*

To solve the LIMITEDLENGTHPARTIALSEEDS problem, we make explicit all nodes corresponding to factors of length $\ell - 1$ and r. This way each edge either contains only nodes at tree levels in $\{\ell, \ldots, r\}$ or none of them. Note that the functions ϕ_v on the subdivided edges stay the same, only the ranges shrink. Consider the following abstract problem.

PROBLEM A1
Input: a linear-oscillation array B of size n and m pairs (ϕ_i, R_i), where ϕ_i is a function $\phi_i(x) = a_i x + b_i + \min(c_i, B[x])$ and $R_i = (\ell_i, r_i] \subseteq [1, n]$ is a non-empty range
Output: the values $x_i = \operatorname{argmax}\{\phi_i(x) : x \in R_i\}$.[1]

Applying Problem A1 for $B = \beta$ and a query for each edge $path(v)$, we obtain $v_j \in path(v)$ maximizing $\mathcal{S}(v_j)$. Taking a global maximum over all edges containing factors of lengths within $\{\ell, \ldots, r\}$, we get the sought factor u, which maximizes $\mathcal{S}(u)$ among all factors of w with $|u| \in \{\ell, \ldots, r\}$. This results in the following lemma.

Lemma 8. *Given $SST(w)$ and an $\mathcal{O}(n + m)$-time off-line solution to Problem A1, the LIMITEDLENGTHPARTIALSEEDS problem can be solved in $\mathcal{O}(n)$ time.*

To solve the PARTIALSEEDS problem, we also apply Problem A1 to compute $\max \mathcal{S}(v_j)$ for each edge $path(v)$ of $SST(w)$ (this time we do not introduce any additional extra nodes). We say that an edge $path(v)$ is *feasible*, if $\max \mathcal{S}(v_j) \geq \alpha$,

---

[1] For a set X and a function $f : X \to \mathbb{R}$, we define $\operatorname{argmax}\{f(x) : x \in X\}$ as the largest argument for which f attains its maximum value, that is, $\max\{x \in X : \forall_{x' \in X} f(x) \geq f(x')\}$ (we use the maximality of x later on). We assume $\max \emptyset = -\infty$ and $\min \emptyset = \infty$.

and *important* if it is feasible and no ancestor edge is feasible. It is easy to see that all shortest α-partial seeds lie on important edges. Also, by Lemma 5, a_v summed over all feasible edges e_v is at most n. Consider the following abstract problem.

PROBLEM A2

Input: a linear-oscillation array B of size n, a positive integer α and m pairs (ϕ_i, R_i), where ϕ_i is a function $\phi_i(x) = a_i x + b_i + \min(c_i, B[x])$, $\sum a_i = \mathcal{O}(n)$, and $R_i = (\ell_i, r_i] \subseteq [1, n]$ is a range

Output: the values $\min\{x \in R_i : \phi_i(x) \geq \alpha\}$

Note that applied for $B = \beta$ and queries for all important edges, it gives the shortest α-partial seed within each important edge. Taking all globally shortest among these candidates, we get all shortest α-partial seeds. This results in the following lemma.

Lemma 9. *Given $SST(w)$ and $\mathcal{O}(n+m)$-time off-line solutions to Problems A1 and A2, the PARTIALSEEDS problem can be solved in $\mathcal{O}(n)$ time.*

5   Solutions to Abstract Problems

We begin with a few auxiliary definitions and lemmas. For a set $X \subseteq \mathbb{Z}$ and an integer $x \in \mathbb{Z}$ we define $\mathrm{pred}(x, X) = \max\{y \in X : y \leq x\}$, $\mathrm{succ}(x, X) = \min\{y \in X : y \geq x\}$, and $\mathrm{rank}(x, X) = |\{y \in X : y \leq x\}|$. The technical proofs of the following two lemmas are left for the full version of the paper.

Lemma 10. *Let $Y[1 \mathinner{.\,.} n]$ be an array of integers of magnitude $\mathcal{O}(n)$. For an integer k let $Y_{\geq k} = \{i : Y[i] \geq k\}$. Given m integer pairs (y_j, k_j) we can compute the values $\mathrm{pred}(y_j, Y_{\geq k_j})$ and $\mathrm{succ}(y_j, Y_{\geq k_j})$ in $\mathcal{O}(n + m)$ time.*

Lemma 11. *Assume we are given a family $\{X_a \subseteq [1 \mathinner{.\,.} n] : a \in [1 \mathinner{.\,.} n]\}$ of sets of total size $\mathcal{O}(n)$ and functions $\psi_a : [1 \mathinner{.\,.} n] \to \mathbb{Z}$, each computable in constant time. Given m pairs (a_j, R_j) or m triplets (a_j, β_j, R_j), where $a_j \in [1 \mathinner{.\,.} n]$, $R_j \subseteq [1 \mathinner{.\,.} n]$ is an interval, and $\beta_j \in \mathbb{Z}$, we can compute*

$$\mathrm{argmax}\{\psi_{a_j}(x) \ : \ x \in X_{a_j} \cap R_j\} \text{ in case of pairs, or}$$

$$\min\{x \in X_{a_j} \cap R_j \ : \ \psi_{a_j}(x) \geq \beta_j\} \text{ in case of triples}$$

offline in $\mathcal{O}(n + m)$ time.

The following simple result is the reason behind the linear-oscillation assumption in both problems.

Observation 12. *Let $F_a = \{x : B[x + 1] < B[x] - a\}$ for a linear-oscillation array B of size n. Then $\sum_{a=1}^{\infty} |F_a| = \mathcal{O}(n)$.*

Proof. Each $x \in \{1, \ldots, n - 1\}$ belongs to F_a if and only if $a < B[x] - B[x + 1]$. The sum is therefore bounded by the total decrease of B, which is $\mathcal{O}(n)$ for a linear-oscillation array. □

The following is a simplified version of Problem A1.

PROBLEM B1
Input: a linear-oscillation array B of size n and m pairs (ϕ_i, R_i), where ϕ_i is a function $\phi_i(x) = a_i x + B[x]$ and $R_i = (\ell_i, r_i] \subseteq [1, n]$ is a non-empty range
Output: the values $x_i = \text{argmax}\{\phi_i(x) : x \in R_i\}$

Lemma 13. *Problem A1 can be reduced to Problem B1 in $\mathcal{O}(n + m)$ time.*

Proof. Let

$$y_i = \max(\{x \in R_i : B[x] \geq c_i\} \cup \{\ell_i\}).$$

Consider any $x \in R_i$ such that $x \leq y_i$. If such x exists then $\phi_i(x) = a_i x + b_i + \min(c_i, B[x]) \leq a_i y_i + b_i + c_i = \phi_i(y_i)$. Consequently, $x_i = \text{argmax}\{\phi_i(x) : x \in R_i\} \geq y_i$. Note that for $x \in R_i$ such that $x > y_i$ we have $\phi_i(x) = a_i x + b_i + B[x]$. Thus, it suffices to solve Problem B1 for $R_i' = (y_i, r_i]$ (if $R_i' \neq \emptyset$). This way we find:

$$x_i' = \text{argmax}\{a_i x + B[x] : x \in R_i'\}.$$

Then x_i is guaranteed to be either x_i' or y_i, and it is easy to check in constant time which of the two it actually is.

The missing step of the reduction is determining y_i's. We claim that these values can be computed off-line in $\mathcal{O}(n + m)$ time. Instead of y_i it suffices to compute $y_i' = \max\{x \leq r_i : B[x] \geq c_i\}$ ($y_i' = -\infty$ if the set is empty). Then y_i can be determined as $\max(y_i', \ell_i)$. Clearly $y_i' = \text{pred}(r_i, B_{\geq c_i})$, and these values can be computed in $\mathcal{O}(n + m)$ time using Lemma 10. $\square$

Now, it remains to solve Problem B1. Recall sets F_a defined in Observation 12. Note that $x_i \in F_{a_i}$ or $x_i = r_i$. Indeed, if $x \notin F_{a_i}$ and $x \neq r_i$, then $x + 1 \in R_i$ and $\phi_i(x + 1) = a_i(x + 1) + B[x + 1] \geq a_i x + a_i + B[x] - a_i = \phi_i(x)$. Consequently, it is enough to compute

$$x_i' = \text{argmax}\{\phi_i(x) : x \in F_{a_i} \cap R_i\}$$

($x_i' = -\infty$ if the set is empty). Then x_i is either x_i' or r_i, and it is easy to check in constant time which of the two it actually is.

By Lemma 11 values x_i' can be computed in $\mathcal{O}(n + m)$ time. This concludes the solution to Problem B1, and together with Lemma 13 implies the following result.

Lemma 14. *Problem A1 can be solved in $\mathcal{O}(n + m)$ time.*

The following is a simplified version of Problem A2.

PROBLEM B2
Input: a linear-oscillation array B of size n and m triples (ϕ_i, α_i, R_i), where $\phi_i(x) = a_i x + B[x]$, $\sum a_i = \mathcal{O}(n)$, α_i is a positive integer and $R_i = (\ell_i, r_i] \subseteq [1, n]$ is a range
Output: the values $\min\{x \in R_i : \phi_i(x) \geq \alpha_i\}$

Lemma 15. *Problem A2 can be reduced to Problem B2 in $\mathcal{O}(n+m)$ time.*

Proof. We set $\alpha_i = \alpha - b_i$. Let $y_i = \left\lceil \frac{\alpha_i - c_i}{a_i} \right\rceil$. If $a_i = 0$, we set $y_i = \infty$ if $c_i < \alpha_i$ and $y_i = -\infty$ otherwise. Note that for $x \in R_i$ such that $x < y_i$ we have $a_i x + c_i < \alpha_i$, so $\phi_i(x) < \alpha$. Therefore $x_i \geq y_i$. On the other hand, if $x \geq y_i$ then $a_i x + c_i \geq \alpha_i$, so $\phi_i(x) \geq \alpha$ if and only if $a_i x + B[x] \geq \alpha_i$. Consequently, it suffices to solve Problem B2, with $R'_i = R_i \cap [y_i, \infty)$. ▢

It remains to solve Problem B2. Recall sets F_a from Observation 12 and set

$$G_a = F_a \cup \{x \in [1..n] \ : \ x \bmod (1 + a^2) = 0\}.$$

It holds that

$$\sum_{a=0}^{\infty} |G_a| \leq \sum_{a=0}^{\infty} |F_a| + n \sum_{a=0}^{\infty} \tfrac{1}{1+a^2} = \mathcal{O}(n)$$

by Observation 12 and since $\sum_{a=0}^{\infty} \frac{1}{1+a^2}$ is $\mathcal{O}(1)$. Note that Lemma 10 applied for an array Y with $Y[x] = B[x] - B[x+1] - 1$ lets us obtain $\mathrm{pred}(x, F_a)$ and $\mathrm{succ}(x, F_a)$. With simple arithmetics we can use these values to compute $\mathrm{pred}(x, G_a)$ and $\mathrm{succ}(x, G_a)$. Assume that x_i exists. Let

$$x'_i = \min(r_i, \mathrm{succ}(x_i, G_{a_i})), \quad x''_i = \max(\ell_i, \mathrm{pred}(x_i - 1, G_{a_i})).$$

Observe ϕ_i is non-decreasing within $R'_i = (x''_i, x'_i]$. Indeed, if $x, x+1 \in R'_i$, then $x \notin F_{a_i}$, so $\phi_i(x+1) \geq \phi_i(x)$, as noted in the solution to Problem B1. We claim that R'_i can be computed in $\mathcal{O}(n+m)$ time. Since $x''_i = \max(\ell_i, \mathrm{pred}(x'_i - 1, G_{a_i}))$ we can compute x''_i once we have x'_i. Thus, the main challenge is to compute x'_i. By monotonicity of ϕ_i in R'_i, we conclude that $\phi_i(x'_i) \geq \alpha_i$. On the other hand, for any $x \in R_i$ such that $x < x_i$ we have $\phi_i(x) < \alpha_i$. Moreover any $x \in R_i \cap G_{a_i}$ smaller than x'_i is smaller than x_i, so $x'_i = \min(\{x \in R_i \cap G_{a_i} : \phi_i(x) \geq \alpha_i\} \cup \{r_i\})$, and such values can be computed off-line in $\mathcal{O}(n+m)$ time by Lemma 11.

Once we have the interval R'_i, by monotonicity of ϕ_i on R'_i, we can find x_i using binary search in $\mathcal{O}(\log a_i)$ time, because $|R'_i| \leq a_i^2 + 1$. The total time complexity is $\mathcal{O}(n + m + \sum_i \log a_i)$, and since $\sum_i a_i = \mathcal{O}(n)$ in Problem B2, this reduces to $\mathcal{O}(n+m)$, which implies the following result.

Lemma 16. *Problem A2 can be solved in $\mathcal{O}(n+m)$ time.*

6   Conclusions

We are now ready to combine all the results obtained so far.

Theorem 2. *Given $CST(w)$, the* LimitedLengthPartialSeeds *problem can be solved in $\mathcal{O}(n)$ time.*

Proof. First, we apply Lemma 5 and construct $SST(w)$. Then, we solve the LimitedLengthPartialSeeds problem using Lemma 8, plugging the algorithm of Lemma 14 for Problem A1. ▢

Theorem 3. *Given $CST(w)$, the* PARTIALSEEDS *problem can be solved in $\mathcal{O}(n)$ time.*

Proof. We proceed much as in Theorem 2: we construct $SST(w)$ and solve PARTIALSEEDS problem using Lemma 9, plugging the algorithms of Lemmas 14 and 16 for Problems A1 and A2, respectively. □

An interesting open question is whether one can compute the shortest α-partial seed for each $\alpha \in \{1, \ldots, n\}$ any faster than applying n times Theorem 3. The corresponding problem for partial covers is known to have an $\mathcal{O}(n \log n)$-time solution [10].

References

1. Apostolico, A., Ehrenfeucht, A.: Efficient detection of quasiperiodicities in strings. Theor. Comput. Sci. 119(2), 247–265 (1993)
2. Apostolico, A., Farach, M., Iliopoulos, C.S.: Optimal superprimitivity testing for strings. Inf. Process. Lett. 39(1), 17–20 (1991)
3. Breslauer, D.: An on-line string superprimitivity test. Inf. Process. Lett. 44(6), 345–347 (1992)
4. Christodoulakis, M., Iliopoulos, C.S., Park, K., Sim, J.S.: Approximate seeds of strings. Journal of Automata, Languages and Combinatorics 10(5/6), 609–626 (2005)
5. Crochemore, M., Ilie, L., Rytter, W.: Repetitions in strings: Algorithms and combinatorics. Theor. Comput. Sci. 410(50), 5227–5235 (2009)
6. Crochemore, M., Rytter, W.: Jewels of Stringology. World Scientific (2003)
7. Iliopoulos, C.S., Moore, D.W.G., Park, K.: Covering a string. Algorithmica 16(3), 288–297 (1996)
8. Kociumaka, T., Kubica, M., Radoszewski, J., Rytter, W., Waleń, T.: A linear time algorithm for seeds computation. In: Rabani, Y. (ed.) SODA, pp. 1095–1112. SIAM (2012)
9. Kociumaka, T., Pissis, S.P., Radoszewski, J., Rytter, W., Waleń, T.: Fast algorithm for partial covers in words. In: Fischer, J., Sanders, P. (eds.) CPM 2013. LNCS, vol. 7922, pp. 177–188. Springer, Heidelberg (2013)
10. Kociumaka, T., Pissis, S.P., Radoszewski, J., Rytter, W., Waleń, T.: Fast algorithm for partial covers in words. In: ArXiv e-prints, arXiv:1401.0163 [cs.DS] (December 2013)
11. Li, Y., Smyth, W.F.: Computing the cover array in linear time. Algorithmica 32(1), 95–106 (2002)
12. Moore, D., Smyth, W.F.: An optimal algorithm to compute all the covers of a string. Inf. Process. Lett. 50(5), 239–246 (1994)
13. Sim, J.S., Park, K., Kim, S., Lee, J.: Finding approximate covers of strings. Journal of Korea Information Science Society 29(1), 16–21 (2002)

Computing k-th Lyndon Word and Decoding Lexicographically Minimal de Bruijn Sequence

Tomasz Kociumaka[1,*], Jakub Radoszewski[1], and Wojciech Rytter[1,2]

[1] Faculty of Mathematics, Informatics and Mechanics,
University of Warsaw, Warsaw, Poland
{kociumaka,jrad,rytter}@mimuw.edu.pl
[2] Faculty of Mathematics and Computer Science,
Copernicus University, Toruń, Poland

Abstract. Let Σ be a finite ordered alphabet. We present polynomial-time algorithms for computing the k-th in the lexicographic order Lyndon word of a given length n over Σ and counting Lyndon words of length n that are smaller than a given word. We also use the connections between Lyndon words and minimal de Bruijn sequences (theorem of Fredricksen and Maiorana) to develop the first polynomial time algorithm for *decoding* minimal de Bruijn sequence of any rank n (it determines the position of an arbitrary word of length n within the de Bruijn sequence). Our tools mostly rely on combinatorics on words and automata theory.

1 Introduction

We consider finite words over a finite ordered alphabet Σ. A *Lyndon word* over Σ is a word that is strictly smaller than all its nontrivial cyclic rotations. Lyndon words have a number of combinatorial properties (see, e.g., [10]) including the famous Lyndon factorization theorem which states that every word can be uniquely written as a concatenation of a lexicographically non increasing sequence of Lyndon words (due to this theorem Lyndon words are also called prime words, see [9]). They are also related to necklaces of n beads in k colors, that is, equivalence classes of k-ary n-tuples under rotation [6,7]. Lyndon words have a number of applications in the field of text algorithms, see e.g. [1,2,3,12].

A *de Bruijn sequence of rank n* is a cyclic sequence of length $|\Sigma|^n$ in which every possible word of length n appears as a subword exactly once. De Bruijn sequences are present in a variety of contexts, such as digital fault testing, pseudo-random number generation, and modern public-key cryptographic schemes; there are numerous algorithms for generating such sequences and their generalizations to other combinatorial structures have been investigated, see [5,9]. Fredricksen and Maiorana [7] have shown a surprising deep connection between de Bruijn sequences and Lyndon words: the lexicographically minimal de Bruijn sequence over Σ is a concatenation, in lexicographic order, of all Lyndon words over Σ whose length is a divisor of n.

---

* Supported by Polish budget funds for science in 2013-2017 as a research project under the 'Diamond Grant' program.

A.S. Kulikov, S.O. Kuznetsov, and P. Pevzner (Eds.): CPM 2014, LNCS 8486, pp. 202–211, 2014.
© Springer International Publishing Switzerland 2014

All Lyndon words of length at most n can be generated in lexicographic order by algorithm of Fredricksen, Kessler and Maiorana (FKM) [6,7] (another algorithm was developed by Duval in [4]). The analysis from [14] shows that the FKM algorithm generates the subsequent Lyndon words in constant amortized time. We give the first polynomial time algorithm for generating Lyndon words of arbitrary rank in lexicographic order. We also generalize the known simple formula for the number of Lyndon words of length n over Σ (see [9,10]) by showing a polynomial time algorithm that computes the number of Lyndon words of length n smaller than a given word.

For several de Bruijn sequences *decoding* algorithms exist which find the position of an arbitrary word of length n in a given de Bruijn sequence in polynomial time [11,15]. Such algorithms prove useful in certain types of position sensing applications of de Bruijn sequences [15]. We obtain the first decoding algorithm for the lexicographically minimal de Bruijn sequence by exploiting its connections with Lyndon words. We also obtain a polynomial-time algorithm for random access of symbols in this sequence and in a related sequence defined in [8]. Note that the FKM algorithm can be used to compute the subsequent symbols of the lexicographically minimal de Bruijn sequence with $\mathcal{O}(n^2)$ time delay (or even with worst-case $\mathcal{O}(1)$ time delay [13]), however it does it only *in order*.

We denote by $\mathcal{L}$ and $\mathcal{L}_n$ the set of all Lyndon words and all Lyndon words of length n, respectively, and define

$$Lynd(w) = \{x \in \mathcal{L}_{|w|} : x \leq w\}.$$

Example 1. For $\Sigma = \{a, b\}$ we have $|Lynd(\text{abbaba})| = 8$ since we have the following Lyndon words of length 6 smaller than abbaba (note that abbaba itself is not a Lyndon word):

aaaaab, aaaabb, aaabab, aaabbb, aababb, aabbab, aabbbb, ababbb.

Let $\mathcal{L}^{(n)} = \bigcup_{d|n} \mathcal{L}_d$. By dB_n we denote the lexicographically first de Bruijn sequence of rank n over the given alphabet Σ. It is the concatenation of all Lyndon words in $\mathcal{L}^{(n)}$ in lexicographic order.

Example 2. For $n = 6$ and binary alphabet we have the following decomposition of dB_6 into Lyndon words:

0 000001 000011 000101 000111 001 001011 001101 001111 01 010111 011 011111 1.

Recently a variant of de Bruijn words was introduced in [8]. Let dB'_n be the concatenation in lexicographic order of Lyndon words of length n over Σ. Then dB'_n is a cyclic sequence containing all primitive words of length n.

Example 3. For $n = 6$ and binary alphabet we have the following decomposition of dB'_6:

000001 000011 000101 000111 001011 001101 001111 010111 011111.

Our Results. We assume that $|\Sigma|$ fits in a single machine word. We present an $\mathcal{O}(n^3)$-time algorithm for computing $|Lynd(w)|$ for a word w of length n. Using binary search this algorithm implies an $\mathcal{O}(n^4 \log |\Sigma|)$-time algorithm for computing the k-th Lyndon word of length n (in the lexicographic order) for a given k. Next we show an $\mathcal{O}(n^3)$-time decoding algorithm that finds the position of an arbitrary $w \in \Sigma^n$ in dB_n. We also obtain $\mathcal{O}(n^4 \log |\Sigma|)$-time algorithms computing the k-th symbol of dB_n and dB'_n for a given k.

2   Preliminaries

Let Σ be a finite ordered alphabet. By Σ^* and Σ^n we denote the set of all words over Σ and the set of all such words of length n. If w is a word then $|w|$ denotes its length, $w[i]$ its i-th letter (for $1 \le i \le |w|$), $w[i,j]$ its factor $w[i]w[i+1]\ldots w[j]$ and $w_{(i)}$ its prefix $w[1,i]$. Additionally w^k is a concatenation of k copies of w and w^∞ is an infinite word composed of an infinite number of copies of w.

By $rot(w, c)$ let us denote a *cyclic rotation* of w obtained by moving $(c \bmod n)$ first letters of w to its end (preserving the order of the letters). We say that the words w and $rot(w, c)$ are *cyclically equivalent* (sometimes called *conjugates*). By $\langle w \rangle$ we denote the lexicographically minimal cyclic rotation of w. We say that w is *primitive* if $w = u^k$ for $k \in \mathbb{Z}_+$ implies that $u = w$, otherwise w is called non-primitive. We say that $\lambda \in \Sigma^*$ is a *Lyndon word* if it is primitive and $\langle \lambda \rangle = \lambda$. All cyclic rotations of a Lyndon word are different primitive words [10]. The technical proofs of the following lemmas are left for the full version.

Lemma 4. *Let $x, y \in \Sigma^*$. Assume $x = \langle x \rangle$ and $x \ge y$. Then $x^\infty \ge y^\infty$.*

Lemma 5. *For a given word $w \in \Sigma^n$ we can compute in $\mathcal{O}(n^2)$ time the lexicographically largest word $w' \in \Sigma^n$ such that $\langle w' \rangle = w' \le w$.*

3   Combinatorial Tools

Our basic goal is to compute $|Lynd(w)|$, that is, the number of Lyndon words in Σ^n not exceeding w ($n = |w|$). It suffices to compute $|Lynd(w)|$ for words w such that $\langle w \rangle = w$. We show how to reduce it to the computation of the cardinality of the following set:

$$CS(v) = \{x \in \Sigma^{|v|} : \langle x \rangle \le v\}$$

for some prefixes v of w. Computation of $|CS(v)|$ is also of independent interest, we apply it in the decoding scheme for minimal de Bruijn sequences.

Let us introduce the following auxiliary sets:

$$CS_\ell(v) = \{x \in \Sigma^\ell : \langle x \rangle^\infty \le v^\infty\}$$
$$CS'_\ell(v) = \{x \in \Sigma^\ell : x \text{ is primitive}, \langle x \rangle^\infty \le v^\infty\}.$$

Note that if $|x| = |v|$ then $\langle x \rangle^\infty \le v^\infty$ is simply equivalent to $x \le v$. Thus $CS(v) = CS_{|v|}(v)$.

Observation 6. $|Lynd(w)| = \frac{1}{n}|CS'_n(w)|$

Proof. Observe that $CS'_n(w)$ is the set of all primitive words of length n that have a cyclic rotation not exceeding w. Each Lyndon word of length n not exceeding w corresponds to n such words: all its cyclic rotations. □

Observation 7. $|CS_\ell(w)| = \sum_{d|\ell}|CS'_d(w)|.$

Proof. For a word x of length ℓ there exists exactly one primitive word y such that $y^k = x$ where $k \in \mathbb{Z}_+$. Thus:

$$CS_\ell(w) = \bigcup_{d|\ell}\left\{y \in \Sigma^d : y \text{ is primitive}, \left\langle y^{\ell/d}\right\rangle^\infty \le w^\infty\right\},$$

and the sum is disjoint. Now $\left\langle y^{\ell/d}\right\rangle^\infty = \langle y\rangle^\infty$ implies the requested formula. □

From Observation 7, using Möbius inversion formula, we obtain:

$$|CS'_\ell(w)| = \sum_{d|\ell}\mu(\tfrac{\ell}{d})|CS_d(w)|.$$

Observation 8. *Let $w \in \Sigma^n$ satisfy $w = \langle w\rangle$. Then $CS_d(w) = CS(w_{(d)})$.*

Proof. If $d = n$ the equality is trivial. Assume $d < n$. Let $y \in \Sigma^d$ be a word. By Lemma 4, $w_{(d)}^\infty \le w^\infty$, so $y^\infty \le w_{(d)}^\infty$ implies $y^\infty \le w^\infty$. On the other hand, if $y^\infty \le w^\infty$, then $y \le w_{(d)}$, so $y^\infty \le w_{(d)}^\infty$. Applying for $y = \langle x\rangle$ we conclude that $\langle x\rangle^\infty \le w_{(d)}^\infty$ if and only if $\langle x\rangle^\infty \le w^\infty$, which proves the claim. □

We conclude with a simple formula for $|Lynd(w)|$ that combines the results of this section.

Lemma 9. *Let $w \in \Sigma^n$ satisfy $\langle w\rangle = w$. Then*

$$|Lynd(w)| = \frac{1}{n}\sum_{d|n}\mu(\tfrac{n}{d})\left|CS(w_{(d)})\right|.$$

Example 10. Let $w = \mathsf{ababbb}$. We have $w_{(1)} = \mathsf{a}$, $w_{(2)} = \mathsf{ab}$, $w_{(3)} = \mathsf{aba}$ and

$$CS(w_{(1)}) = \{\mathsf{a}\}, \qquad\qquad CS(w_{(2)}) = \{\mathsf{aa}, \mathsf{ab}, \mathsf{ba}\},$$
$$CS(w_{(3)}) = \{\mathsf{aaa}, \mathsf{aab}, \mathsf{aba}, \mathsf{baa}\}, \qquad |CS(w)| = 54,$$

$$|Lynd(w)| = \tfrac{1}{6}\cdot\big(\mu(1)|CS(w)| + \mu(2)\left|CS(w_{(3)})\right| + \mu(3)\left|CS(w_{(2)})\right|$$
$$+ \mu(6)\left|CS(w_{(1)})\right|\big) = \tfrac{1}{6}\cdot(54 - 4 - 3 + 1) = 8.$$

The set $Lynd(w)$ contains the following words:

$$\mathsf{aaaaab},\ \mathsf{aaaabb},\ \mathsf{aaabab},\ \mathsf{aaabbb},\ \mathsf{aababb},\ \mathsf{aabbab},\ \mathsf{aabbbb},\ \mathsf{ababbb}.$$

4 Automata-Theoretic Tools: Computing CS

In this section we design an algorithm computing $|CS(w)|$ for a word $w \in \Sigma^n$. Note that we may assume that $\langle w\rangle = w$, since $CS(w) = CS(w')$ where $w' \in \Sigma^n$ is the largest word such that $\langle w'\rangle = w' \le w$.

Let $\mathrm{Pref}_-(w) = \{w_{(i)}s : i \in [0, n-1], s \in \Sigma, s < w[i+1]\} \cup \{w\}$. Consider a language $L(w)$ containing words that have a factor $y \in \mathrm{Pref}_-(w)$. Equivalently, $x \in L(w)$ if there exists a factor of x which is smaller than or equal to w, but is not a proper prefix of w. For a language $L \subseteq \Sigma^*$ let $\sqrt{L} = \{x : x^2 \in L\}$.

Fact 11. $CS(w) = \sqrt{L(w)} \cap \Sigma^n$

Proof. Consider a word $x \in \Sigma^n$. If $x \in CS(w)$ then $\langle x \rangle \leq w$. Take $y = \langle x \rangle$, which is a factor of x^2. Note that $y \leq w$, so some prefix of y belongs to $\mathrm{Pref}_-(w)$. This prefix is a factor of x^2, so $x^2 \in L(w)$. Consequently, $x \in \sqrt{L(w)}$.

On the other hand, assume $x \in \sqrt{L(w)}$, so x^2 contains a factor $y \in \mathrm{Pref}_-(w)$. Let us fix the first occurrence of y in x^2. Observe that y can be extended to a cyclic rotation x' of x. Note that $y \in \mathrm{Pref}_-(w)$ implies that $x' \leq w$, hence $\langle x \rangle \leq x' \leq w$ and $x \in CS(w)$. $\qquad\square$

We construct a deterministic finite automaton A recognizing $L(w)$. It has $n+1$ states: one for each proper prefix of w, and an auxiliary accepting state AC. The transitions are defined as follows: we set $\delta(AC, c) = AC$ for any $c \in \Sigma$ and

$$\delta(w_{(i)}, c) = \begin{cases} w_{(0)} & \text{if } c > w[i+1], \\ w_{(i+1)} & \text{if } c = w[i+1] \text{ and } i \neq n-1, \\ AC & \text{otherwise.} \end{cases}$$

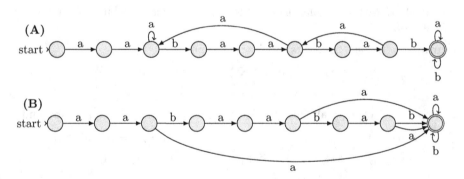

Fig. 1. Automata for a word $w = \mathbf{aabaabab}$: **(A)** accepts all words containing w as a factor, **(B)** accepts all words in $L(w)$, i.e. containing a factor $y \in \mathrm{Pref}_-(w)$. Missing links lead to the initial state.

The following fact implies that $L(A) = L(w)$.

Fact 12. *Let $x \in \Sigma^*$ and let q be the state of A after reading x. If $x \in L(w)$ then $q = AC$. Otherwise q corresponds to the longest prefix of w which is a suffix of x.*

Proof. The proof goes by induction on $|x|$. If $|x| = 0$ the statement is clear. Consider a word x of length $|x| \geq 1$. Let $x = x'c$ where $c \in \Sigma$. If $x' \in L(w)$ then clearly $x \in L(w)$. By inductive assumption after reading x' the automaton is in AC, and A is constructed so that it stays in AC once it gets there. Thus the conclusion holds in this case. From now on we assume that $x' \notin L(w)$.

Let $w_{(i)}$ be the state of A after reading x'. If $c < w[i+1]$, clearly $x \in L(w)$ ($y = w_{(i)}c \in \mathrm{Pref}_-(w)$), and the automaton proceeds to AC as desired. Similarly, it behaves correctly if $i = n-1$ and $c = w[i+1]$. Consequently we may assume that $c \geq w[i+1]$ and that w is not a suffix of x.

Take any j such that $w_{(j)}$ is a suffix of x' (possibly empty). Note that then $w_{(j)}$ is a border of $w_{(i)}$. Consequently $w_{(j)}w[i+1,n]w_{(i-j)}$ is a cyclic rotation of w, so $w_{(j)}w[i+1,n]w_{(i-j)} \geq \langle w \rangle = w = w_{(j)}w[j+1,n]$, hence $c \geq w[i+1] \geq w[j+1]$. This implies that $w_{(j)}c$ could be a prefix of w only if $c = w[i+1] = w[j+1]$. In particular, A indeed shifts to the longest prefix of w being a suffix of x. Now we only need to prove that $x \notin L(w)$. For a proof by contradiction, choose a factor y of x such that $y \in \mathrm{Pref}_-(w)$ and $|y|$ is minimal. Note that y is a suffix of x (since $x' \notin L(w)$). We have $y = w_{(j)}c$ for some $j \leq n-1$ and $c < w[j+1]$. As we have already noticed, such a word cannot be a suffix of x. □

We say that an automaton with the set of states Q is *sparse* if the underlying directed graph has $\mathcal{O}(|Q|)$ edges counting parallel edges as one. Note that the transitions from any state q of A lead to at most 3 distinct states, so A is sparse.

The following corollary summarizes the construction of A.

Corollary 13. *Let $w \in \Sigma^n$ satisfy $\langle w \rangle = w$. One can construct a sparse automaton A with $\mathcal{O}(n)$ states recognizing $L(w)$.*

For a deterministic automaton $A = (Q, q_0, F, \delta)$ and $q, q' \in Q$ let us define $L_A(q, q') = \{x \in \Sigma^* : \delta(q, x) = q'\}$. Then $L_A(q, q')$ is recognized by $(Q, q, \{q'\}, \delta)$. Note that $L(A) = \bigcup_{q \in F} L_A(q_0, q)$ where the sum is disjoint. The proof of the following lemma is based on matrix multiplication, we omit it in this version.

Lemma 14. *Let $A = (Q, q_0, F, \delta)$ be a deterministic automaton with n states, and let $m \in \mathbb{Z}_{\geq 0}$.*

(a) In poly$(n+m)$ time, one can compute $|L_A(q, q') \cap \Sigma^k|$ for all $q, q' \in Q$ and $k \leq m$.

(b) If A is sparse, it takes $\mathcal{O}(m^2 n)$ time to compute all values $|L_A(q, q') \cap \Sigma^k|$, $k \leq m$, for a fixed state q or q'.

Observation 15. *Let $A = (Q, q_0, F, \delta)$ be a deterministic automaton. Then*

$$\sqrt{L(A)} = \bigcup_{q \in Q, q' \in F} L_A(q_0, q) \cap L_A(q, q')$$

and the sum is disjoint.

Proof. Consider a word $x \in \Sigma^*$. Note that $x \in L_A(q_0, q) \cap L_A(q, q')$ for $q = \delta(q_0, x)$ and $q' = \delta(q, x)$ and no other pair of states q, q'. Clearly $x \in \sqrt{L(A)}$ if and only if $q' \in F$. □

Corollary 16. *If $w \in \Sigma^n$ satisfies $\langle w \rangle = w$, then one can compute $|CS(w)|$ in poly(n) time.*

Proof. We construct the automaton A with $L(A) = L(w)$ as in Corollary 13. By Fact 11, it suffices to compute $|\sqrt{L(A)} \cap \Sigma^n|$. Observation 15 reduces this to computing $|L_A(w_{(0)}, q) \cap L_A(q, AC) \cap \Sigma^n|$ for all states q of A. One can construct an automaton with $\mathcal{O}(n^2)$ states representing pairs of states of A that recognizes $L_A(w_{(0)}, q) \cap L_A(q, AC)$. Now it is enough to apply Lemma 14(a) to determine $|L_A(w_{(0)}, q) \cap L_A(q, AC) \cap \Sigma^n|$. □

Corollary 16 already gives a polynomial-time algorithm to compute $|CS(w)|$, however, the approach it takes is rather inefficient. Below, we give an algorithm working with a more reasonable $\mathcal{O}(n^3)$ bound for the running time, which exploits the structure of both the automaton A and the language $L(w)$.

Lemma 17. *If $w \in \Sigma^n$ satisfies $\langle w \rangle = w$, then one can compute $|CS(w)|$ in $\mathcal{O}(n^3)$ time.*

Proof. As before we apply Fact 11 with Corollary 13 and actually compute $|\{x \in \Sigma^n : x^2 \in L(A)\}|$. If $x \in L(A)$, then obviously $x^2 \in L(A)$. Moreover, Lemma 14(b) lets us compute $|\Sigma^n \cap L(A)|$ in $\mathcal{O}(n^3)$ time. Thus, it suffices to count $x \in \Sigma^n$ such that $x^2 \in L(A)$ but $x \notin L(A)$. This is done based on the following claim, see Fig. 2.

Claim. Assume $|x| = n$ and $x^2 \in L(A)$ but $x \notin L(A)$. Then there is a unique decomposition $x = x_1 x_2 x_3$ such that $x_1, x_3 \neq \varepsilon$, $x_3 x_1 \in \text{Pref}_-(w)$ and $x_1 x_2 \in L_A(w_{(0)}, w_{(0)})$.

Proof (of the claim). Let va (for $v \in \Sigma^*, a \in \Sigma$) be the shortest prefix of x^2 which belongs to $L(A)$. Let $w_{(i)} = \delta(w_{(0)}, v)$ be the state of A after reading v. Also, let u be the prefix of v of length $|v| - i$. The structure of the automaton implies that $\delta(w_{(0)}, u) = w_{(0)}$, actually u is the longest prefix of x^2 which belongs to $L_A(w_{(0)}, w_{(0)})$. Note that $v = u w_{(i)}$ and $w_{(i)} a \in \text{Pref}_-(w)$, so $x \notin L(A)$ implies $|u| < n \leq |v|$. We set the decomposition so that $x_1 x_2 = u$ and $x_3 x_1 = w_{(i)} a$. Uniqueness follows from deterministic behaviour of the automaton. □

The algorithm considers all n^2 choices of $|x_1|$ and $|x_3|$, and counts the number of x's conditioned on these values. Let $x_1 = x_1' a$ where $x_1' \in \Sigma^*$, $a \in \Sigma$. Note that

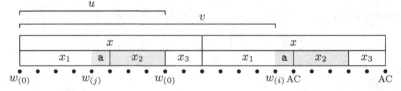

Fig. 2. Illustration of the claim. Both lines represent different factorizations of the same word x^2. Black circles represent states of the automaton. Only shaded letters are not necessarily uniquely determined by $|x_3|$ and $|x_1|$ for a fixed w.

$x_3 x_1 = x_3 x_1' a \in \mathrm{Pref}_-(w)$, so $x_3 x_1'$ is a prefix $w_{(i)}$ of w and $\delta(w_{(i)}, a) = AC$. Hence, i is uniquely determined by $|x_1|$ and $|x_3|$. In particular x_3 and x_1' are uniquely determined, the latter lets us compute $w_{(j)} = \delta(w_{(0)}, x_1')$. If we obtain $\delta(w_{(0)}, x_1') = AC$, we would have $x_1 \in \mathcal{L}(w)$, a contradiction, which implies that no x's for our choice of $|x_1|$ and $|x_3|$ exist.

We need to count ax_2 such that $|x_2| = n - i - 1$, $ax_2 \in L_A(w_{(j)}, w_{(0)})$ and $\delta(w_{(i)}, a) = AC$. Note that $\delta(w_{(j)}, a) \in \{w_{(0)}, w_{(j+1)}\}$, since $\delta(w_{(j)}, a) = AC$ would imply that $ax_2 \in L_A(w_{(j)}, AC)$ rather than $ax_2 \in L_A(w_{(j)}, w_{(0)})$. Thus the number of words ax_2 is equal to $|L(q, w_{(0)}) \cap \Sigma^{n-i-1}|$ summed for $q \in \{w_{(0)}, w_{(j+1)}\}$ taken with coefficients $|\{a \in \Sigma : \delta(w_{(j)}, a) = q \wedge \delta(w_{(i)}, a) = AC\}|$. Lemma 14(b) lets us compute all the values of the type $|L(q, w_{(0)}) \cap \Sigma^{n-i-1}|$ that we may need altogether in $\mathcal{O}(n^3)$ time. The coefficient can be computed for any j, i and q in $\mathcal{O}(n)$ time, and in total we need to sum $\mathcal{O}(n^2)$ integers fitting into $\mathcal{O}(n)$ machine words each. This concludes the $\mathcal{O}(n^3)$-time algorithm. $\square$

Lemma 18. *For an arbitrary word $w \in \Sigma^n$ one can compute $|CS(w)|$ in $\mathcal{O}(n^3)$ time.*

Proof. Let $w' \in \Sigma^n$ be the largest word such that $\langle w' \rangle = w' \leq w$. Note that $CS(w') = CS(w)$. By Lemma 5 we can compute w' in $\mathcal{O}(n^2)$ time. $\square$

5   Ranking Lyndon Words and De Bruijn Sequences

Fact 19. *Let $\alpha > 1$ be a real number. Then $\sum_{d|n} d^\alpha = \mathcal{O}(n^\alpha)$.*

Proof. Recall that for $\alpha > 1$ we have $\sum_{n=1}^\infty \frac{1}{n^\alpha} = \mathcal{O}(1)$. Consequently

$$\sum_{d|n} d^\alpha = \sum_{d|n} \left(\frac{n}{d}\right)^\alpha \leq \sum_{d=1}^\infty \left(\frac{n}{d}\right)^\alpha = n^\alpha \sum_{d=1}^\infty \frac{1}{d^\alpha} = \mathcal{O}(n^\alpha). \qquad \square$$

Theorem 20. *We can compute $|Lynd(w)|$ in $\mathcal{O}(n^3)$ time.*

Proof. We use the formula given by Lemma 9 and the algorithm of Lemma 18. The time complexity is $\mathcal{O}(\sum_{d|n} d^3)$ which, by Fact 19, reduces to $\mathcal{O}(n^3)$. $\square$

Theorem 21. *The k-th Lyndon word of length n can be found in $\mathcal{O}(n^4 \log |\Sigma|)$ time.*

Proof. By definition we look for the smallest $w \in \Sigma^n$ such that $|Lynd(w)| \geq k$. We binary search Σ^n with respect to the lexicographic order, using the algorithm of Theorem 20 to check whether $|Lynd(w)| \geq k$. The size of the search space is $|\Sigma|^n$, which gives an additional $n \log |\Sigma|$-time factor. $\square$

The proof of the theorem of Fredricksen and Maiorana [7] is constructive, i.e. for any word w of length n it shows the concatenation of a constant number of consecutive Lyndon words of length dividing n that contain w. This, together with the following lemma which relates dB_n to CS, lets us compute the exact position where w occurs in dB_n. Recall that $\mathcal{L}^{(n)}$ is the set of Lyndon words whose length is a divisor of n.

Lemma 22. *Let $w \in \Sigma^n$ and $\mathcal{L}(w) = \{\lambda \in \mathcal{L}^{(n)} : \lambda^\infty \leq w^\infty\}$. Then the concatenation, in lexicographic order, of words $\lambda \in \mathcal{L}(w)$ forms a prefix of dB_n and its length, $\sum_{\lambda \in \mathcal{L}(w)} |\lambda|$, is equal to $|CS(w)|$. Moreover, if $w = \lambda^d$ for some $\lambda \in \mathcal{L}$ and $d \in \mathbb{Z}_+$, then λ is the lexicographically largest element of $\mathcal{L}(w)$.*

Proof. First, observe that by Lemma 4 the lexicographic order on $\mathcal{L}$, the set of all Lyndon words, coincides with the lexicographic order of the infinite powers of these words. In particular, this remains true in $\mathcal{L}^{(n)}$ and shows that concatenation of elements of $\mathcal{L}(w)$ indeed forms a prefix of dB_n, and that if $w = \lambda^d$, then λ is the lexicographically largest element of $\mathcal{L}(w)$.

It remains to show that $\sum_{\lambda \in \mathcal{L}(w)} |\lambda| = |CS(w)|$. We shall build a mapping $\phi : \Sigma^n \to \mathcal{L}^{(n)}$ such that $|\phi^{-1}(\lambda)| = |\lambda|$ and $\langle x \rangle \leq w$ if and only if $\phi(x) \in \mathcal{L}(w)$.

Let $x \in \Sigma^n$. There is a unique primitive word y and a positive integer k such that $x = y^k$. We set $\phi(x) = \langle y \rangle$, note that it indeed belongs to $\mathcal{L}^{(n)}$. Moreover, to each Lyndon word λ of length $d \mid n$ we have assigned $v^{\frac{n}{d}}$ for each cyclic rotation v of λ. Also, $\langle x \rangle = \langle y \rangle^{\frac{n}{d}}$, so $\langle x \rangle \leq w$ if and only if $\phi(x)^{\frac{n}{d}} \leq w$, i.e. $\phi(x)^\infty \leq w^\infty$, i.e. $\phi(x) \in \mathcal{L}(w)$. □

Theorem 23. *Given a word $w \in \Sigma^n$, its position in the de Bruijn sequence dB_n can be found in $\mathcal{O}(n^3)$ time.*

Proof. Let $\lambda_1 < \lambda_2 < \ldots < \lambda_p$ be all Lyndon words in $\mathcal{L}^{(n)}$ (we have $\lambda_1 \lambda_2 \ldots \lambda_p = dB_n$). The proof of theorem of Fredricksen and Maiorana [7,9] describes the location of w in dB_n which can be stated succinctly as follows.

Claim ([7,9]). Assume that $w = (\alpha\beta)^d$, where $d \in \mathbb{Z}_+$ and $\beta\alpha = \lambda_k \in \mathcal{L}^{(n)}$. Denote $a = \min \Sigma$ and $z = \max \Sigma$.

(a) If $w = z^i a^{n-i}$ for $i \geq 1$, then w occurs at position $|\Sigma|^n - i$.
(b) If $\alpha \neq z^{|\alpha|}$ then w is a factor of $\lambda_k \lambda_{k+1}$.
(c) If $\alpha = z^{|\alpha|}$ and $d > 1$ then w is a factor of $\lambda_{k-1} \lambda_k \lambda_{k+1}$.
(d) If $\alpha = z^{|\alpha|}$ and $d = 1$ then w is a factor of $\lambda_{k'-1} \lambda_{k'} \lambda_{k'+1}$, where $\lambda_{k'}$ is the largest $\lambda \in \mathcal{L}^{(n)}$ such that $\lambda < \beta$.

In case (a) it is easy to locate w in dB_n, we omit it from further considerations. Observe that λ_k can be retrieved as the primitive root of $\langle w \rangle$. Also note that, by Lemma 4, $\lambda_{k'}$ is the primitive root of the largest $w' \in \Sigma^n$ such that $w' = \langle w' \rangle$ and $w' < \beta a^{|\alpha|}$, and thus it can be computed in $\mathcal{O}(n^2)$ time using Lemma 5.

Once we know $\lambda_{k'}$ and λ_k, depending on the case, we need to find the successor in $\mathcal{L}^{(n)}$ and possibly the predecessor in $\mathcal{L}^{(n)}$ of one of them. For any $\lambda \in \mathcal{L}^{(n)}$ the successor in $\mathcal{L}^{(n)}$ can be generated by iterating a single step of the FKM algorithm at most $(n-1)/2$ times [6], i.e. in $\mathcal{O}(n^2)$ time. For the predecessor in $\mathcal{L}^{(n)}$, a version of the FKM algorithm that visits the Lyndon words in reverse lexicographic order can be used [9], it also takes $\mathcal{O}(n^2)$ time to find the predecessor. In all cases we obtain in $\mathcal{O}(n^2)$ time the Lyndon words whose concatenation contains w.

Then we perform a pattern matching for w in the concatenation. This gives us a relative position of w in dB_n with respect to the position of the canonical occurrence of λ_k or $\lambda_{k'}$ in dB_n. Lemma 22 proves that such an occurrence of $\lambda \in \mathcal{L}^{(n)}$ ends at position $|CS(\lambda^{\frac{n}{|\lambda|}})|$, which can be computed in $\mathcal{O}(n^3)$ time by Lemma 18. Applied to λ_k or $\lambda_{k'}$ this concludes the proof. $\square$

To compute the k-th symbol of dB_n we have to locate the Lyndon word from $\mathcal{L}^{(n)}$ containing the k-th position of dB_n. We apply binary search as in Theorem 21. The k-th symbol of dB'_n is much easier to find due to a simpler structure of the sequence. This gives a rough idea of the proof of the following theorem. We omit the full proof in this version of the paper.

Theorem 24. *Given integers n and k, the k-th symbol of dB_n and dB'_n can be computed in $\mathcal{O}(n^4 \log |\Sigma|)$ time.*

References

1. Bonomo, S., Mantaci, S., Restivo, A., Rosone, G., Sciortino, M.: Suffixes, conjugates and Lyndon words. In: Béal, M.-P., Carton, O. (eds.) DLT 2013. LNCS, vol. 7907, pp. 131–142. Springer, Heidelberg (2013)
2. Crochemore, M., Iliopoulos, C.S., Kubica, M., Radoszewski, J., Rytter, W., Waleń, T.: Extracting powers and periods in a word from its runs structure. Theor. Comput. Sci. (2013), doi:10.1016/j.tcs.2013.11.018
3. Crochemore, M., Rytter, W.: Text Algorithms. Oxford University Press (1994)
4. Duval, J.-P.: Génération d'une section des classes de conjugaison et arbre des mots de Lyndon de longueur bornée. Theor. Comput. Sci. 60, 255–283 (1988)
5. Chung, R.G.F., Diaconis, P.: Universal cycles for combinatorial structures. Discrete Mathematics 110, 43–59 (1992)
6. Fredricksen, H., Kessler, I.J.: An algorithm for generating necklaces of beads in two colors. Discrete Mathematics 61(2-3), 181–188 (1986)
7. Fredricksen, H., Maiorana, J.: Necklaces of beads in k colors and k-ary de Bruijn sequences. Discrete Mathematics 23(3), 207–210 (1978)
8. Hin Au, Y.: Shortest sequences containing primitive words and powers. ArXiv e-prints (April 2009)
9. Knuth, D.E.: The Art of Computer Programming, vol. 4, Fascicle 2. Addison-Wesley (2005)
10. Lothaire, M.: Combinatorics on Words. Addison-Wesley, Reading (1983)
11. Mitchell, C.J., Etzion, T., Paterson, K.G.: A method for constructing decodable de Bruijn sequences. IEEE Transactions on Information Theory 42(5), 1472–1478 (1996)
12. Mucha, M.: Lyndon words and short superstrings. In: Khanna, S. (ed.) SODA, pp. 958–972. SIAM (2013)
13. Radoszewski, J.: Generation of lexicographically minimal de Bruijn sequences with prime words. Master's thesis, University of Warsaw (2008) (in Polish)
14. Ruskey, F., Savage, C.D., Wang, T.M.Y.: Generating necklaces. J. Algorithms 13(3), 414–430 (1992)
15. Tuliani, J.: De Bruijn sequences with efficient decoding algorithms. Discrete Mathematics 226(1-3), 313–336 (2001)

Searching of Gapped Repeats and Subrepetitions in a Word[*]

Roman Kolpakov[1], Mikhail Podolskiy[1], Mikhail Posypkin[2],
and Nickolay Khrapov[2]

[1] Lomonosov Moscow State University,
Leninskie Gory, Moscow, 119992 Russia
[2] Institute for Information Transmission Problems
Bolshoy Karetny per., Moscow, 127994 Russia

Abstract. A gapped repeat is a factor of the form uvu where u and
v are nonempty words. The period of the gapped repeat is defined as
$|u| + |v|$. The gapped repeat is maximal if it cannot be extended to
the left or to the right by at least one letter with preserving its period. The gapped repeat is called α-gapped if its period is not greater
than $\alpha|u|$. A δ-subrepetition is a factor which exponent is less than 2
but is not less than $1 + \delta$ (the exponent of the factor is the quotient of
the length and the minimal period of the factor). The δ-subrepetition
is maximal if it cannot be extended to the left or to the right by at
least one letter with preserving its minimal period. We obtain that in a
word of length n the number of maximal α-gapped repeats is bounded
by $O(\alpha^2 n)$ and the number of maximal δ-subrepetitions is bounded by
$O(n/\delta^2)$. Using the obtained upper bounds, we propose algorithms for
finding all maximal α-gapped repeats and all maximal δ-subrepetitions
in a word of length n. The algorithm for finding all maximal α-gapped repeats has $O(\alpha^2 n)$ time complexity for the case of constant alphabet size
and $O(n \log n + \alpha^2 n)$ time complexity for the general case. For finding
all maximal δ-subrepetitions we propose two algorithms. The first algorithm has $O(\frac{n \log \log n}{\delta^2})$ time complexity for the case of constant alphabet
size and $O(n \log n + \frac{n \log \log n}{\delta^2})$ time complexity for the general case. The
second algorithm has $O(n \log n + \frac{n}{\delta^2} \log \frac{1}{\delta})$ expected time complexity.

1 Inroduction

Let $w = w[1]w[2] \ldots w[n]$ be an arbitrary word. A fragment $w[i] \cdots w[j]$ of w,
where $1 \leq i \leq j \leq n$, is called a *factor* of w and is denoted by $w[i..j]$. By
positions in w we mean the order numbers $1, 2, \ldots, n$ of letters of the word w.
For any factor $v = w[i..j]$ of w the positions i and j are called *start position*
of v and *end position* of v and denoted by $\mathrm{beg}(v)$ and $\mathrm{end}(v)$ respectively. The
factor v *covers* a letter $w[k]$ if $\mathrm{beg}(v) \leq k \leq \mathrm{end}(v)$. For any two factors u, v
of w the factor u *is contained* (*is strictly contained*) in v if $\mathrm{beg}(v) \leq \mathrm{beg}(u)$ and

---

[*] This work is partially supported by Russian Foundation for Fundamental Research
(Grant 12-07-00216).

A.S. Kulikov, S.O. Kuznetsov, and P. Pevzner (Eds.): CPM 2014, LNCS 8486, pp. 212–221, 2014.
© Springer International Publishing Switzerland 2014

$end(u) \leq end(v)$ (if $beg(v) < beg(u)$ and $end(u) < end(v)$). Let u, v be two factors of w such that $beg(v) = end(u) + 1$. In this case we say that v *follows* u. The number $end(u)$ is called *the frontier* between the factors u and v. A factor v *contains* a frontier j if $beg(v) - 1 \leq j \leq end(v)$.

We denote by $p(w)$ the minimal period of w and by $e(w)$ the ratio $|w|/p(w)$ which is called the *exponent* of w. A word is called *primitive* if its exponent is not an integer greater than 1. By repetition in a word we mean any factor of exponent greater than or equal to 2. Repetitions are fundamental objects, due to their primary importance in word combinatorics [18] as well as in various applications, such as string matching algorithms [8,3], molecular biology [9], or text compression [19]. The simplest and best known example of repetitions is factors of the form uu which are called *squares*. A square uu is called *primitive* if u is primitive. The questions concerned to squares are well studied in the literature. In particular, it is known (see, e.g., [3]) that a word of length n contains no more than $\log_\varphi n$ primitive squares. In [2] an $O(n \log n)$-time algorithm for finding of all primitive squares in a word of length n is proposed. In [10] an algorithm for finding of all primitive squares in a word of length n with time complexity $O(n + S)$ where S is the size of output is proposed for the case of constant alphabet size.

A repetition $r = w[i..j]$ in w is called *maximal* if it satisfies the following conditions: if $i > 1$, then $w[i - 1] \neq w[i - 1 + p(r)]$, and, if $j < n$, then $w[j + 1 - p(r)] \neq w[j + 1]$. Maximal repetitions are usually called *runs* in the literature. Since runs contain all the other repetitions in a word, the set of all runs can be considered as a compact encoding of all repetitions in the word which has many useful applications (see, for example, [6]). It was proved in [12] that the number of runs in w is $O(n)$. Moreover, in [12] an $O(n)$ time algorithm for finding of all runs in a word of length n is proposed for the case of constant alphabet size (in the case of arbitrary alphabet size all runs in a word of length n can be found in $O(n \log n)$ time). Further many papers were devoted to obtaining more precise upper bounds on the number of runs in words. In our knowledge, the best upper bound was obtained in [5].

A natural generalization of squares is factors of the form uvu where u and v are nonempty words. We call such factors *gapped repeats*. In the gapped repeat uvu the first (second) factor u is called *the left (right) copy*, and v is called *the gap*. By *the period* of this gapped repeat we will mean the value $|u| + |v|$. For a gapped repeat σ we denote the length of copies of σ by $c(\sigma)$ and the period of σ by $p(\sigma)$. By (u', u'') we will denote the gapped repeat with the left copy u' and the right copy u''. Note that gapped repeats with distinct periods can be the same factor, i.e. can have the same both start and end positions in the word. In this case, for convenience, we will consider this repeats as different ones, i.e. a gapped repeat is not determined uniquely by its start and end positions in the word. For any real $\alpha > 1$ a gapped repeat σ is called α-*gapped* if $p(\sigma) \leq \alpha c(\sigma)$. A gapped repeat $(w[i'..j'], w[i''..j''])$ in w is called *maximal* if it satisfies the following conditions: if $i' > 1$, then $w[i' - 1] \neq w[i'' - 1]$, and, if $j'' < n$, then $w[j' + 1] \neq w[j'' + 1]$. In other words, a gapped repeat in a word is maximal if its copies cannot be extended to the left or to the right in the word by at least one letter with preserving its period. Note that any α-gapped repeat is contained either in a determined uniquely maximal α-gapped repeat with the same period or, otherwise, in a determined uniquely maximal repetiton which minimal period is a divisor of the period of the repeat. Therefore, for computing all α-gapped repeats in a given word it is enough to find all maximal α-gapped repeats and all maximal repetitions in this word. Thus, we can conclude that the problem of computing all α-gapped repeats in a word is reduced to the problem of finding all maximal α-gapped repeats in a word. The set of all maximal α-gapped repeats in w will be denoted by $\mathcal{GR}_\alpha(w)$. The problem of finding gapped repeats in a word was investigated before. In particular, it is shown in [1] that all maximal gapped repeats with a gap length belonging to a specified interval can be found in a word of length n with time complexity $O(n \log n + S)$ where S is the size of output. An algorithm for finding in a word all gapped repeats with a fixed gap length is proposed in [13]. The proposed algorithm has time complexity $O(n \log d + S)$ where d is the gap length, n is the word length, and S is the size of output.

Another natural generalization of repetitions is factors with exponents strictly less than 2. We will call such factors *subrepetitions*. More precisely, for any δ such that $0 < \delta < 1$ by δ-subrepetition we mean a factor v such that $1 + \delta \leq e(v) < 2$. Note that the notion of maximal repetition is directly generalized to the case of subrepetitions: maximal subrepetitions are defined exactly in the same way as maximal repetitions. Further we reveal a close relation between maximal subrepetitions and maximal gapped repeats. Some results concerning the possible number of maximal subrepetitions in words were obtained in [15]. In particular, it was proved that the number of maximal δ-subrepetitions in a word of length n is bouned by $O(\frac{n}{\delta} \log n)$.

The aim of our research is to develop effective algorithms of finding maximal gapped repeats and maximal subrepetitions in a given word. We show that in the case of constant alphabet size all maximal α-gapped repeats in a word of length n can be found in $O(\alpha^2 n)$ time. For finding all maximal δ-subrepetitions in the word we propose two algorithms. The first algorithm has time complexity $O(\frac{n \log \log n}{\delta^2})$ in the case of constant alphabet size and $O(n \log n + \frac{n \log \log n}{\delta^2})$ in the general case. The second algorithm has $O(n \log n + \frac{n}{\delta^2} \log \frac{1}{\delta})$ expected time complexity.

2   Auxiliary Definitions and Results

Further we will consider an arbitrary word $w = w[1]w[2] \ldots w[n]$ of length n. Let $\delta < 1$ and r be a maximal δ-subrepetition in w. Then we can consider in w the repeat $\sigma = (w[\mathrm{beg}(r)..\mathrm{end}(r) - p(r)], w[\mathrm{beg}(r) + p(r)..\mathrm{end}(r)])$. It follows from $e(r) < 2$ that σ is gapped. Moreover, $p(\sigma) = p(r)$ and, since r is maximal, it is obvious that σ is maximal. Since r is a δ-subrepetition, we have also that $|r| - p(r) \geq \delta p(r)$, so $c(\sigma) = |r| - p(r) \geq \delta p(r) = \delta p(\sigma)$, i.e. $p(\sigma) \leq \frac{1}{\delta} c(\sigma)$. Thus, σ is a maximal $\frac{1}{\delta}$-gapped repeat in w. We will call the subrepetition r and the repeat σ

respective to each other. Note that for each maximal δ-subrepetition r there exists a maximal $\frac{1}{\delta}$-gapped repeat σ respective to r. Moreover, the subrepetition r is determined uniquely by the repeat σ, so the same repeat can not be respective to different subrepetitions. Thus we have

Proposition 1. *Let $0 < \delta < 1$. Then in any word the number of maximal δ-subrepetitions is no more then the number of maximal $1/\delta$-gapped repeats.*

On the other hand, it is easy to see that a maximal gapped repeat can have no a respective maximal subrepetition. Maximal gapped repeats which have respective maximal subrepetitions will be called *principal*. Thus we have the one-to-one correspondence between maximal δ-subrepetitions and principal $\frac{1}{\delta}$-gapped repeats in a word. It is easy to check the following fact.

Proposition 2. *A maximal gapped repeat σ in w is principal if and only if $p(w[\text{beg}(\sigma)..\text{end}(\sigma)]) = p(\sigma)$.*

Let σ be a maximal gapped repeat, and r be a maximal repetition or subrepetition. We will say that σ *is stretched* by r if σ is contained in r and $p(r) < p(\sigma)$ and call σ *stretchable* if σ is stretched by some maximal repetition or subrepetition. It follows from Proposition 2 that σ is not principal if and only if $p(w[\text{beg}(\sigma)..\text{end}(\sigma)]) < p(\sigma)$, i.e. σ is contained in some maximal repetition or subrepetition with minimal period less than $p(\sigma)$. So we obtain

Proposition 3. *A maximal gapped repeat is principal if and only if it is not stretchable.*

We will say that a gapped repeat σ *is stretched* by a gapped repeat σ' if σ is contained in σ' and $p(\sigma') < p(\sigma)$. It is easy to see that a gapped repeat is stretched by a subrepetition if and only if this repeat is stretched by the gapped repeat respective to this subrepetition. Using this observation, we can derive the following

Proposition 4. *A maximal δ-gapped repeat is stretchable if and only if it is stretched by either a maximal repetition or a maximal δ-gapped repeat.*

For the time complexity analysis of the proposed algorithms we use the following upper bound on the number of maximal α-gapped repeats in a word[1]

Lemma 1. *For any $\alpha > 1$ the number of maximal α-gapped repeats in w is $O(\alpha^2 n)$.*

From Lemma 1, using the Proposition 1, one can easily derive the following upper bound for maximal δ-subrepetitions.

Corollary 1. *Let $0 < \delta < 1$. Then the number of maximal δ-subrepetitions in w is $O(n/\delta^2)$.*

---

[1] By reason of space limitations we omit the proof of this lemma. The proof is available in the full preprint [17] of this paper.

3   Computing of Maximal Gapped Repeats

In this section we propose an algorithm for finding of all maximal α-gapped repeats in the given word w for a fixed value of α. The proposed algorithm is actually a modification of the algorithm described in [14] for finding repeats with a fixed gap in a given word. In particular, the two following basic tools are used in this modification.

The first tool is special functions which are defined as follows. Let u, v be two arbitrary words. For each $i = 2, 3, \ldots, |u|$ we define $\mathrm{LP}_u(i)$ as the length of the longest common prefix of u and $u[i..|u|]$. For each $i = 1, 2, \ldots, |u| - 1$ we define $\mathrm{LS}_u(i)$ as the length of the longest common suffix of u and $u[1..|u| - i]$. For each $i = 0, 1, \ldots, |u| - 1$ we define $\mathrm{LP}_{u|v}(i)$ as the length of the longest common prefix of $u[|u| - i..|u|]v$ and v. For each $i = 1, 2, \ldots, |v|$ we define $\mathrm{LS}_{u|v}(i)$ as the length of the longest common suffix of u and $uv[1..i]$. The functions LP_u and LS_u can be computed in $O(|u|)$ time and the functions $\mathrm{LP}_{u|v}$ and $\mathrm{LS}_{u|v}$ can be computed in $O(|u| + |v|)$ time (see, e.g., [14]).

The second tool is a factorization $f = f_1 f_2 \ldots f_t$ of the word w which is called *non-overlapping s-factorization* and defined inductively as follows:

- $f_1 = w[1]$.
- Let for $i > 1$ the factors $f_1, \ldots, f_{i-1}$ are already computed, and $w[j]$ be the letter which follows the factor f_{i-1} in w. Then $f_i = w[j]$ if the letter $w[j]$ has no occurences in $f_1 f_2 \ldots f_{i-1}$; otherwise f_i is the longest factor in w which follows f_{i-1} and has an occurence in $f_1 f_2 \ldots f_{i-1}$.

The factorization f can be computed in $O(n)$ time for the case of constant alphabet size and in $O(n \log n)$ time for the general case (see, e.g., [14]). By a_i (b_i) we denote the start (end) position of the factor f_i. The length of f_i is denoted by l_i. For $i = 1, 2, \ldots, t - 1$ we will consider also the factor $w[a_i..b_i + 1]$ which is denoted by f'_i.

For convenience sake we consider the case when α is integer, i.e. for any integer $k \geq 2$ we describe the algorithm of finding in w all repeats from $\mathcal{GR}_k(w)$. To this purpose we divide the set $\mathcal{GR}_k(w)$ into the following two nonoverlapping subsets: $\mathcal{FGR}$ is the set of all repeats from $\mathcal{GR}_k(w)$ which are not strictly contained in any factor f_i of the factorization f, and $\mathcal{SGR}$ is the set of all repeats from $\mathcal{GR}_k(w)$ which are strictly contained in factors of the factorization f. To compute the set $\mathcal{GR}_k(w)$, we compute separately the sets $\mathcal{FGR}$ and $\mathcal{SGR}$. For each $i = 2, 3, \ldots, t$ we define in the set $\mathcal{FGR}$ the following subsets: $\mathcal{FGR}'_i$ is the set of all repeats σ from $\mathcal{FGR}$ such that $b_{i-1} < \mathrm{end}(\sigma) \leq b_i$ and $\mathrm{beg}(\sigma) \leq a_i$, and $\mathcal{FGR}''_i$ is the set of all repeats σ from $\mathcal{FGR}$ such that $\mathrm{end}(\sigma) = b_i$ and $\mathrm{beg}(\sigma) > a_i$. It is easy to see that all the subsets $\mathcal{FGR}'_i$ and $\mathcal{FGR}''_i$ are nonoverlapping. Moreover, taking into account that the factor f_1 consists of only one letter, we have that $\mathcal{FGR} = \bigcup_{i=2}^{t} \mathcal{FGR}'_i \cup \bigcup_{i=2}^{t} \mathcal{FGR}''_i$. To compute the set $\mathcal{FGR}$, we compute separately the sets $\mathcal{FGR}'_i$ and $\mathcal{FGR}''_i$ for $i = 2, 3, \ldots, t$.

To compute $\mathcal{FGR}'_i$, we consider in this set the following nonoverlapping subsets: $\mathcal{FGR}_i^{\mathrm{lrt}}$ is the set of all repeats from $\mathcal{FGR}'_i$ which left copies contain the frontier between the factors f_{i-1} and f_i, $\mathcal{FGR}_i^{\mathrm{rrt}}$ is the set of all repeats from

$\mathcal{FGR}'_i$ which right copies contain the frontier between the factors f_{i-1} and f_i, and $\mathcal{FGR}^{\mathrm{mid}}_i$ is the set of all repeats σ from $\mathcal{FGR}'_i$ such that neither left nor right copies of σ contain the frontier between the factors f_{i-1} and f_i. It is obvious that $\mathcal{FGR}'_i = \mathcal{FGR}^{\mathrm{lrt}}_i \cup \mathcal{FGR}^{\mathrm{rrt}}_i \cup \mathcal{FGR}^{\mathrm{mid}}_i$. We compute separately the considered subsets of $\mathcal{FGR}'_i$.

1. **Computing the set $\mathcal{FGR}^{\mathrm{lrt}}_i$.** Let $\sigma = (w[i'..j'], w[i''..j''])$ be a repeat from $\mathcal{FGR}^{\mathrm{lrt}}_i$ with a period p. Note that in this case $p \leq l_i$, so $c(\sigma) < l_i$. Thus, σ is strictly contained in the factor $w[a_i - l_i..a_{i+1}] = g_i f'_i$ where $g_i = w[a_i - l_i..b_{i-1}]$. Since $w[i'..j']$ contains the frontier between f_{i-1} and f_i, we have $i' - 1 \leq b_{i-1} \leq j'$. Note that $|w[a_i..j']| = |w[a_i + p..j'']| = \mathrm{LP}_{f'_i}(p+1)$. Note also that $|w[i'..b_{i-1}]| = |w[i''..b_{i-1} + p]| = \mathrm{LS}_{g_i|f_i}(p)$. Thus,

$$\sigma = (w[a_i - \hat{\mathrm{LS}}(p)..b_{i-1} + \hat{\mathrm{LP}}(p)], w[a_i + p - \hat{\mathrm{LS}}(p)..b_{i-1} + p + \hat{\mathrm{LP}}(p)]) \quad (1)$$

where $\hat{\mathrm{LP}}(p) = \mathrm{LP}_{f'_i}(p+1)$ and $\hat{\mathrm{LS}}(p) = \mathrm{LS}_{g_i|f_i}(p)$. Since $\hat{\mathrm{LP}}(p) + \hat{\mathrm{LS}}(p) = c(\sigma)$ and σ is a k-gapped repeat, we have the following restrictions for $\hat{\mathrm{LP}}(p)$ and $\hat{\mathrm{LS}}(p)$:

$$p/k \leq \hat{\mathrm{LP}}(p) + \hat{\mathrm{LS}}(p) < p. \quad (2)$$

Moreover, from the condition $\mathrm{end}(\sigma) \leq b_i$ we have the restriction

$$\hat{\mathrm{LP}}(p) \leq l_i - p. \quad (3)$$

On the other hand, if for some p such that $p \leq l_i$ the conditions (2) and (3) hold, in the set $\mathcal{FGR}^{\mathrm{lrt}}_i$ there exists the maximal k-gapped repeat (1) with the period p. Thus, to compute $\mathcal{FGR}^{\mathrm{lrt}}_i$, for each $p = 1, 2, \ldots, l_i$ we compute the values $\hat{\mathrm{LP}}(p)$ and $\hat{\mathrm{LS}}(p)$ and check the conditions (2) and (3). If these conditions are valid we add the corresponding repeat (1) to $\mathcal{FGR}^{\mathrm{lrt}}_i$. As noted above, all the values $\hat{\mathrm{LP}}(p)$ and $\hat{\mathrm{LS}}(p)$ can be computed in $O(|g_i| + |f'_i|) = O(l_i)$ time, and all the conditions (2) and (3) can be checked in $O(l_i)$ time. Thus, the set $\mathcal{FGR}^{\mathrm{lrt}}_i$ can be computed in $O(l_i)$ time.

2. **Computing the set $\mathcal{FGR}^{\mathrm{rrt}}_i$.** Let $\sigma = (w[i'..j'], w[i''..j''])$ be a repeat from $\mathcal{FGR}^{\mathrm{rrt}}_i$ with a period p. Then for σ we have the following[2]

Proposition 5. *The right copy of σ doesn't contain the frontier between the factors f_{i-2} and f_{i-1}.*

Corollary 2. $c(\sigma) < l_{i-1} + l_i$.

Thus, $p \leq kc(\sigma) < k(l_{i-1} + l_i)$ and $i'' > a_{i-1}$ by Proposition 5. Therefore, $i' = i'' - p > a_{i-1} - k(l_{i-1} + l_i)$, i.e. σ is strictly contained in the factor $g'_i f'_i$ where $g'_i = w[a_{i-1} - k(l_{i-1} + l_i)..b_{i-1}]$. Note that $|w[a_i..j'']| = |w[a_i - p..j']| = \mathrm{LP}_{g'_i|f'_i}(p-1)$. Note also that $|w[i''..b_{i-1}]| = |w[i'..b_{i-1} - p]| = \mathrm{LS}_{g'_i}(p)$. Hence

$$\sigma = (w[a_i - p - \hat{\mathrm{LS}}(p)..b_{i-1} - p + \hat{\mathrm{LP}}(p)], w[a_i - \hat{\mathrm{LS}}(p)..b_{i-1} + \hat{\mathrm{LP}}(p)]) \quad (4)$$

---

[2] The proofs of Propositions 5 and 6 can be found in [17].

where $\hat{\mathrm{LP}}(p) = \mathrm{LP}_{g'_i|f'_i}(p-1)$ and $\hat{\mathrm{LS}}(p) = \mathrm{LS}_{g'_i}(p)$. As in the the case of computing $\mathcal{FGR}_i^{\mathrm{lrt}}$, we have for the period p the restrictions (2). Moreover, since $b_{i-1} < j'' \le b_i$, we have the following additional restriction:

$$0 < \hat{\mathrm{LP}}(p) \le l_i. \tag{5}$$

On the other hand, if for some p such that $p < k(l_{i-1} + l_i)$ the conditions (2) and (5) hold, in the set $\mathcal{FGR}_i^{\mathrm{rrt}}$ there exists the k-gapped repeat (4) with the period p. Thus, to compute $\mathcal{FGR}_i^{\mathrm{rrt}}$, for each $p < k(l_{i-1} + l_i)$ we check the conditions (2) and (5) for the values $\hat{\mathrm{LP}}(p)$ and $\hat{\mathrm{LS}}(p)$. If these conditions hold we add the corresponding repeat (4) to $\mathcal{FGR}_i^{\mathrm{rrt}}$. Note that all the values $\hat{\mathrm{LP}}(p)$ and $\hat{\mathrm{LS}}(p)$ can be computed in $O(|g'_i| + |f'_i|) = O(k(l_{i-1} + l_i))$ time, and all the conditions (2) and (5) can be checked in $O(k(l_{i-1} + l_i))$ time. Thus, the set $\mathcal{FGR}_i^{\mathrm{rrt}}$ can be computed in $O(k(l_{i-1} + l_i))$ time.

3. **Computing the set $\mathcal{FGR}_i^{\mathrm{mid}}$.** Note that the right copies of all repeats from $\mathcal{FGR}_i^{\mathrm{mid}}$ are strictly contained in f'_i. Let $q = \lfloor \log k/(k-1)l_i \rfloor$. We denote by d_s the position $\lfloor ((k-1)/k)^s l_i \rfloor + 1$ for $s = 0, 1, \ldots, q$ and divide the set $\mathcal{FGR}_i^{\mathrm{mid}}$ into nonoverlapping subsets $MP_1, MP_2, \ldots, MP_q$ where MP_s is the set of all repeats from $\mathcal{FGR}_i^{\mathrm{mid}}$ which right copies cover the letter $f'_i[d_s]$ but don't cover the letter $f'_i[d_{s-1}]$.

Proposition 6. $\mathcal{FGR}_i^{\mathrm{mid}} = \bigcup_{s=1}^{q} MP_s$.

Using Proposition 6, for computing $\mathcal{FGR}_i^{\mathrm{mid}}$ we compute separately the sets $MP_1, MP_2, \ldots, MP_q$. In order to compute the set MP_s, consider an arbitrary repeat $\sigma = (w[i'..j'], w[i''..j''])$ with a period p in this set. Note that in this case the right copy of σ is strictly contained in $f'_i[1..d_{s-1}]$, so $c(\sigma) < d_{s-1}$. Thus, $j < kd_{s-1}$ and σ is strictly contained in the factor $h_{is}h'_{is}$ where $h_{is} = w[a_i - kd_{s-1}..b_{i-1}]f'_i[1..d_s - 1]$ and $h'_{is} = f'_i[d_s..d_{s-1}]$. Note that

$$|w[i'..b_{i-1} + d_s - p]| = |w[i''..b_{i-1} + d_s]| = \mathrm{LS}_{h_{is}}(p),$$

$$|w[a_{i-1} + d_s - p..j']| = |w[a_{i-1} + d_s..j'']| = \mathrm{LP}_{h_{is}|h'_{is}}(p-1).$$

Thus, σ is defined uniquely by the period p as

$$(w[a_i+d_s-p-\hat{\mathrm{LS}}(p)..b_{i-1}+d_s-p+\hat{\mathrm{LP}}(p)], w[a_i+d_s-\hat{\mathrm{LS}}(p)..b_{i-1}+d_s+\hat{\mathrm{LP}}(p)]) \tag{6}$$

where $\hat{\mathrm{LS}}(p) = \mathrm{LS}_{h_{is}}(p)$ and $\hat{\mathrm{LP}}(p) = \mathrm{LP}_{h_{is}|h'_{is}}(p-1)$. Since the repeat σ is k-gapped, the conditions (2) have to be valid for the period p. Moreover, p has to satisfy the additional restrictions

$$\hat{\mathrm{LP}}(p) \le p - d_s, \tag{7}$$

$$0 < \hat{\mathrm{LP}}(p) \le d_{s-1} - d_s, \tag{8}$$

$$\hat{\mathrm{LS}}(p) < d_s - 1, \tag{9}$$

following from the definition of the set MP_s. On the other hand, for each p satisfying the inequality $p < kd_{s-1}$ and the conditions (2), (7), (8), and (9), there exists the k-gapped repeat (6) with the period p in the set MP_s. Thus, to compute MP_s, we check the conditions (2), (7), (8), and (9) for each p such that $p < kd_{s-1}$. If for some p these conditions hold we add the corresponding repeat (6) to MP_s. Note that the time required for computing the involved values $\hat{LP}(p)$ and $\hat{LS}(p)$ is bounded by $O(|h_{is}| + |h'_{is}|) = O(kd_{s-1})$ and the total time required for checking these conditions is bounded by $O(kd_{s-1})$. Thus, MP_s can be computed in $O(kd_{s-1}) = O(((k-1)/k)^{s-1} kl_i)$ time. Hence by Proposition 6 we obtain that $\mathcal{FGR}_i^{\mathrm{mid}}$ can be computed in $O(k^2 l_i)$ time.

Summing up the obtained time bounds for computing the sets $\mathcal{FGR}_i^{\mathrm{lrt}}$, $\mathcal{FGR}_i^{\mathrm{rrt}}$ and $\mathcal{FGR}_i^{\mathrm{mid}}$, we conclude that $\mathcal{FGR}'_i$ can be computed in $O(kl_{i-1} + k^2 l_i)$ time. It is easy to note that the set $\mathcal{FGR}''_i$ can also be computed in $O(l_i)$ time by a simplified version of the described above algorithm for computing $\mathcal{FGR}_i^{\mathrm{rrt}}$. The set $\mathcal{SGR}$ can be computed analogously to the computation of the set $\mathcal{GR}''$ in the algorithm from [14] for finding repeats with a fixed gap (the details can be found in [17]). Thus, provided by the factorization f, all repeats from $\mathcal{GR}_k(w)$ can be computed in $O(k^2 n + |\mathcal{SGR}|)$ time. Sinse $|\mathcal{SGR}| < |\mathcal{GR}_k(w)| = O(k^2 n)$ by Lemma 1, taking into account the time for constructing the factorization f, we conclude that $\mathcal{GR}_k(w)$ can be computed in $O(k^2 n)$ time for the case of constant alphabet size and in $O(n \log n + k^2 n)$ time for the general case. It is easy to see that the proposed algorithm can be directly generalized to the case of maximal α-gapped repeats for real $\alpha > 1$ with preserving the upper bound for time complexity. Thus we have

Theorem 1. *For any real $\alpha > 1$ all maximal α-gapped repeats in w can be computed in $O(\alpha^2 n)$ time for the case of constant alphabet size and in $O(n \log n + \alpha^2 n)$ time for the general case.*

Now consider the problem of finding all maximal δ-subrepetitions in a word for a fixed δ. Because of the established above one-to-one correspondence between maximal δ-subrepetitions and principal $\frac{1}{\delta}$-gapped repeats, this problem is reduced to computing all principal $\frac{1}{\delta}$-gapped repeats in a word. We propose the following algorithm for computing all principal $\frac{1}{\delta}$-gapped repeats in the word w. Further, for convenience, by the period of a repetition we will mean its minimal period. First we compute the ordered set $\mathcal{OSR}_\delta$ of all maximal repetitions and all maximal $\frac{1}{\delta}$-gapped repeats in w such that all elements of $\mathcal{OSR}_\delta$ are ordered in non-decreasing order of their start positions and, furthermore, elements of $\mathcal{OSR}_\delta$ with the same start position are ordered in increasing order of their periods (it is easy to note that any element of $\mathcal{OSR}_\delta$ determined uniquely by its start position and its period, so the introduced order in $\mathcal{OSR}_\delta$ is uniquely defined). To compute $\mathcal{OSR}_\delta$, we find in w all maximal repetitions and all maximal $\frac{1}{\delta}$-gapped repeats. Using Theorem 1 and the algorithm for finding maximal repetitions proposed in [12], it can be done in $O(n/\delta^2)$ time for the case of constant alphabet size and in $O(n \log n + n/\delta^2)$ time for the general case. Then we arrange the found repetitions and repeats in the order required for $\mathcal{OSR}_\delta$. By Lemma 1 the number of the maximal $\frac{1}{\delta}$-gapped repeats is $O(n/\delta^2)$, so $|\mathcal{OSR}_\delta| = O(n/\delta^2)$. Therefore, using backet sort, the required arrangement can be done in $O(n + |\mathcal{OSR}_\delta|)$ time which is bounded by $O(n/\delta^2)$. Thus, $\mathcal{OSR}_\delta$ can be computed in $O(n/\delta^2)$ time for the case of constant alphabet size and in $O(n \log n + n/\delta^2)$ time for the general case. Note that by Proposition 3 for discovering all principal repeats from the maximal $\frac{1}{\delta}$-gapped repeats it is enough to compute all stretchable $\frac{1}{\delta}$-gapped repeats in w. To compute stretchable $\frac{1}{\delta}$-gapped repeats, we maintain an auxiliary two-way queue SRQ consisting of elements from $\mathcal{OSR}_\delta$. Elements from $\mathcal{OSR}_\delta$ are presented by pairs (p, q) where p and q are respectively the period and the end position of the presented element (it is easy to note that any element of $\mathcal{OSR}_\delta$ determined uniquely by its period and its start position, so two different elements can not be presented by the same pair in SRQ). At any time the queue SRQ has a form:

$$(p_1, q_1), (p_2, q_2), \dots, (p_s, q_s) \tag{10}$$

where $p_1 < p_2 < \dots < p_s$ and $q_1 < q_2 < \dots < q_s$. Starting from empty SRQ, we try to insert in SRQ each element of $\mathcal{OSR}_\delta$ in the prescribed order by the following way. Let an element τ with period p and end position q be the next candidate for insertion in the queue SRQ presented in (10). Firstly we find the periods p_i and p_{i+1} such that $p_i \le p < p_{i+1}$ and[3] compare q with q_i. If $q \le q_i$ we establish that τ is a stretchable repeat[4] and don't insert τ in SRQ. Otherwise we insert τ in SRQ and remove from SRQ all pairs (p_j, q_j) such that $j > i$ and $q_j \le q$ in order to preserve SRQ in the proper form. Using Proposition 4, one can check that the described procedure compute correctly all stretchable repeats from $\mathcal{OSR}_\delta$ which allows to compute all principal $\frac{1}{\delta}$-gapped repeats in w. For effective execution of operations required in this procedure we use the data structure proposed in [7]. This data structure can be constructed in $O(n \log \log n)$ time and allows to execute the operations of finding p_i, inserting an element to SRQ and removing an element from SRQ in $O(\log \log n)$ time. Note that in the described procedure no more than one of each of these three operations is required for treating any element from $\mathcal{OSR}_\delta$. Thus, the time required for computing all stretchable repeats in $\mathcal{OSR}_\delta$ is $O(n \log \log n + |\mathcal{OSR}_\delta|) \log \log n)$, so can be bounded by $O(n \log \log n/\delta^2)$. Summing up this time bound with the time bound for computing the set $\mathcal{OSR}_\delta$, we obtain

Theorem 2. *Let $0 < \delta < 1$. Then all maximal δ-subrepetitions in w can be computed in $O(\frac{n \log \log n}{\delta^2})$ time for the case of constant alphabet size and in $O(n \log n + \frac{n \log \log n}{\delta^2})$ time for the general case.*

Another algorithm for computing all principal $\frac{1}{\delta}$-gapped repeats in a word is based on Proposition 2. By this proposition, in order to check if a maximal gapped repeat σ in w is principal we can compute the minimal period of $w[\text{beg}(\sigma)..\text{end}(\sigma)]$ and compare this period with $p(\sigma)$: if these periods are equal then σ is principal; otherwise σ is not principal. The problem of effective answering to queries related to minimal periods of factors in a word is studied in [11]. In particular, in [11] a hash table data structure is proposed for resolving this problem. Using the data structure from [11], we can prove

---

[3] We describe our algorithm for the general case when both p_i and p_{i+1} are exist. The cases when eigther p_i or p_{i+1} does not exist are easily derived from this general case.

[4] It is easy to check that in this case τ can not be a repetition.

Theorem 3. *Let* $0 < \delta < 1$. *Then all maximal δ-subrepetitions in w can be computed in* $O(n \log n + \frac{n}{\delta^2} \log \frac{1}{\delta})$ *expected time.*

References

1. Brodal, G., Lyngso, R., Pedersen, C., Stoye, J.: Finding Maximal Pairs with Bounded Gap. J. of Discrete Algorithms 1(1), 77–104 (2000)
2. Crochemore, M.: An optimal algorithm for computing the repetitions in a word. Information Processing Letters 12, 244–250 (1981)
3. Crochemore, M., Rytter, W.: Squares, cubes, and time-space efficient string searching. Algorithmica 13, 405–425 (1995)
4. Crochemore, M., Hancart, C., Lecroq, T.: Algorithms on Strings. Cambridge University Press (2007)
5. Crochemore, M., Ilie, L., Tinta, L.: Towards a solution to the "runs" conjecture. In: Ferragina, P., Landau, G.M. (eds.) CPM 2008. LNCS, vol. 5029, pp. 290–302. Springer, Heidelberg (2008)
6. Crochemore, M., Iliopoulos, C., Kubica, M., Radoszewski, J., Rytter, W., Waleń, T.: Extracting powers and periods in a string from its runs structure. In: Chavez, E., Lonardi, S. (eds.) SPIRE 2010. LNCS, vol. 6393, pp. 258–269. Springer, Heidelberg (2010)
7. van Emde Boas, P., Kaas, R., Zulstra, E.: Design and Implementation of an Efficient Priority Queue. Mathematical Systems Theory 10, 99–127 (1977)
8. Galil, Z., Seiferas, J.: Time-space optimal string matching. J. of Computer and System Sciences 26(3), 280–294 (1983)
9. Gusfield, D.: Algorithms on Strings, Trees, and Sequences. Cambridge University Press (1997)
10. Gusfield, D., Stoye, J.: Linear time algorithms for finding and representing all the tandem repeats in a string. J. of Computer and System Sciences 69(4), 525–546 (2004)
11. Kociumaka, T., Radoszewski, J., Rytter, W., Waleń, T.: Efficient Data Structures for the Factor Periodicity Problem. In: Calderón-Benavides, L., González-Caro, C., Chávez, E., Ziviani, N. (eds.) SPIRE 2012. LNCS, vol. 7608, pp. 284–294. Springer, Heidelberg (2012)
12. Kolpakov, R., Kucherov, G.: On Maximal Repetitions in Words. J. of Discrete Algorithms 1(1), 159–186 (2000)
13. Kolpakov, R., Kucherov, G.: Finding Repeats with Fixed Gap. In: 7th International Symposium on String Processing and Information Retrieval (SPIRE 2000), pp. 162–168 (2000)
14. Kolpakov, R., Kucherov, G.: Periodic structures in words. Chapter for the 3rd Lothaire volume Applied Combinatorics on Words. Cambridge University Press (2005)
15. Kolpakov, R., Kucherov, G., Ochem, P.: On maximal repetitions of arbitrary exponent. Information Processing Letters 110(7), 252–256 (2010)
16. Kolpakov, R.: On primary and secondary repetitions in words. Theoretical Computer Science 418, 71–81 (2012)
17. Kolpakov, R., Podolskiy, M., Posypkin, M., Khrapov, N.: Searching of gapped repeats and subrepetitions in a word, http://arxiv.org/abs/1309.4055
18. Lothaire, M.: Combinatorics on Words. Encyclopedia of Mathematics and Its Applications, vol. 17. Addison-Wesley (1983)
19. Storer, J.: Data compression: Methods and theory. Computer Science Press, Rockville (1988)

Approximate String Matching
Using a Bidirectional Index

Gregory Kucherov[1,2], Kamil Salikhov[1,3], and Dekel Tsur[2]

[1] CNRS/LIGM, Université Paris-Est Marne-la-Vallée, France
[2] Department of Computer Science, Ben-Gurion University of the Negev, Israel
[3] Mechanics and Mathematics Department, Lomonosov Moscow State University,
Russia

Abstract. We study strategies of approximate pattern matching that
exploit bidirectional text indexes, extending and generalizing ideas of [5].
We introduce a formalism, called search schemes, to specify search strategies of this type, then develop a probabilistic measure for the efficiency
of a search scheme, prove several combinatorial results on efficient search
schemes, and finally, provide experimental computations supporting the
superiority of our strategies.

1 Introduction

Approximate string matching has numerous practical applications and long been
a subject of extensive studies by algorithmic researchers. If errors are allowed
in a match between a pattern string and a text string, most of the fundamental
ideas behind exact string search algorithms become inapplicable. The Approximate string matching problem comes in different variants. In this paper, we
are concerned with the *indexing* variant, when a static text is available for preprocessing and storing in a data structure, before any matching query is made.

The quest for efficient approximate string matching algorithms has been
boosted by a new generation of DNA sequencing technologies, able to produce
huge quantities of short DNA *reads*. Those reads can then be mapped to a given
genomic sequence, which requires very fast and accurate approximate string
matching algorithms. Many algorithms and associated software programs have
been designed for this task, we refer to [9] for a survey, and many of them rely
on full-text indexes.

The classical indexing paradigm consists in building a text index in order to
quickly identify pattern occurrences, preferably within a worst-case time weakly
dependent on the text length. In the context of approximate matching, even the
case of one error turned out to be highly nontrivial and gave rise to a series of
works (see [6] and references therein). In the case of k errors, existing solutions
generally have time or space complexity that are exponential in k, see [14] for a
survey.

Some of the existing algorithms use standard text indexes, such as suffix tree
or suffix arrays. However, for large datasets occurring in modern applications,

A.S. Kulikov, S.O. Kuznetsov, and P. Pevzner (Eds.): CPM 2014, LNCS 8486, pp. 222–231, 2014.

these indexes are known to take too much memory. Suffix arrays and suffix trees typically require at least 4 or 10 *bytes* per character respectively. The last years saw the emergence of *succinct* or *compressed full-text indexes* that occupy virtually as much memory as the sequence itself and yet provide very powerful functionalities [10]. For example, the FM-index [4], based on the Burrows-Wheeler Transform [2], may occupy 2–4 *bits* of memory per character for DNA texts. FM-index has now been used in many practical bioinformatics software programs, e.g. [7, 8, 13]. Even if succinct indexes are primarily designed for exact string search, using them for approximate matching naturally became an attractive opportunity. This direction has been taken in several papers, see [11], as well as in practical implementations [13].

Interestingly, succinct indexes can provide even more functionalities than classical ones. In particular, several succinct indexes can be made *bidirectional*, i.e. can perform pattern search in both directions [1, 5, 11, 12]. Lam et al. [5] showed how a bidirectional FM-index can be used to efficiently search for strings up to a small number (one or two) errors. The idea is to partition the pattern into $k + 1$ equal parts, where k is the number of errors, and then perform multiple searches on the FM-index, where each search assumes a different distribution of mismatches among the pattern parts. Lam et al. implemented the proposed algorithm and reported that it outperforms in speed the best existing read alignment software [5]. Related algorithmic ideas appear also in [11].

In this paper, we extend the search strategy of [5] in two main directions. We consider the case of arbitrary k and propose to partition the pattern into more than $k + 1$ parts that can be of *unequal* size. To demonstrate the benefit of both ideas, we first introduce a general formal framework for this kind of algorithm, called *search scheme*, that allows us to easily specify them and to reason about them (Section 2). Then, in Section 3 we perform a probabilistic analysis that provides us with a quantitative measure of performance of a search scheme, and give an efficient algorithm for obtaining the optimal pattern partition for a given scheme. Furthermore, we prove several combinatorial results on the design of efficient search schemes (Section 4). Finally, Section 5 contains comparative analytical estimations, based on our probabilistic analysis, that demonstrate the superiority of our search strategies for many practical parameter ranges. We further report on large-scale experiments on genomic data supporting this analysis.

2   Bidirectional Search

In the framework of text indexing, pattern search is usually done by scanning the pattern online and recomputing *index points* referring to the occurrences of the scanned part of the pattern. With classical text indexes, such as suffix trees or suffix arrays, the pattern is scanned left-to-right (*forward search*). However, some compact indexes such as FM-index provide a search algorithm that scans the pattern right-to-left (*backward search*).

Consider now approximate string matching, where k letter mismatches are allowed between a pattern P and a text T. In this paper, we present our ideas for the case of Hamming distance. However, they also apply to the edit distance, see Conclusions. Both forward and backward search can be extended to approximate search in a straightforward way, by exploring all possible mismatches along the search, as long as their number does not exceed k and the current pattern still occurs in the text. For the forward search, for example, the algorithm enumerates all substrings of T with Hamming distance at most k to a *prefix* of P. Starting with the empty string, the enumeration is done by extending the current string with the corresponding letter of P, and with all other letters provided that the number of accumulated mismatches has not yet reached k. For each extension, its positions in T are computed using the index. Note that the set of enumerated strings is closed under prefixes and therefore can be represented by the nodes of a trie. Similarly to forward search, *backward search* enumerates all substrings of T with Hamming distance at most k to a *suffix* of P.

Clearly, backward and forward search are symmetric and, once we have an implementation of one, the other can be implemented similarly by constructing an index for the reversed text. However, combining both forward and backward search within one algorithm results in a more efficient search. To illustrate this, consider the case $k = 1$. Partition P into two equal length parts $P = P_1 P_2$. The idea is to perform two complementary searches: forward search for occurrences of P with a mismatch in P_2 and backward search for occurrences with a mismatch in P_1. In both searches, branching is performed only after $|P|/2$ characters are matched. Then, the number of strings enumerated by the two searches is much less than the number of strings enumerated by a single standard forward search, even though two searches are performed instead of one.

A *bidirectional index* of a text allows one to extend the current string S both left and right, that is, compute the positions of both cS or Sc from the positions of S. Note that a bidirectional index allows forward and backward searches to alternate, which will be crucial for our purposes. Lam et al. [5] showed how the FM-index can be made bidirectional. Other succinct bidirectional indexes were given in [1, 11, 12]. Using a bidirectional index, such as FM-index, forward and backward searches can be performed in time linear in the number of enumerated strings. Therefore, our main goal is to organize the search so that the number of enumerated strings is minimized.

Lam et al. [5] gave a new search algorithm, called *bidirectional search*, that utilizes the bidirectional property of the index. Consider the case $k = 2$, studied in [5]. In this case, the pattern is partitioned into three equal length parts, $P = P_1 P_2 P_3$. There are now 6 cases to consider according to the placement of mismatches within the parts: 011 (i.e. one mismatch in P_2 and one mismatch in P_3), 101, 110, 002, 020, and 200. The algorithm of Lam et al. [5] performs three searches:

(1) A forward search that allows no mismatches when processing characters of P_1, and 0 to 2 accumulated mismatches when processing characters of P_2 and P_3. This search handles the cases 011, 002, and 020 above.

(2) A backward search that allows no mismatches when processing characters of P_3, 0 to 1 accumulated mismatches when processing characters of P_2, and

0 to 2 accumulated mismatches when processing characters of P_1. This search handles the cases 110 and 200 above.

(3) The remaining case is 101. This case is handled using a *bidirectional search*. It starts with a forward search on string $P' = P_2 P_3$ that allows no mismatches when processing characters of P_2, and 0 to 1 accumulated mismatches when processing the characters of P_3. For each string S of length $|P'|$ enumerated by the forward search whose Hamming distance from P' is exactly 1, a backward search for P_1 is performed by extending S to the left, allowing one additional mismatch. In other words, the search allows 1 to 2 accumulated mismatches when processing the characters of P_1.

We now give a formal definition for the above. Suppose that pattern P is partitioned into p parts. A *search* is a triplet of strings $S = (\pi, L, U)$ where π is a permutation string of length p over $\{1, \ldots, p\}$, and L, U are strings of length p over $\{0, \ldots, k\}$. String π indicates the order in which the parts of P are processed, and thus it must satisfy the following property: For every $i > 1$, $\pi(i)$ is either $(\min_{j<i} \pi(j)) - 1$ or $(\max_{j<i} \pi(j)) + 1$. Strings U and L give upper and lower bounds on the number of mismatches: When the j-th part is processed, the number of accumulated mismatches between the active strings and the corresponding substring of P must be between $L[j]$ and $U[j]$. Formally, for a string M over integers, the *weight* of M is $\sum_i M[i]$. A search $S = (\pi, L, U)$ *covers* a string M if $L[i+1] \le \sum_{j=1}^{i} M[j] \le U[i]$ for all i (assuming $L[p+1] = 0$). A k-*mismatch search scheme* S is a collection of searches such that for every string M of weight k, there is a search in S that covers M. For example, the 2-mismatch scheme of Lam et al. consists of searches $S_f = (123, 000, 022)$, $S_b = (321, 000, 012)$, and $S_{bd} = (231, 001, 012)$. We denote this scheme by S_{LLTWWY}.

In this work, we introduce two types of improvements over the search scheme of Lam et al.

Uneven partition. In S_{LLTWWY}, search S_f enumerates more strings than the other two searches, as it allows 2 mismatches on the second processed part of P, while the other two searches allow only one mismatch. If we increase the length of P_1 in the partition of P, the number of strings enumerated by S_f will decrease, while the number of strings enumerated by the two other searches will increase. We show that for some typical parameters of the problem, the decrease in the former number is larger than the increase of the latter number, leading to a more efficient search.

More parts. Another improvement can be achieved using partitions with $k + 2$ or more parts, rather than $k + 1$ parts.

3   Analysis of Search Schemes

In this section we show how to estimate the performance of a given search scheme S. Using this technique, we present a dynamic programming algorithm for designing an optimal partition of a pattern.

3.1   Estimating the Efficiency of a Search Scheme

To measure the efficiency of a search scheme, we estimate the number of strings enumerated by all the searches of $\mathcal{S}$. We assume that performing single steps of forward, backward, or bidirectional searches takes the same amount of time. It is fairly straightforward to extend the method of this section to the case when these times are not equal. Note that the bidirectional index of Lam et al. [5] reportedly spends slightly more time (order of 10%) on forward search than on backward search.

For the analysis, we assume that characters of T and P are randomly chosen uniformly and independently from the alphabet. We note that it is possible to extend the method of this section to a non-uniform distribution. For more complex distributions, a Monte Carlo simulation can be applied which, however, requires much more time than the method of this section.

Let $\#\mathrm{str}(S, X, \sigma, n)$ denote the expected number of strings enumerated when performing a search $S = (\pi, L, U)$ on a random text of length n and random pattern of length m, where X is a partition of the pattern and σ is the size of the alphabet (note that m is not a parameter for $\#\mathrm{str}$ since the value of m is implied from X). For a search scheme $\mathcal{S}$, $\#\mathrm{str}(\mathcal{S}, X, \sigma, n) = \sum_{S \in \mathcal{S}} \#\mathrm{str}(S, X, \sigma, n)$.

Fix S, X, σ, and n. Let $\mathcal{A}_l$ be the set of enumerated strings of length l when performing the search S on a random pattern of length m, partitioned by X, and a text $\hat{T}$ containing all the strings of length at most m as substrings. Select a random order on the elements of $\mathcal{A}_l$, and let $A_{l,i}$ be the i-th element of $\mathcal{A}_l$. By the linearity of the expectation,

$$\#\mathrm{str}(S, X, \sigma, n) = \sum_{l \geq 1} \sum_{i=1}^{n_l} \Pr_{T \in \Sigma^n}[A_{l,i} \text{ is a substring of } T],$$

where $n_l = |\mathcal{A}_l|$. For any l and i, the string $A_{l,i}$ is a random string with uniform distribution over Σ^l. Therefore, the probability that $A_{l,i}$ is a substring of T can be approximated by $1 - e^{-n/\sigma^l}$ using the Chen-Stein method [3]. Therefore,

$$\#\mathrm{str}(S, X, \sigma, n) \approx \sum_{l=1}^{m} n_l(1 - e^{-n/\sigma^l}). \qquad (1)$$

In order to compute the values of n_l, we give some definitions. Let $n_{l,d}$ be the number of strings in $\mathcal{A}_l$ of length l with Hamming distance d to the prefix of P of length l. Let U' be a string obtained from U by replacing each character $U[i]$ of U by a run of $U[i]$ of length $x_{\pi(i)}$, where x_j is the length of the j-th part in the partition X. The string L' is defined analogously. In other words, the values $L'[i], U'[i]$ give a lower and upper bounds on the number of allowed mismatches for an enumerated string of length i. The values of n_l are given by the following recurrence.

$$n_l = \sum_{d=L'[l]}^{U'[l]} n_{l,d}, \quad n_{l,d} = \begin{cases} n_{l-1,d} + (\sigma - 1)n_{l-1,d-1} & \text{if } l \geq 1 \text{ and } L'[l] \leq d \leq U'[l] \\ 1 & \text{if } l = 0 \text{ and } d = 0 \\ 0 & \text{otherwise} \end{cases} \tag{2}$$

For a specific search, a closed form formula can be given for n_l.

Consider equation (1). The value of the term $1 - e^{-n/\sigma^l}$ is very close to 1 for $l \leq \log_\sigma n - O(1)$. When $l \geq \log_\sigma n$, the value of this term decreases exponentially. Note that n_l increases exponentially, but the base of the exponent of n_l is $\sigma - 1$ whereas the base of $1 - e^{-n/\sigma^l}$ is $1/\sigma$. We can then approximate $\#str(S, X, \sigma, n)$ with function $\#str'(S, X, \sigma, n)$ defined by

$$\#str'(S, X, \sigma, n) = \sum_{l=1}^{\lceil \log_\sigma n \rceil + c_\sigma} n_l(1 - e^{-n/\sigma^l}), \tag{3}$$

where c_σ is a constant chosen so that $((\sigma - 1)/\sigma)^{c_\sigma}$ is sufficiently small.

If a search scheme S contains two or more searches with the same π-strings, these searches can be merged in order to eliminate the enumeration of the same string twice or more. It is straightforward to modify the computation of $\#str(S, X, \sigma, n)$ to account for this optimization.

3.2   Computing an Optimal Partition

Let p be the number of parts. An optimal partition can be naively found by enumerating all $\binom{m-1}{p-1}$ possible partitions, and for each partition X, computing $\#str'(S, X, \sigma, n)$. We now describe a more efficient dynamic programming algorithm that computes an optimal partition for a given search scheme S.

The algorithm takes advantage of the fact that the value of $\#str'(S, X, \sigma, n)$ does not depend on the entire partition X, but only on the partition of a substring of P of length $N = \lceil \log_\sigma n \rceil + c_\sigma$ induced by X. We first give some definitions. Define $S(i, j)$ to be the set containing every search $S \in S$ such that i appears before j in the π-string of S (we assume $S(i, p+1) = S$). A partition of P is a partition of the set $\{1, 2, \ldots, m\}$ into disjoint sets of the form $\{i, i+1, \ldots, j\}$ which will be called *intervals*. A partition of a substring $P[i..j]$ is a partition of $\{i, i+1, \ldots, j\}$ into disjoint intervals.

The algorithm builds *partial partitions* of prefixes of P, incrementally from left to right. That is, the partitions of $P[1..m'']$ are built from partitions of $P[1..m']$ for $m' < m''$ by extending them with interval $\{m'+1, \ldots, m''\}$. Consider a partition X' of $P[1..m']$, $m' \geq N$, into p' parts, and let p'_2 be the left-to-right rank of the interval in X' containing the position $m' - N + 1$. For a search $S \in S(p'_2, p' + 1)$, the value of $\#str'(S, X, \sigma, n)$ is the same for every extension of X' to a partition X of P, since $\#str'(S, X, \sigma, n)$ is determined by the partition of a substring of P

of length N contained in $P[1..m']$. Thus, for every partial partition X', the algorithm computes the value of $\sum_{S \in \mathcal{S}(p'_2, p'+1)} (S, X, \sigma, n)$ for some extension of X' to a partition X. This value will be denoted $v(X')$. Note that for a partition X of P, $v(X) = \#\mathrm{str}'(\mathcal{S}, X, \sigma, n)$. When the algorithm extends the partial partition X' to a partial partition $X'' = X' \cup \{m'+1, \ldots, m''\}$, it has to compute $v(X'')$. This is done by adding $v(X'') - v(X')$ to the already computed $v(X')$. By definition, $v(X'') - v(X') = \sum_{S \in \mathcal{S}(p''_2, p''+1) \setminus \mathcal{S}(p'_2, p'+1)} \#\mathrm{str}'(S, X, \sigma, n)$, where $p'' = p'+1$ and p''_2 is the left-to-right rank of the interval in X' containing the position $m'' - N + 1$. If X'_1 and X'_2 are partitions of $P[1..m']$ with the same number of parts and these partitions induce the same partition on $P[m' - N + 1..m']$, then for every partition X'' of $P[m'+1..m'']$ we have $v(X'_1 \cup X'') - v(X'_1) = v(X'_2 \cup X'') - v(X'_2)$. Thus, for every $p' \le p$ and every partition X' of $P' = P[m' - N + 1..m']$, we need to store only one partition X^* of $P[1..m']$ into p' parts that induces the partition X' on P'. The partition X^* is chosen such that $v(X^*)$ is minimum among all partitions of $P[1..m']$ into p' parts that induce the partition X' on P'. We obtain an algorithm whose time complexity is $O(m^2 + (|\mathcal{S}|Nk + mp)\sum_{p'=1}^{p}\binom{N-1}{p'-1})$. Further details are omitted due to space limitations.

4   Properties of Optimal Search Schemes

Designing an efficient search scheme for a given set of parameters consists of (1) choosing the number of parts, (2) choosing the searches, (3) choosing the partition of the pattern. While it is possible to enumerate all possible choices, and evaluate the efficiency of the resulting scheme using Section 3.1, this is generally infeasible due to a large number of possibilities. It is therefore desirable to have a combinatorial characterization of optimal search schemes.

The *critical string* of a search scheme $\mathcal{S}$ is the lexicographically maximal U-string of a search in $\mathcal{S}$. A search of $\mathcal{S}$ is *critical* if its U-string is equal to the critical string of $\mathcal{S}$. For example, the critical string of $\mathcal{S}_{\mathrm{LLTWWY}}$ is 022, and S_f is the critical search. For typical parameters, critical searches of a search scheme constitute the bottleneck. Consider a search scheme $\mathcal{S}$, and assume that the L-strings of all searches contain only zeros. Assume further that the pattern is partitioned into equal-size parts. Let ℓ be the maximum index such that for every search $S \in \mathcal{S}$ and every $i \le \ell$, $U[i]$ of S is no larger than the number in position i in the critical string of $\mathcal{S}$. From Section 3, the number of strings enumerated by a search $S \in \mathcal{S}$ depends mostly on the prefix of the U-string of S of length $\lceil \lceil \log_\sigma n \rceil / (m/p) \rceil$. Thus, if $\lceil \lceil \log_\sigma n \rceil / (m/p) \rceil \le \ell$, a critical search enumerates an equal or greater number of strings than a non-critical search.

We now consider the problem of designing a search scheme whose critical string is minimal. Let $\alpha(k, p)$ denote the lexicographically minimal critical string of a k-mismatch search scheme that partitions the pattern into p parts. The next theorem gives the values of $\alpha(k, k+2)$ and $\alpha(k, k+1)$. We omit the proof due to space limitations.

Theorem 1. $\alpha(k, k+1) = 013355 \cdots kk$ *for every odd* k, *and* $\alpha(k, k+1) = 02244 \cdots kk$ *for every even* k. $\alpha(k, k+2) = 0123 \cdots (k-1)kk$ *for every* $k \ge 1$.

Table 1. Values of $\#\mathrm{str}(\mathcal{S}, X, 4, 4^{16})$ for 2-mismatch search schemes (recall that 4 is the size of the alphabet, and 4^{16} is the length of the text). The second column gives #str values for the 3-part search scheme with equal-size parts. The other columns give #str values for different search schemes using an optimal partition of the pattern. For each search scheme, the optimal value of #str is shown in the first sub-column, and the optimal partition in the second sub-column.

m	3 equal	3 unequal		4 unequal		5 unequal	
24	1197	1077	9,7,8	959	7,4,4,9	939	7,1,6,1,9
36	241	165	15,10,11	140	12,5,7,12	165	11,1,9,1,14
48	53	53	16,16,16	51	16,7,9,16	53	16,1,15,1,15

5   Case Studies

In this section, we examine the efficiency of several 2-mismatch and 3-mismatch search schemes. The search schemes were generated by a greedy algorithm. At each step, the algorithm considers the uncovered string M of weight k such that the lexicographically minimal U-strings of searches that covers M is maximal. Among the searches that cover M with minimal U-string, a search that covers the maximum number of uncovered strings of weight k is chosen. The L-string of the search is chosen to be lexicographically maximal among all possible L-string that do not decrease the number of uncovered strings. For each search scheme and each choice of parameters, we obtained an optimal partition and computed the efficiency of the scheme according to Section 3.

Results for 2 mismatches are given in Table 1 and Table 2, for 4-letter and 30-letter alphabets respectively, and results for 3 mismatches are given in Table 3.

Our theoretical analysis indicates that an increased number of parts combined with uneven partitioning improves over a scheme with $k+1$ equal sized parts when $m/(k+1)$ is smaller than $\log_\sigma n$ (details omitted due to lack of space). This is confirmed by the numerical results in Tables 1, 2, and 3, which show significant improvement when $m/(k+1)$ is small.

For big alphabets (Table 2), we observe a larger gain in efficiency. The is due to the fact that the values of n_l (see equation (2)) increase more rapidly when the alphabet is large, and thus a change in the size of parts can have a bigger influence on these values.

For 3 mismatches (Table 3), we observe smaller gain. This is partly explained by Theorem 1, as with 3 mismatches and 4 parts, the critical string starts with 01 (compared to 02 for 2 mismatches and 3 parts), therefore 4 parts provide already a competitive search. Another interesting observation is that with 4 parts, the optimal partition is an even one, as the U-strings in all searches in the 4-part scheme are the same.

We implemented our method using the *2BWT* library, provided by [5] (available at http://i.cs.hku.hk/2bwt-tools/) and experimentally compared different search schemes. The experiments were done on the sequence of human chromosome 14. The sequence is 88M long, with nucleotide distribution 29%, 21%, 21%, 29%. Searched patterns were obtained from Next-Generation Sequencing reads by

Table 2. Values of #str($\mathcal{S}, X, 30, 30^7$) for 2-mismatch search schemes

m	3 equal	3 unequal		4 unequal		5 unequal	
15	846	286	6,4,5	231	5,2,3,5	286	5,1,3,1,5
18	112	111	7,6,5	81	6,2,4,6	111	6,1,4,1,6
21	24	24	7,7,7	23	7,3,4,7	24	7,1,6,1,6

Table 3. Values of #str($\mathcal{S}, X, 4, 4^{16}$) for 3-mismatch search schemes

m	4 unequal		5 unequal	
24	11222	6,6,6,6	8039	4,6,5,1,8
36	416	9,9,9,9	549	6,11,5,1,13
48	185	12,12,12,12	213	11,11,11,1,14

Table 4. Total time (in seconds) of searching for one million patterns in human chromosome 14, up to 2 mismatches

m	3 equal	3 unequal		4 equal	4 unequal	
18	230	230 (100%)	6,6,6	209 (91%)	203 (88%)	5,5,1,7
21	175	161 (92%)	8,6,7	147 (84%)	150 (86%)	6,4,3,8
24	142	120 (85%)	10,7,7	117 (82%)	107 (75%)	7,4,4,9
27	119	99 (83%)	12,7,8	96 (81%)	82 (69%)	9,4,5,9
30	101	84 (83%)	12,9,9	66 (65%)	68 (67%)	10,4,6,10
33	83	70 (84%)	13,10,10	53 (64%)	58 (70%)	11,5,6,11
36	68	66 (97%)	13,11,12	49 (72%)	50 (74%)	12,5,7,12
39	56	56 (100%)	13,13,13	48 (86%)	45 (80%)	13,6,7,13
42	45	45 (100%)	14,14,14	44 (98%)	38 (84%)	14,6,8,14

cutting out strings of required length. Both the sequence and the reads were downloaded from http://gage.cbcb.umd.edu/data/. For every pattern length and every search scheme, 10^6 patterns were searched and the average number of enumerated strings was computed.

For the case of 2 mismatches, we implemented the 3-part and 4-part schemes from Section 5, as well as their equal part versions for comparison. For each pattern length, we computed an optimal partition, taking into account a non-identical distribution of nucleotides. Results are presented in Table 4. Using unequal parts for 3-part schemes yields a notable time decrease (8–17%) for patterns of length between 21 and 33. Furthermore, using 4 parts leads to an even more important improvement, showing a significantly better results for all pattern length compared to the 3-equal-parts scheme of [5]. For pattern lengths from 21 to 42 we observe 16–36% improvement in running time and 12–35% improvement in the number of enumerated strings. For pattern lengths 24, 27, 39, 42, we observe that using unequal part lengths for 4-part schemes is beneficial. Overall, the experimental results are consistent with numerical estimations of Section 5. However, for pattern lengths 30-36, the 4-equal-parts scheme performs better than the 4-unequal-parts one, which illustrates that the optimal partition found for random texts may not be the best one for genomic sequences.

6   Conclusions

This paper can be seen as the first step towards an automated design of efficient search schemes for approximate string matching, based on bidirectional indexes.

More research has to be done in order to allow an automated design of optimal search schemes. It would be very interesting to study an approach when a search scheme is designed simultaneously with the partition, rather than independently as it was done in our work. The results of this paper can be extended to approximate string matching under edit distance. The estimation of n_l (Section 3) becomes more complicated though.

Acknowledgements. GK has been supported by the ABS2NGS grant of the French government (program *Investissement d'Avenir*) as well as by a EU Marie-Curie Intra-European Fellowship for Carrier Development. KS has been supported by the *co-tutelle* PhD fellowship of the French government. DT has been supported by ISF grant 981/11.

References

1. Belazzougui, D., Cunial, F., Kärkkäinen, J., Mäkinen, V.: Versatile succinct representations of the bidirectional burrows-wheeler transform. In: Bodlaender, H.L., Italiano, G.F. (eds.) ESA 2013. LNCS, vol. 8125, pp. 133–144. Springer, Heidelberg (2013)
2. Burrow, M., Wheeler, D.: A block-sorting lossless data compression algorithm. Technical report 124, Digital Equipment Corporation, California (1994)
3. Chen, L.H.Y.: Poisson approximation for dependent trials. The Annals of Probability, 534–545 (1975)
4. Ferragina, P., Manzini, G.: Opportunistic data structures with applications. In: Proc. 41st Symposium on Foundation of Computer Science (FOCS), pp. 390–398 (2000)
5. Lam, T.W., Li, R., Tam, A., Wong, S.C.K., Wu, E., Yiu, S.-M.: High throughput short read alignment via bi-directional BWT. In: Proc. IEEE International Conference on Bioinformatics and Biomedicine (BIBM), pp. 31–36 (2009)
6. Lam, T.-W., Sung, W.-K., Wong, S.-S.: Improved approximate string matching using compressed suffix data structures. In: Deng, X., Du, D.-Z. (eds.) ISAAC 2005. LNCS, vol. 3827, pp. 339–348. Springer, Heidelberg (2005)
7. Langmead, B., Trapnell, C., Pop, M., Salzberg, S.: Ultrafast and memory-efficient alignment of short DNA sequences to the human genome. Genome Biology 10(3), R25 (2009)
8. Li, H., Durbin, R.: Fast and accurate short read alignment with Burrows-Wheeler transform. Bioinformatics 25(14), 1754–1760 (2009)
9. Li, H., Homer, N.: A survey of sequence alignment algorithms for next-generation sequencing. Briefings in Bioinformatics 11(5), 473–483 (2010)
10. Navarro, G., Mäkinen, V.: Compressed full-text indexes. ACM Computing Surveys 39(1) (2007)
11. Russo, L.M.S., Navarro, G., Oliveira, A.L., Morales, P.: Approximate string matching with compressed indexes. Algorithms 2(3), 1105–1136 (2009)
12. Schnattinger, T., Ohlebusch, E., Gog, S.: Bidirectional search in a string with wavelet trees and bidirectional matching statistics. Information and Computation 213, 13–22 (2012)
13. Simpson, J.T., Durbin, R.: Efficient de novo assembly of large genomes using compressed data structures. Genome Research 22(3), 549–556 (2012)
14. Sung, W.-K.: Indexed approximate string matching. In: Kao, M.-Y. (ed.) Encyclopedia of Algorithms, pp. 1–99. Springer, US (2008)

String Range Matching[*]

Juha Kärkkäinen, Dominik Kempa, and Simon J. Puglisi

Department of Computer Science, University of Helsinki, and
Helsinki Institute for Information Technology HIIT, Finland
{firstname.lastname}@cs.helsinki.fi

Abstract. Given strings X and Y the *exact string matching* problem is
to find the occurrences of Y as a substring of X. An alternative formulation asks for the lexicographically consecutive set of suffixes of X that
begin with Y. We introduce a generalization called *string range matching* where we want to find the suffixes of X that are in an arbitrary
lexicographical range bounded by two strings Y and Z. The problem has
applications in distributed suffix sorting, where Y and Z are themselves
suffixes of X. Exact string matching can be solved in linear time and
constant extra space under the standard comparison model. Our conjecture is that string range matching is a harder problem and cannot be
solved within the same time–space complexity. In this paper, we trace
the upper bound on the complexity of string range matching by describing algorithms that are within a logarithmic factor of the time–space
complexity of exact string matching, as well as variants of the problem
and the model that can be solved in linear time and constant extra space.

1 Introduction

Exact string matching, the problem of finding all the occurrences of a string
$Y[0..m)$ (the pattern) in a larger string $X[0..n)$ (the text) is a foundational problem in computer science, and has applications throughout modern computer software. Among the numerous algorithms for exact string matching [5], there are
several that are optimal in space ($O(1)$ extra space) as well as in time ($O(n+m)$)
under the comparison model [8,9,3,2,4]. We have recently considered a generalization called *longest prefix matching*, where we want to find the occurrences of
the longest prefix of the pattern that occurs in the text, and have shown that
(at least) one of these algorithms can be generalized to longest prefix matching
within the same optimal time–space complexity [11].

In this paper we introduce a further generalization based on an alternative
view of the above problems. If we regard the text as a collection of its suffixes, the
above problems can be stated as reporting the starting positions of all suffixes
that begin with the pattern or with the longest prefix of the pattern producing a
non-empty result. The resulting subset of suffixes is lexicographically consecutive
and the query could be expressed as a lexicographical range query on the set of

---

[*] This research is partially supported by the Academy of Finland through grant 118653
(ALGODAN) and grant 250345 (CoECGR).

A.S. Kulikov, S.O. Kuznetsov, and P. Pevzner (Eds.): CPM 2014, LNCS 8486, pp. 232–241, 2014.
© Springer International Publishing Switzerland 2014

suffixes. A natural generalization, which we call *string range matching*, then asks for suffixes in an arbitrary lexicographical range: Given strings $X[0..n]$, $Y[0..m_1]$ and $Z[0..m_2]$ report all i such that $Y \leq X[i..n) < Z$.

We describe two basic algorithms for string range matching. One is based on Crochemore's string matching algorithm [2] (or its simplification in [11]) and solves the problem in $O(n \log(m_1 + m_2))$ time using constant extra space. For certain important special cases the algorithm can be made to run in linear time, in particular when the strings Y and Z share a prefix of length $\epsilon(m_1 + m_2)$ for any constant $\epsilon > 0$. The second algorithm is based on the string matching algorithm by Galil and Seiferas [8] and solves the *counting* version of the problem in linear time using $O(\log(m_1 + m_2))$ extra space. In both cases, the modification to solve string range matching is non-trivial.

Furthermore, we show that the problem can be solved in linear time using just constant extra space by slightly cheating about the extra space aspect. We describe two distinct "cheats": (i) The algorithm can overwrite the strings Y and Z with other data but must restore the strings to their original state at the end. The overwritten memory does not count as extra space. (ii) The algorithm has read access to its own output, which does not count as extra space. The algorithm works for most reasonable output formats but only for the reporting version of the problem, not for the counting version.

We conjecture that string range matching is a harder problem than exact string matching and cannot be solved in linear time and constant extra space in the comparison model without cheating, i.e., when the only access to the input is by character comparisons and the output is write-only.

Applications to Suffix Sorting. The suffix array [13] and the Burrows–Wheeler transform [1] are central to modern string processing and the key task in their construction is sorting the suffixes of a string. For long strings we may want to split the task into smaller subtasks in order to distribute the work load or reduce the memory requirements. One approach partitions the lexicographical space [10]. For each partition, we then need to collect all suffixes within the given lexicographical range. The collecting problem is exactly string range matching.

Another approach to suffix sorting for large strings is to split the string into smaller blocks, sort the suffixes of each block separately, and then merge [6]. Assume the string is ABC, where B is the current block. When sorting the suffixes of B, what we really want is the lexicographical ordering of the $|B|$ longest suffixes of BC, which can differ significantly from the ordering of the suffixes of B as a standalone string. The correct ordering can be obtained without accessing C if we know which of the $|B|$ longest suffixes of BC are smaller than C. This information can be obtained using a one-sided string range matching query.

In both of the papers mentioned above [10,6], the string range matching problem is solved using an algorithm introduced in [10], which is based on the Knuth–Morris–Pratt [12] exact string matching algorithm, and is the only prior work on string range matching that we are aware of. However, the $O(m_1 + m_2)$ extra space needed by the algorithm is a problem since Y and Z are potentially very long suffixes of X. The techniques used in those papers to reduce the space requirement (a global data structure [10] and sequential processing of the blocks [6]) are not well suited for distributed computation. Thus our small space algorithms open up new possibilities for distributed suffix sorting.

2   General Framework

Consider a string $X = X[0..n) = X[0]X[1]\ldots X[n-1]$ of $|X| = n$ symbols drawn from an ordered alphabet. Here and elsewhere we use $[i..j)$ as a shorthand for $[i..j-1]$. For $i \in [0..n)$ we write X_i to denote the *suffix* of X of length $n-i$, that is $X_i = X[i..n) = X[i]X[i+1]\ldots X[n-1]$. We also generalize the notation to sets of suffixes: for any $S \subseteq [0..n)$, $X_S = \{X_i \mid i \in S\}$. The *empty string* is denoted by ε. For strings X and Y, we use $\mathsf{lcp}(X,Y)$ to denote the length of the *longest common prefix* of X and Y. If $X \le Y$, we write $[X,Y)$ to denote the *lexicographical range* of strings between X and Y, i.e., the set $\{Z \mid X \le Z < Y\}$. For $X \le Y \le Z$, we have $[X,Z) = [X,Y) \cup [Y,Z)$, where $\cup$ denotes the disjoint union. A positive integer p is a *period* of X if $X[i] = X[i+p]$ for all $i \in [0..n-p)$. If p and q are periods of X and $p + q \le |X|$, then $\gcd(p,q)$ is a period of X too (Weak Periodicity Lemma [7]). The smallest period of X is denoted $\mathsf{per}(X)$. A string X is called *primitive* if it cannot be written as $X = Y^k$ for an integer $k > 1$.

We focus on the one-sided string range matching problem of computing the set $X_{[0..n)} \cap [\varepsilon, Y)$, to which the two-sided version can be reduced since $[Y,Z) = [\varepsilon, Z) \setminus [\varepsilon, Y)$. All our algorithms use a similar basic approach. In a generic step, they compute $\ell = \mathsf{lcp}(X_i, Y)$ and then i is incremented by h, where either $h = p = \mathsf{per}(Y[0..\ell))$ or $h < p$ and $h = \Theta(\ell)$. The efficient computation of ℓ and h is based on well-known exact string matching algorithms, which need no further information as none of the skipped suffixes X_{i+j}, $j \in [1..h)$, can have Y as a prefix. Our contribution is to show how to add order comparisons between Y and the skipped suffixes based on the following lemma.

Lemma 1. *Let $\ell = \mathsf{lcp}(X_i, Y)$ and $p = \mathsf{per}(Y[0..\ell))$. Then $X_{i+j} < Y$ iff $Y_j < Y$ for any $j \in [1..p)$.*

Proof. Fix $j \in [1..p)$. If we had $\mathsf{lcp}(X_{i+j}, Y) \ge \ell - j$, this would imply that j is a period of $Y[0..\ell)$, which violates the assumption that $p = \mathsf{per}(Y[0..\ell))$. Thus $\mathsf{lcp}(X_{i+j}, Y) < \ell - j$ and the claim follows since $\mathsf{lcp}(X_{i+j}, Y_j) = \ell - j$. □

In other words, the status of the suffixes in the skipped segment depends only on Y enabling the use of precomputed information.

3   Linear Time and Logarithmic Extra Space

Our first algorithm solves the one-sided counting variant of the string range matching problem, that is, computes the value $|X_{[0..n)} \cap [\varepsilon, Y)|$. The two-sided problem needs two calls to the algorithm. The algorithm is built on the exact string matching algorithm of Galil and Seiferas that uses $O(\log m)$ extra space [8] (or more precisely on the cleaner formulation due to Crochemore and Rytter [4]).

Assume that $k \geq 3$ is an integer constant. A prefix P of Y is called a *k-highly repeating prefix* (*k*-hrp) of Y if P is primitive and P^k is a prefix of Y. With each *k*-hrp P of Y we associate the interval of positions $[2|P|, |P| + \text{lcp}(Y, Y_{|P|}))$ called the *scope* of P. Scopes have a number of useful properties:

Lemma 2 ([4]). *A string* Y *of length* m *has* $O(\log m)$ *scopes and they can be computed in* $O(m)$ *time and* $O(\log m)$ *extra space. Two distinct scopes do not overlap and for any prefix* $Y[0..\ell]$ *of* Y *it holds that:*

$$\text{per}(Y[0..\ell)) = b/2 \qquad \text{if } \ell \in [b, e) \text{ and } [b, e) \text{ is a scope of some } k\text{-hrp of } Y; \text{ and}$$
$$\text{per}(Y[0..\ell)) > \ell/k \qquad \text{otherwise.}$$

Using the scopes, we can compute the skip value h in the generic matching step as either $b/2$ or $\lfloor \ell/k \rfloor + 1$, which is sufficient for exact string matching [4]. For string range matching, we need additional precomputed information. First, for each k-hrp P of Y we precompute, in addition to its scope $[b, e)$, the value $c = |Y_{[0..|P|)} \cap [\varepsilon, Y)|$. We store all triples (b, e, c) in a sorted list $\mathcal{S}_p$, which allows us to count all suffixes that are skipped-over whenever $h = \text{per}(Y[0..\ell))$. If the match length ℓ does not belong to a scope, the Galil–Seiferas algorithm would use a skip value $h = \lfloor \ell/k \rfloor + 1$. Our strategy in this case is to do a shorter (by a constant factor) skip h for which we know the value $|Y_{[0..h)} \cap [\varepsilon, Y)|$. To be able to do that, we will precompute a second list $\mathcal{S}_n = ((b_1, c_1), \ldots, (b_t, c_t))$ of length $O(\log m)$ that satisfies: $c_i = |Y_{[0..b_i)} \cap [\varepsilon, Y)|$ for $i \in [1..t]$ and $2b_i \leq b_{i+1} < 4b_i$ for $i \in [1..t-1]$. Moreover $4b_t > m$.

Algorithm OneSidedStringRangeCounting(X, Y)

Input: strings $X[0..n)$, $Y[0..m)$.
Output: $|X_{[0..n)} \cap [\varepsilon, Y)|$.
1: $(\mathcal{S}_p, \mathcal{S}_n) \leftarrow$ Precompute(Y)
2: $count \leftarrow i \leftarrow \ell \leftarrow 0$
3: **while** $i < n$ **do**   // Invariant: $count = |X_{[0..i)} \cap [\varepsilon, Y)|$
4:      **while** $i + \ell < n$ and $\ell < m$ and $X[i + \ell] = Y[\ell]$ **do** $\ell \leftarrow \ell + 1$
5:      $(b, e, c) \leftarrow$ contains$(\mathcal{S}_p, \ell)$
6:      **if** $\ell < m$ and $(i + \ell = n$ or $X[i + \ell] < Y[\ell])$ **then**
7:          $count \leftarrow count + 1$
8:      **if** $b \neq 0$ **then**   // $\text{per}(Y[0..\ell)) = b/2$
          // $c = |Y_{[1..b/2)} \cap [\varepsilon, Y)| = |X_{[i+1..i+b/2)} \cap [\varepsilon, Y)|$
9:          $count \leftarrow count + c$
10:         $i \leftarrow i + b/2; \ell \leftarrow \ell - b/2$
11:     **else**   // $\text{per}(Y[0..\ell)) > \ell/k$
12:         $(b, c) \leftarrow$ pred$(\mathcal{S}_n, \lfloor \ell/k \rfloor + 1)$ // $(\lfloor \ell/k \rfloor + 1)/4 < b$
          // $c = |Y_{[1..b)} \cap [\varepsilon, Y)| = |X_{[i+1..i+b)} \cap [\varepsilon, Y)|$
13:         $count \leftarrow count + c$
14:         $i \leftarrow i + b; \ell \leftarrow 0$
15: **return** $count$

Fig. 1. String range counting in linear time and logarithmic space

---

Algorithm Precompute(Y)

Input: a string $Y[0..m]$.
Output: lists $\mathcal{S}_p$ and $\mathcal{S}_n$.
1: $\mathcal{S}_p \leftarrow ()$; $\mathcal{S}_n \leftarrow ((1,0))$
2: $i \leftarrow last \leftarrow 1$; $\ell \leftarrow count \leftarrow 0$
3: **while** $i < m$ **do**   // Invariant: $count = |Y_{[0..i)} \cap [\varepsilon, Y)|$
4:      **while** $i + \ell < m$ **and** $Y[i + \ell] = Y[\ell]$ **do** $\ell \leftarrow \ell + 1$
5:      $(b, e, c) \leftarrow \text{contains}(\mathcal{S}_p, \ell)$
6:      **if** $k \cdot i \leq i + \ell$ **and** $b = 0$ **then**
7:          $(b, e, c) \leftarrow (2i, i + \ell + 1, count)$
8:          $\text{add}(\mathcal{S}_p, (b, e, c))$ // $Y[0..i)$ is a new k-hrp
9:      **if** $2 \cdot last \leq i$ **then**
10:          $\text{add}(\mathcal{S}_n, (i, count))$; $last \leftarrow i$
11:      **if** $i + \ell = m$ **or** $Y[i + \ell] < Y[\ell]$ **then** $count \leftarrow count + 1$
12:      **if** $b \neq 0$ **then**
13:          $count \leftarrow count + c$
14:          $i \leftarrow i + b/2$; $\ell \leftarrow \ell - b/2$
15:      **else**
16:          $(b, c) \leftarrow \text{pred}(\mathcal{S}_n, \lfloor \ell/k \rfloor + 1)$
17:          $count \leftarrow count + c$
18:          $i \leftarrow i + b$; $\ell \leftarrow 0$
19: **return** $(\mathcal{S}_p, \mathcal{S}_n)$

Fig. 2. Computation of lists $\mathcal{S}_p$ and $\mathcal{S}_n$

The main algorithm is given in Fig. 1 and the algorithm for precomputing the lists $\mathcal{S}_p$ and $\mathcal{S}_n$ in Fig. 2. The latter is essentially the same as the former except it matches the pattern against itself. The construction of $\mathcal{S}_n$ is possible, because i never more than doubles. The algorithms use the following additional procedures: $\text{add}(L, x)$ adds an element x at the end of a list L, $\text{pred}(\mathcal{S}_n, x)$ returns $(b, c) \in \mathcal{S}_n$, $b \leq x$ with the largest b, and $\text{contains}(\mathcal{S}_p, x)$ returns $(b, e, c) \in \mathcal{S}_p$ such that $b \leq x < e$ or $(0, \cdot, \cdot)$ if no such triple belongs to $\mathcal{S}_p$.

Theorem 1. *The algorithm in Fig. 1 solves the counting version of the string range matching problem in linear time and* $O(\log(m_1 + m_2))$ *extra space.*

Proof. The correctness of the algorithm follows from Lemmas 1 and 2 as explained above and detailed in the invariants included in the code. The time complexity is linear by the same arguments as in [4]. In particular, the shifts made by our algorithm may be shorter but only by a constant factor. □

4 $O(n \log(m_1 + m_2))$ Time and Constant Extra Space

In this section, we first describe an algorithm solving the string range matching problem in linear time and constant extra space for the special case, where the query range is of the form $[Y[0..r), Y)$ with $2\lfloor m/3 \rfloor \leq r < m = |Y|$. For simplicity we assume $m \geq 3$. We will then show how the general string range matching problem can be solved using $O(\log m)$ calls to the special case algorithm.

Algorithm RestrictedOneSidedStringRangeReporting(X, Y, r)

Input: strings $X[0..n)$, $Y[0..m)$ and an integer $r \in [2\lfloor m/3 \rfloor, m)$.
Output: the set $\mathcal{R} = \{j \in [0..n) \mid Y[0..r) \leq X_j < Y\}$.
1: $(q, e) \leftarrow$ Precompute(Y, r)
2: $\mathcal{R} \leftarrow \emptyset$
3: $i \leftarrow \ell \leftarrow s \leftarrow p \leftarrow 0$
4: **while** $i < n$ **do**   // Invariant: $\mathcal{R} = \{j \in [0..i) \mid Y[0..r) \leq X_j < Y\}$
5:      **while** $i + \ell < n$ **and** $\ell < m$ **and** $X[i+\ell] = Y[\ell]$ **do**
6:          $(\ell, s, p) \leftarrow$ UpdateMS(Y, ℓ, s, p) // Increments ℓ by one
7:      **if** $r \leq \ell < m$ **and** $(i + \ell = n$ **or** $X[i+\ell] < Y[\ell])$ **then** $\mathcal{R} \leftarrow \mathcal{R} \cup \{i\}$
8:      **if** $0 < p \leq \ell/3$ **and** $Y[0..s) = Y[p..p+s)$ **then**   // per$(Y[0..\ell)) = p$
9:          $h \leftarrow p; \ell \leftarrow \ell - p$
10:     **else**   // per$(Y[0..\ell)) > \ell/3$
11:         $h \leftarrow \lfloor \ell/3 \rfloor + 1; (\ell, s, p) \leftarrow (0, 0, 0)$
12:     **if** $e = m$ **or** $Y[e] < Y[e - q]$ **then**   // $Y_q < Y$
13:         $g \leftarrow \min(\lfloor \frac{h-1}{q} \rfloor, \lfloor \frac{e-r}{q} \rfloor)$
14:         $\mathcal{R} \leftarrow \mathcal{R} \cup \{i + q, i + 2q, \ldots, i + gq\}$
15:     $i \leftarrow i + h$
16: **return** $\mathcal{R}$

Fig. 3. Reporting suffixes $X_j \in [Y[0..r), Y)$ in linear time and constant extra space. Due to lack of space, we omit the code for UpdateMS(Y, ℓ, s, p), which can be found in [11], and the precomputation of q and e.

The backbone of the algorithm is the exact string matching algorithm of Crochemore [2] (see [11] for a simplified version, which we use here). Similarly to the Galil–Seiferas algorithm, it computes the skip value h in the generic step either as $h = \mathsf{per}(Y[0..\ell)) \leq \ell/3$ or $h = \lfloor \ell/3 \rfloor + 1 \leq \mathsf{per}(Y[0..\ell))$, but now using only $O(1)$ extra space. Our addition is two precomputed values $q = \mathsf{per}(Y[0..r))$ and $e = q + \mathsf{lcp}(Y, Y_q)$, i.e., $Y[0..e)$ is the longest prefix of Y having period q. Both values can be computed in linear time using constant extra space [2]. Determining the status of the skipped suffixes is based on Lemma 1 and the following result. The pseudo-code is given in Figure 3.

Lemma 3. Let $h \leq \lfloor m/3 \rfloor + 1$ be a positive integer and $g = \min(\lfloor \frac{h-1}{q} \rfloor, \lfloor \frac{e-r}{q} \rfloor)$. Then:
$$\{j \in [1..h) \mid Y[0..r) \leq Y_j < Y\} = \begin{cases} \{q, 2q, \ldots, gq\} & \text{if } Y_q < Y \\ \emptyset & \text{otherwise} \end{cases}$$

Proof. First note that $Y[0..r) \leq Y_j < Y$ iff $Y_j < Y$ and $\mathsf{lcp}(Y_j, Y) \geq r$. If $\mathsf{lcp}(Y_j, Y) \geq r$, then j is a period of $Y[0..r)$. Since $q = \mathsf{per}(Y[0..r))$ and $j + q \leq 2j \leq 2\lfloor m/3 \rfloor \leq r$, the Weak Periodicity Lemma implies that j must be a multiple of q. Thus $j = kq$ for some integer $k > 0$ such that $kq \leq h - 1 \leq e$. Then $\mathsf{lcp}(Y_j, Y) = e - j$ and $Y_j < Y$ iff $Y_q < Y$. Thus, if $Y_q \geq Y$, the set in the lemma is empty. Otherwise, the set includes all $j = kq$ such that $kq \leq h - 1$ and $\mathsf{lcp}(Y_j, Y) \geq r$, which give the limits $k \leq \lfloor (h-1)/q \rfloor$ and $k \leq \lfloor (e-r)/q \rfloor$, respectively.   □

Let us now consider more general forms of queries. First, consider the query $[Y[0..\ell), Y)$ with $\ell < r = 2\lfloor |Y|/3 \rfloor$. We can break the query into two distinct subqueries: $[Y[0..\ell), Y) = [Y[0..\ell), Y[0..r)) \cup [Y[0..r), Y)$. This leads to a tail-recursive algorithm that makes $O(\log(|Y|/\ell))$ calls to the algorithm in Fig. 3 and thus runs in $O(n \log(|Y|/\ell))$ time still using only constant extra space. An alternative formulation for the query $[Y[0..\ell), Y)$ is to compute the set $\{X_j \mid X_j < Y$ and $\mathsf{lcp}(X_j, Y) \geq \ell\}$. By a symmetrical procedure, we can also compute the set $\{X_j \mid Y \leq X_j$ and $\mathsf{lcp}(X_j, Y) \geq \ell\}$ in $O(n \log(|Y|/\ell))$ time and constant extra space.

Consider now the fully general query $[Y, Z)$ and let $\ell = \mathsf{lcp}(Y, Z)$. We can partition the result set $X_{[0..n)} \cap [Y, Z)$ into three disjoint sets $\mathcal{R}_1 = \{X_j \mid X_j < Z$ and $\mathsf{lcp}(X_j, Z) > \ell\}$, $\mathcal{R}_2 = \{X_j \mid Y \leq X_j$ and $\mathsf{lcp}(X_j, Y) > \ell\}$ and $\mathcal{R}_3 = \{X_j \mid Y \leq X_j < Z$ and $\mathsf{lcp}(X_j, Y) = \mathsf{lcp}(X_j, Z) = \ell\}$. The sets $\mathcal{R}_1$ and $\mathcal{R}_2$ can be computed as described above. $\mathcal{R}_3$ can be found using any time–space optimal exact string matching algorithm (e.g. [2]). We have proven the following result.

Theorem 2. *It is possible to solve the string range matching problem using constant extra space in* $O(n \log((m_1 + m_2)/(1 + \mathsf{lcp}(Y, Z))))$ *time.*

The worst case time complexity is $O(n \log(m_1 + m_2))$. If $\mathsf{lcp}(Y, Z) \geq \epsilon(m_1 + m_2)$ for some constant $\epsilon > 0$, the algorithm runs in linear time.

5   Linear Time and Constant Extra Space

In this section, we describe two algorithms for string range matching that run in linear time and constant extra space by taking advantage of the space reserved for input or output, which we do not count as extra space. Both algorithms are based on Crochemore's string matching algorithm.

We will describe the algorithms for one-sided queries $[\varepsilon, Y)$. Since these algorithms produce their output in a sequential order, we can answer two-sided queries by running two one-sided queries in parallel: $[Y, Z) = [\varepsilon, Z) \setminus [\varepsilon, Y)$. To ease the description of the algorithms, we assume that the output is produced as a bitvector $B[0..n)$ satisfying, for all $i \in [0..n)$, $B[i] = 1$ iff $X_i < Y$. We will later describe how to produce output in other formats.

Copying Output. Consider the actions of Crochemore's algorithm as it compares the pattern Y against the text suffix X_i. It computes $\ell = \mathsf{lcp}(X_i, Y)$ and then shifts the pattern forward by h positions. With one character comparison we can determine $B[i]$, so the problem now is computing the output for the skipped-over positions $B[i + 1..i + h)$.

Let $\ell_{\max}$ be the longest previously found match, which occurred at position j, i.e., $\ell_{\max} = \mathsf{lcp}(X_j, Y)$. If $\ell \leq \ell_{\max}$, then by Lemma 1 we have $B[i + 1..i + h) = B[j + 1..j + h)$. Thus, if we have access to the previous output, we can simply copy the missing output. Note that this works even if the two intervals $[i + 1..i + h)$ and $[j + 1..j + h)$ overlap.

Algorithm OneSidedStringRangeReporting(X, Y)

Input: strings $X[0..n)$ and $Y[0..m)$.
Output: bitvector $B[0..n)$, where $B[i] = 1$ iff $X[i..n) < Y$.
1: $B \leftarrow (0, 0, \ldots, 0)$
2: $i \leftarrow \ell \leftarrow p \leftarrow s \leftarrow 0$
3: $i_{max} \leftarrow \ell_{max} \leftarrow p_{max} \leftarrow s_{max} \leftarrow 0$
4: **while** $i < n$ **do**
5:       **while** $i + \ell < n$ **and** $\ell < m$ **and** $X[i + \ell] = Y[\ell]$ **do**
6:            $(\ell, s, p) \leftarrow$ UpdateMS(Y, ℓ, s, p) // Increments ℓ by one
7:       **if** $\ell < m$ **and** $(i + \ell = n$ **or** $X[i + \ell] < Y[\ell])$ **then** $B[i] = 1$
8:       $j \leftarrow i_{max}$
9:       **if** $\ell > \ell_{max}$ **then**
10:           swap (ℓ, s, p) and $(\ell_{max}, s_{max}, p_{max})$
11:           $i_{max} \leftarrow i$
12:      **if** $0 < p \le \ell/3$ **and** $Y[0..s) = Y[p..p + s)$ **then**   // per$(Y[0..\ell)) = p$
13:           $B[i + 1..i + p) \leftarrow B[j + 1..j + p)$
14:           $i \leftarrow i + p$
15:           $\ell \leftarrow \ell - p$
16:      **else**   // per$(Y[0..\ell)) > \ell/3$
17:           $h \leftarrow \lfloor \ell/3 \rfloor + 1$
18:           $B[i + 1..i + h) \leftarrow B[j + 1..j + h)$
19:           $i \leftarrow i + h$
20:           $(\ell, s, p) \leftarrow (0, 0, 0)$
21: **return** B

Fig. 4. Linear time, constant extra space algorithm

If $\ell > \ell_{max}$, we cannot obtain all the missing values by copying. In this case, we set ℓ to ℓ_{max} (after computing $B[i]$) and continue as if the match had been a shorter one. This shorter match length leads to a shorter shift and allows copying the output for all the skipped-over positions. Shortening the match length does not violate the correctness of the algorithm as long as the state of algorithm is adjusted accordingly. The full match length becomes the new value of ℓ_{max}, i.e., we are actually swapping the values of ℓ and ℓ_{max}.

The pseudocode for the algorithm is given in Fig. 4. The key changes to the original algorithm in [11, Figure 2] are the computation of B on lines 7, 13 and 18, and the swap of ℓ and ℓ_{max} on line 10. The two other variables, s and p, swapped together with ℓ represent the rest of the state of the algorithm that needs to change when ℓ changes. We refer to [11] for their explanation. Here it suffices to know that they are functions of ℓ (and Y).

Theorem 3. *The algorithm in Fig. 4 solves the string range reporting problem in linear time using constant extra space.*

Proof. The correctness of the algorithm has been established above, and the algorithm clearly uses only constant extra space. In [11], it is shown that the time spent in any round of the original algorithm is proportional to the increase of the value $6i + \ell + s$ during that round. Because of the swap on line 10, this might not hold for the modified algorithm, but this can be corrected by using the value $6i + \ell + s + \ell_{\max} + s_{\max}$ instead. The copying of the output does not add more than $O(n)$ time. Thus the time complexity is still $O(n + m)$.   □

The algorithm can be easily modified to handle other output formats such as a sequence of starting positions. On the other hand, the algorithm cannot be used if the output is streamed or compressed in such a way that it cannot be quickly decompressed from an arbitrary position. Also, the counting version of the problem cannot be solved this way without extra space.

Overwriting Input. Suppose we have separate storage space of m bits and consider the following variant of the algorithm in Fig. 4. Whenever $i_{\max}$ changes we write $B[i_{\max} + 1..i_{\max} + m)$ to our separate storage at the same time as it is being written to the output. Whenever we need to copy a part of the output, we can copy from our separate storage instead of from the output itself. Now the algorithm never needs to access the output, and so can stream the output, encode it in any way, or just count the number of occurrences without producing any output until the end.

Let $\hat{\ell}$ be the length of the longest prefix of Y that also occurs elsewhere in Y. We will use the space occupied by $Y[0..\hat{\ell})$ as a separate storage of at least $\hat{\ell}$ bits. Any access to the characters of $Y[0..\hat{\ell})$ are redirected to that other occurrence. Clearly this can be made to work even if that other occurrence overlaps $Y[0..\hat{\ell})$. The value $\hat{\ell}$ is easily computed in $O(m)$ time using $O(1)$ extra space.

The largest number of bits the algorithm may need to copy from the separate storage at a given stage is at most $\ell_{\max}/3$, since the longest possible shift is $\lfloor \ell_{\max}/3 \rfloor + 1$. Thus as long as $\ell_{\max} \leq 3\hat{\ell}$, the algorithm can operate as described.

Now consider a round i of the algorithm with $\ell = \text{lcp}(X_i, Y) > 3\hat{\ell}$. We call this a *long match*. In this case, the algorithm behaves as it would for a shorter match except the value of $\ell_{\max}$ (and the associated $s_{\max}$ and $p_{\max}$) is not modified. Thus $\ell_{\max}$ never grows too large for the separate storage. The extra time spent in round i because of this is at most $O(\ell)$, and the total extra time is bounded by the total length of the long matches. Thus the following lemma shows that the time complexity of the algorithm is still $O(n)$.

Lemma 4. *The total length of all long matches is at most* $1.5n$.

Proof. Consider a long match of length ℓ at position i, i.e., $X[i..i + \ell) = Y[0..\ell)$, and assume that the next long match starts at $j > i$ and has length ℓ'. Then $\min\{\ell, \ell'\} > 3\hat{\ell}$. First observe that we must have $j + \ell' > i + \ell$. Otherwise, $Y[0..\ell')$ has another occurrence as $Y[j - i..j - i + \ell')$ and $\hat{\ell} \geq \ell'$. Furthermore, we must have $j > i + \ell - \lceil \ell/3 \rceil$. Otherwise, $Y[0..\lceil \ell/3 \rceil)$ has another occurrence as $Y[j - i..j - i + \lceil \ell/3 \rceil)$ and $3\hat{\ell} \geq \ell$. Thus at most one third of a long match can be overlapped by a later long match, which proves the lemma.   □

We have proven the following.

Theorem 4. *The string range matching problem can be solved in linear time and constant extra space if each character of the range boundary strings occupies at least one bit and that bit can be overwritten by the algorithm.*

Note that the algorithm can restore the overwritten part of Y at the end as long as equal characters are always represented by identical bit patterns. The restoration might not be possible in some cases, for example, when upper and lower case letters are considered equal in comparisons.

6 Concluding Remarks

This article is the first to directly consider the string range matching problem. Many interesting avenues for future work remain, but perhaps the most challenging open problem is to establish if the algorithms described in this paper for the general setting (read-only input, inaccessible output) are optimal, or if linear-time constant extra-space methods indeed exist.

References

1. Burrows, M., Wheeler, D.J.: A block sorting lossless data compression algorithm. Technical Report 124, Digital Equipment Corporation, Palo Alto, California (1994)
2. Crochemore, M.: String-matching on ordered alphabets. Theor. Comp. Sci. 92, 33–47 (1992)
3. Crochemore, M., Perrin, D.: Two-way string matching. J. ACM 38(3), 651–675 (1991)
4. Crochemore, M., Rytter, W.: Squares, cubes, and time-space efficient string searching. Algorithmica 13(5), 405–425 (1995)
5. Faro, S., Lecroq, T.: The exact online string matching problem: A review of the most recent results. ACM Comp. Surv. 45(2), 13 (2013)
6. Ferragina, P., Gagie, T., Manzini, G.: Lightweight data indexing and compression in external memory. Algorithmica 63(3), 707–730 (2012)
7. Fine, N.J., Wilf, H.S.: Uniqueness theorems for periodic functions. Proc. Amer. Math. Soc. 16(1), 109–114 (1965)
8. Galil, Z., Seiferas, J.: Saving space in fast string-matching. SIAM J. Comp. 9(2), 417–438 (1980)
9. Galil, Z., Seiferas, J.: Time-space optimal string matching. J. Comp. Sys. Sci. 26, 280–294 (1983)
10. Kärkkäinen, J.: Fast BWT in small space by blockwise suffix sorting. Theor. Comp. Sci. 387(3), 249–257 (2007)
11. Kärkkäinen, J., Kempa, D., Puglisi, S.J.: Crochemore's string matching algorithm: Simplification, extensions, applications. In: Proc. PSC 2013, pp. 168–175. Czech Technical University (2013)
12. Knuth, D., Morris, J.H., Pratt, V.: Fast pattern matching in strings. SIAM J. Comp. 6(2), 323–350 (1977)
13. Manber, U., Myers, G.W.: Suffix arrays: a new method for on-line string searches. SIAM J. Comp. 22(5), 935–948 (1993)

On Hardness of Several String Indexing Problems[⋆]

Kasper Green Larsen[1], J. Ian Munro[2],
Jesper Sindahl Nielsen[1], and Sharma V. Thankachan[2]

[1] Aarhus University, Denmark
{larsen,jasn}@cs.au.dk
[2] University of Waterloo, Canada
{imunro,thanks}@uwaterloo.ca

Abstract. Let $\mathcal{D} = \{d_1, d_2, ..., d_D\}$ be a collection of D string documents of n characters in total. The two-pattern matching problems ask to index $\mathcal{D}$ for answering the following queries efficiently.

- report/count the unique documents containing P_1 *and* P_2.
- report/count the unique documents containing P_1, *but not* P_2.

Here P_1 and P_2 represent input patterns of length p_1 and p_2 respectively. Linear space data structures with $O(p_1 + p_2 + \sqrt{nk} \log^{O(1)} n)$ query cost are already known for the reporting version, where k represents the output size. For the counting version (i.e., report the value k), a simple linear-space index with $O(p_1 + p_2 + \sqrt{n})$ query cost can be constructed in $O(n^{3/2})$ time. However, it is still not known if these are the best possible bounds for these problems. In this paper, we show a strong connection between these string indexing problems and the boolean matrix multiplication problem. Based on this, we argue that these results cannot be improved significantly using purely combinatorial techniques. We also provide an improved upper bound for a related problem known as *two-dimensional substring indexing*.

1 Introduction

Document listing is a fundamental problem in information retrieval, where the task is to index a collection of documents, such that whenever a pattern P comes as a query, we can efficiently find the unique documents containing P as a substring. This problem was introduced by Matias et al. and they provide a linear space and near-optimal time solution [15]. Later Muthukrishnan improved the result by providing a linear-space and optimal query time index [16]. The counting case asks to find the number of documents containing the query pattern. See [17] for an excellent survey on more results and extensions of document retrieval problems. In this paper, our focus is on the case where the query consists of two patterns (known as *two-pattern query problems*). The formal definitions of the problems under consideration are given below.

[⋆] Work supported in part by the Danish National Research Foundation grant DNRF84 through Center for Massive Data Algorithmics (MADALGO).

A.S. Kulikov, S.O. Kuznetsov, and P. Pevzner (Eds.): CPM 2014, LNCS 8486, pp. 242–251, 2014.
© Springer International Publishing Switzerland 2014

Problem 1 Given a set of strings $\mathcal{D} = \{d_1, d_2, \ldots, d_D\}$ with $\sum_{i=1}^{D} |d_i| = n$, preprocess $\mathcal{D}$ to answer queries: given two strings P_1 and P_2 report all i's where both P_1 and P_2 occur in d_i.

Problem 2 Given a set of strings $\mathcal{D} = \{d_1, d_2, \ldots, d_D\}$ with $\sum_{i=1}^{D} |d_i| = n$, preprocess $\mathcal{D}$ to answer queries: given two strings P^+ and P^- report all i's where P^+ occurs in string d_i and P^- does not occur in string d_i.

Problem 3 Let $\mathcal{D} = \{(d_{1,1}, d_{1,2}), (d_{2,1}, d_{2,2}), \ldots, (d_{D,1}, d_{D,2})\}$, be a set of pairs of strings with $\sum_{i=1}^{D} |d_{i,1}| + |d_{i,2}| = n$. Preprocess $\mathcal{D}$ to answer queries: given two strings P_1 and P_2 report all i's where P_1 occurs in $d_{i,1}$ and P_2 occurs in $d_{i,2}$.

Problem 1 was introduced by Muthukrishnan [16]. He presented a data structure using $O(n^{1.5} \log^{O(1)} n)$-space (in words) with $O(p_1 + p_2 + \sqrt{n} + k)$ time for query processing, where $p_1 = |P_1|$ and $p_2 = |P_2|$ and k is the output size[1]. Later Cohen and Porat [6] presented a space efficient structure of $O(n \log n)$-space, but with a higher query time of $O(p_1 + p_2 + \sqrt{nk \log n} \log^2 n)$. The space and the query time of was improved by Hon et al. [11] to $O(n)$ words and $O(p_1 + p_2 + \sqrt{nk \log n} \log n)$ time. See [13] for a succinct space solution for this problem. Problem 2 is known as the forbidden (or excluded) pattern query problem. This was introduced by Fischer et al. [8], where they presented an $O(n^{3/2})$-bit solution with query time $O(p_1 + p_2 + \sqrt{n} + k)$. Immediately, Hon et al. [12] improved its space occupancy to $O(n)$ words, but with a higher query time of $O(p_1 + p_2 + \sqrt{nk \log n} \log^2 n)$. They presented an $O(n)$-space and $O(p_1 + p_2 + \sqrt{n} \log \log n)$ query time structure for the counting version of Problem 2 (i.e., just report the value k). We remark that the same framework can be adapted to handle the counting version of Problem 1 as well. Also the $O(\log \log n)$ term in the query time can be removed by replacing predecessor search queries in their algorithm by range emptiness queries. In summary, we have $O(n)$-space and $\tilde{\Omega}(\sqrt{n})$ query time solutions for the reporting/counting versions of these problems. However, the question whether these are the best possible bounds remains unanswered.

Problem 3 is a generalization of Problem 1 known as the *two-dimensional substring indexing* problem, and was introduced by Ferragina et al. [7]. They reduced it to another problem known as the *common colors query* problem, where the task is to preprocess an array of colors and maintain a data structure, such that whenever two ranges comes as a query, we can output the unique colors which are common to both ranges. Based on their solution for this new problem, they presented an $O(n^{2-\epsilon})$ space and $O(n^\epsilon + k)$ query time solution for Problem 3, where ϵ is any constant in $(0, 1]$. Later Cohen and Porat [6] presented a space efficient solution for the common colors query problem of space $O(n \log n)$ words and query time $O(\sqrt{nk \log n} \log^2 n)$. Therefore, the current best data structure for *two-dimensional substring indexing* problem occupies $O(n \log n)$ space and processes a query in $O(p_1 + p_2 + \sqrt{nk \log n} \log^2 n)$ time.

---

[1] Specifically k is the maximum of 1 and the output size.

1.1 Our Results

In this paper, we use the Word-RAM model of computation with word size $w = \Omega(\log n)$. The following summarizes our main results.

- We present a strong connection between the counting versions of the string indexing problems (Problem 1, 2, and 3) and the boolean matrix multiplication problem. Specifically, we show that multiplying two $\sqrt{n} \times \sqrt{n}$ boolean matrices can be reduced to the problem of indexing $\mathcal{D}$ (in Problem 1, Problem 2, or Problem 3) and answering n counting queries. However, matrix multiplication is a well known hard problem and this connection gives us a hardness result for the pattern matching problems under considerations.
- We present an improved upper bound for the common colors query problem, where the space and query time are $O(n)$ and $O(\sqrt{nk} \log n)$ respectively. Therefore, we now have a linear-space and $O(p_1 + p_2 + \sqrt{nk} \log n)$ query time index for the two-dimensional substring indexing problem (Problem 3).

2 Hardness Results

The hardness results are reductions from *boolean matrix multiplication*. Through this section we use similar techniques to [3,4,5]. In the boolean matrix multiplication problem we are given two $n \times n$ matrices A and B with $\{0,1\}$ entries. The task is to compute the boolean product of A and B, that is replace multiplication by logical and, and replace addition by logical or. Letting $a_{i,j}, b_{i,j}, c_{i,j}$ denote entry i, j of respectively A, B and C the task is to compute for all i, j

$$c_{i,j} = \bigvee_{k=1}^{n} (a_{i,k} \wedge b_{k,j}).$$

The asymptotically fastest algorithm known for matrix multiplication currently uses $O(n^{2.3728639})$ time [9]. This bound is achieved using algebraic techniques (like in Strassen's matrix multiplication algorithm) and the fastest combinatorial algorithm is still cubic divided by some poly-logarithmic factor [1].

In this section, we prove that the problem of multiplying two $\sqrt{n} \times \sqrt{n}$ boolean matrices A and B can be reduced to the problem of indexing $\mathcal{D}$ (in Problem 1 or 2) and answering n counting queries. This is evidence that unless better matrix multiplication algorithms are discovered we should not expect to be able to preprocess the data and answer the queries much faster than $\Omega((\sqrt{n})^{\omega}) = \Omega(n^{\omega/2})$ (ignoring poly-logarithmic factors) where ω is the matrix multiplication exponent. In other words one should not expect to be able to have small preprocessing and query time simultaneously. Currently we cannot achieve better than $\Omega(n^{1.18635})$ preprocessing time and $\Omega(n^{0.18635})$ query time simultaneously.

We start the next section with a brief discussion on how to view boolean matrix multiplication as solving many set intersection problems. Then we give the reductions from the matrix multiplication problem to Problem 2 and describe how to adapt it for Problem 1.

2.1   Boolean Matrix Multiplication

A different way to phrase the boolean matrix multiplication problem is that entry $c_{i,j} = 1$ if and only if $\exists k : a_{i,k} = b_{k,j} = 1$. For any two matrices A and B let $A_i = \{k \mid a_{i,k} = 1\}$ and similarly let $B_j = \{k \mid b_{k,j} = 1\}$. It follows that $c_{i,j} = 1$ if and only if $A_i \cap B_j \neq \emptyset$. In this manner we view each row of matrix A as a set containing the elements corresponding to the indices where there is a 1, and similarly for columns in B. For completeness we also use $\overline{A_i} = \{k \mid a_{i,k} = 0\}$ and $\overline{B_i} = \{k \mid b_{k,j} = 0\}$.

A naive approach to solving Problem 1 would be to index the documents such that we can find all documents containing a query pattern fast. This way a query would be to find all the documents that P_1 occurs in and the documents that P_2 occurs in separately and then return the intersection of the two result sets. This is obviously not a good solution in the worst case, but it illustrates that the underlying challenge is to solve set intersection.

We observed that boolean matrix multiplication essentially solves set intersection between rows of A and columns of B, so the idea for the reductions is to use the fact that queries for Problems 1 and 2 essentially also solve set intersection. We now give the reductions.

2.2   The Reductions

We first relax the data structure problems. Instead of returning a list of documents satisfying the criteria we just want to know whether the list is empty or not, i.e. return 0 if empty and 1 if nonempty.

Let A and B be two $\sqrt{n} \times \sqrt{n}$ boolean matrices and suppose we have an algorithm for building the data structure for the relaxed Problem 2. We now wish to create a set of strings $\mathcal{D}$ based on A and B, build the data structure on $\mathcal{D}$ and do n queries, one for each entry in the product of A and B. In the following we need to represent a subset of $\{0, 1, \ldots, 2\sqrt{n}\}$ as a string, which we do in the following manner.

Definition 1. *Let $X = \{x_1, x_2, \ldots, x_\ell\} \subseteq \{0, 1, \ldots, 2\sqrt{n}\}$, then we represent the set X as $\mathrm{str}(X) = \mathrm{bin}(x_1)\#\mathrm{bin}(x_2)\# \cdots \#\mathrm{bin}(x_\ell)\#$ where $\mathrm{bin}(\cdot)$ gives the binary representation of the number $\cdot$ using $\lceil \frac{1}{2}\log n + 1\rceil$ bits.*

For the matrix A we define $\sqrt{n}$ strings: $d_1^A, d_2^A, \ldots, d_{\sqrt{n}}^A$ and similarly for B we define $d_1^B, d_2^B, \ldots, d_{\sqrt{n}}^B$. Construct $d_j^B = \mathrm{str}(\{k + \sqrt{n} \mid k \in \overline{B_j}\})$ and $d_i^A = \mathrm{str}(A_i)$. We construct the $\sqrt{n}$ strings in $\mathcal{D}$ as: $d_\ell = d_\ell^A d_\ell^B$ for $1 \leq \ell \leq \sqrt{n}$.

Lemma 1. *Each string in $\mathcal{D}$ is at most $O(\sqrt{n}\log n)$ characters long.*

Proof. There are at most $\sqrt{n}$ elements in A_ℓ and at most $\sqrt{n}$ elements in $\overline{B_\ell}$. Each element in A_ℓ and $\overline{B_\ell}$ contributes exactly one number to the string d_ℓ and one '#'. Each number uses exactly $\lceil \frac{1}{2}\log n + 1\rceil$ characters. In total we get $(|A_\ell + |\overline{B_\ell}|)(\lceil \frac{1}{2}\log n + 1\rceil + 1) \leq 2\sqrt{n}\lceil \frac{1}{2}\log n + 1\rceil + 2\sqrt{n} = O(\sqrt{n}\log n)$ $\square$

Corollary 1. *The total length of the strings in $\mathcal{D}$ is $\sum_{i=1}^{\sqrt{n}} d_i = O(n \log n)$.*

Proof. Follows since there are at most $\sqrt{n}$ strings in $\mathcal{D}$ and by Lemma 1 each string is at most $O(\sqrt{n} \log n)$ characters long. □

We have now created the set of strings, $\mathcal{D}$, that we wish to build the data structure on. We now specify the queries and prove that using these queries we can solve the matrix multiplication problem.

Lemma 2. *The entry $c_{i,j} = 1$ if and only if the query $P^+ = \text{bin}(i)$, $P^- = \text{bin}(\sqrt{n} + j)$ returns 1.*

Proof. Suppose that $c_{i,j} = 1$. Then by a previous discussion there must exist a k such that $a_{i,k} = b_{k,j} = 1$. By construction the string $\text{bin}(i)$ occurs as a substring in d_k^A since $a_{i,k} = 1$. If $b_{k,j} = 1$ we know that $k \in B_j$ and therefore $k \notin \overline{B_j}$ so the string $\text{bin}(\sqrt{n} + j)$ is not in d_k^B. It follows that the string d_k satisfies the conditions and the query returns 1.

Suppose the query (P^+, P^-) returns 1, then by definition there exists a $d_\ell \in \mathcal{D}$ such that $\text{bin}(i)$ occurs in d_ℓ and $\text{bin}(\sqrt{n} + j)$ does not occur in d_ℓ. All numbers in d_ℓ but the first are surrounded by '#' and all numbers are the same length as $|P^+|$. Furthermore any number in d_ℓ less than $\sqrt{n}$ is only there because of a 1 in column ℓ of A. In particular $a_{i,\ell} = 1$, otherwise d_ℓ would not satisfy the conditions. Additionally by construction the binary representation of any number $2\sqrt{n} \geq m > \sqrt{n}$ appears in d_ℓ if and only if $b_{\ell,m} = 0$. In particular $\text{bin}(j + \sqrt{n})$ did not occur in d_ℓ, therefore $b_{\ell,j} = 1$. We now have a witness (ℓ) where $a_{i,\ell} = b_{\ell,j} = 1$, and we conclude $c_{i,j} = 1$. □

We are now able to give the following theorems:

Theorem 1. *Let $P(n)$ be the preprocessing time for building the data structure for Problem 1 on a set of strings of total length n and let $Q(n)$ be the query time. In time $O(P(n \log n) + n \cdot Q(n \log n) + n \log n)$ we can compute the product of two $\sqrt{n} \times \sqrt{n}$ boolean matrices.*

Proof. Follows by the lemmas and the discussion above. □

Similarly for Problem 2 we obtain:

Theorem 2. *Let $P(n)$ be the preprocessing time for building the data structure for Problem 2 on a set of strings of total length n and let $Q(n)$ be the query time. In time $O(P(n \log n) + n \cdot Q(n \log n) + n \log n)$ we can compute the product of two boolean matrices.*

Proof. In the discussion above substitute $\overline{B_j}$ with B_j, P^+ and P^- with P_1 and P_2 respectively. □

As a side note, observe that if we replace the problems by their counting version (i.e. count the number of strings in $\mathcal{D}$ that satisfy the condition) then these problems solve matrix multiplication with 0/1 matrices and the regular addition and multiplication operations using the same reductions.

3   The Common Colors Query Problem

In this section, we present an improved upper bound for the common colors query problem and the main result is summarized below.

Theorem 3. *An array E of n colors can be indexed in $O(n)$-word space so that the following query can be answered in $O(\sqrt{nk} \log n)$ time: report the unique colors appearing in both $E[a...b]$ and $E[c...d]$, where a, b, c and d are input parameters and k is the output size.*

Corollary 2. *The two dimensional substring indexing problem (problem 3) can be solved in linear space and $O(p_1 + p_2 + \sqrt{nk} \log n)$ query time.*

Proof. Ferragina et al. showed that Problem 3 can be reduced to the common colors problem [7]. When combined with Theorem 3, we achieve the result.   □

First we give an overview, and then present the details of the proposed data structure. Let $\Sigma = \{\sigma_1, \sigma_2, \sigma_3, ..., \sigma_{|\Sigma|}\}$ be the set of colors appearing in E. Without loss of generality we assume $|\Sigma| \leq n$. The main structure is a binary tree Δ (not necessarily balanced) of $|\Sigma|$ nodes, where each color is associated with a unique node in Δ. Specifically, the color associated with a node u is given by $\sigma_{ino(u)}$, where $ino(u)$ is the in-order rank of u. Also we use $\Sigma(u)$ to represent the set of colors associated with the nodes in the subtree of u. Let $[q, r]$ and $[s, t]$ be two given ranges in E, then $Out_{q,r,s,t}$ denotes the set of colors present in both $[q, r]$ and $[s, t]$. We maintain auxiliary data structures for answering the following subqueries efficiently.

Subquery 1. *Given $i \in [1, |\Sigma|]$ and ranges $[q, r]$ and $[s, t]$ is $\sigma_i \in Out_{q,r,s,t}$?*

Subquery 2. *Given $u \in \Delta$ and ranges $[q, r]$ and $[s, t]$ is $\Sigma(u) \cap Out_{q,r,s,t}$ empty?*

3.1   Query Algorithm

To answer the query (i.e., find $Out_{a,b,c,d}$, where $[a, b]$ and $[c, d]$ are the input ranges), we perform a *preorder* traversal of Δ. Upon reaching a node u, we issue a Subquery 2: Is $\Sigma(u) \cap Out_{a,b,d,d}$ empty?

- If the answer is *yes* (i.e., empty), we can infer that none of the color associated with any node in the subtree of u is an output. Therefore, we skip the subtree of u and move to the next node in the preorder traversal.
- On the other hand if Subquery 2 at u returns *no*, there exists at least one node v in the subtree of u, where $\sigma_{ino(v)}$ is an output. Notice that v can be the node u itself. Therefore, we first check if $\sigma_{ino(u)} \in Out_{a,b,c,d}$ using Subquery 1. If the query returns *yes*, we report $\sigma_{ino(u)}$ as an output and continue the preorder traversal.

By the end of this procedure, all colors in $Out_{a,b,c,d}$ has been reported.

3.2   Details of the Data Structure

We now present the details. For any node u in Δ, we use n_u to represent the number of elements in E with colors in $\Sigma(u)$. i.e., $n_u = |\{i | E[i] \in \Sigma(u)\}|$. Then, we construct Δ as follows, maintaining the invariant:

$$n_u \le n \left(\frac{1}{2}\right)^{depth(u)}$$

Here $depth(u) \le \log n$ is the number of ancestors of u. (We remark that this property is essential to achieve the result in Lemma 4 stated below). The following recursive algorithm can be used for constructing Δ. Let f_i be the number of occurrences of σ_i in E. Initialize u as the root node and $\Sigma(u) = \Sigma$. Then, find the color $\sigma_z \in \Sigma(u)$, where

$$\sum_{i<z, \sigma_i \in \Sigma(u)} f_i \le \frac{1}{2} \sum_{\sigma_i \in \Sigma(u)} f_i \quad \text{and} \quad \sum_{i>z, \sigma_i \in \Sigma(u)} f_i \le \frac{1}{2} \sum_{\sigma_i \in \Sigma(u)} f_i$$

Partition $\Sigma(u)$ into three disjoint subsets $\Sigma(u_L), \Sigma(u_R)$ and $\{\sigma_z\}$, where

$$\Sigma(u_L) = \{\sigma_i | i < z, \sigma_i \in \Sigma(u)\}$$

$$\Sigma(u_R) = \{\sigma_i | i > z, \sigma_i \in \Sigma(u)\}$$

If $\Sigma(u_L)$ is not empty, then we add a left child u_L for u and recurse further from u_L. Similarly, if $\Sigma(u_R)$ is not empty, we add a right child u_R for u and recurse on u_R. This completes the construction of Δ. Since Δ is a tree of $O(|\Sigma|) = O(n)$ nodes, it occupies $O(n)$ words. The following lemmas summarize the results on the structures for handling Subquery 1 and Subquery 2.

Lemma 3. *Subquery 1 can be answered in $O(\log \log n)$ time using an $O(n)$-word structure.*

Proof. The array E can be stored using $n \log |\Sigma| (1 + o(1))$-bits (or $O(n)$-word of space), and supporting $rank_E(j, \sigma_i)$: find the number of occurrences of σ_i in $E[1...j]$ in $O(\log \log |\Sigma|) = O(\log \log n)$ time for any color $\sigma_i \in \Sigma$ [10]. The answer to Subquery 1 is *yes* if both $(rank_E(r, \sigma_i) - rank_E(q - 1, \sigma_i))$ and $(rank_E(t, \sigma_i) - rank_E(s - 1, \sigma_i))$ are nonzero and *no* otherwise.   □

Lemma 4. *There exists an $O(n)$-word structure for handling Subquery 2 in the following manner:*

- *If $\Sigma(u) \cap Out_{q,r,s,t} = \emptyset$, then return* yes *in time $O\left(\frac{\log n}{\log \log n} + \sqrt{n_u} \log^\epsilon n\right)$*
- *Otherwise, one of the following will happen*
  - *Return* no *in $O\left(\frac{\log n}{\log \log n}\right)$ time*
  - *Return the set $\Sigma(u) \cap Out_{q,r,s,t}$ in $O\left(\frac{\log n}{\log \log n} + \sqrt{n_u} \log^\epsilon n\right)$ time*

Proof. See Section 3.4.

By putting all the pieces together, the total space becomes $O(n)$ words.

3.3   Analysis of Query Algorithm

The structures described in Lemma 3 and Lemma 4 can be used as black boxes to support our query algorithm. However, we slightly optimize the algorithm as follows: when we issue Subquery 2 at a node u, and the structure in Lemma 4 returns the set $\Sigma(u) \cap Out_{a,b,c,d}$ (this includes the case where Subquery 2 returns yes), we do not recurse further in the subtree of u. Next we bound the total time for all Subqueries.

Let $k = |Out_{a,b,c,d}|$ be the output size and Δ' be the subtree of Δ consisting only of those nodes which we visited processing the query. Then we can bound the size of Δ':

Lemma 5. *The number of nodes in Δ' is $O(k \log(n/k))$*

Proof. The parent of any node in Δ' must be a node on the path from root to some node u, where $\sigma_{ino(u)} \in Out_{a,b,c,d}$. Since the height of Δ' is at most $\log n$, the number of nodes with depth at least $\log k$ on any path is at most $\log n - \log k = \log(n/k)$. Therefore, number of nodes in Δ' with depth at least $\log k$ is $O(k \log(n/k))$. Also the total number of nodes in Δ with depth at most $\log k$ is $O(k)$. $\square$

We spend $O(\log \log n)$ time for Subquery 1 in every node in Δ'. If a node u is an internal node in Δ', then Subquery 2 in u must have returned yes in $O\left(\frac{\log n}{\log \log n}\right)$ time (otherwise, the algorithm does not explore its subtree). On the other hand, if a node v is a leaf node in Δ', we spend a lot more time. Thus by combining all, the overall query processing time can be bounded as follows.

$$O\left(k \log(n/k) \frac{\log n}{\log \log n} + \sqrt{n} \log^\epsilon n \sum_{v \in leaves} 2^{-depth(v)/2} \right)$$

$$= O\left(k \log(n/k) \frac{\log n}{\log \log n} + \sqrt{n} \log^\epsilon n \sqrt{\sum_{v \in leaves} 1^2 \sum_{v \in leaves} 2^{-depth(v)}} \right) \quad (1)$$

$$= O\left(k \log(n/k) \frac{\log n}{\log \log n} + \sqrt{n} \log^\epsilon n \sqrt{k \log(n/k) \times 1} \right) \quad (2)$$

$$= O\left(\sqrt{nk} \log n \right),$$

Here Equation (1) is by Cauchy-Schwarz's inequality,[2] while Equation (2) is using Kraft's inequality: for any binary tree, $\sum_{\ell \in leaves} 2^{-depth(\ell)} \leq 1$. This completes the proof of Theorem 3.

3.4   Proof of Lemma 4

The following are the building blocks of our $O(n)$-word structure.

---

[2] $\sum_{i=1}^n x_i y_i \leq \sqrt{\sum_{i=1}^n x_i^2} \sqrt{\sum_{i=1}^n y_i^2}$.

1. For every node u in Δ, we define (but not store) E_u as an array of length n_u, where $E_u[i]$ represents the ith leftmost color in E among all colors in $\Sigma(u)$. Thus for any given range $[x, y]$, the list of colors in $E[x...y]$, which are from $\Sigma(u)$ appears in a contiguous region $[x_u, y_u]$ in E_u, where
   - $x_u = 1+$ the number of elements in $E[1...x-1]$, which are from $\Sigma(u)$.
   - $y_u =$ the number of elements in $E[1...y]$, which are from $\Sigma(u)$.

   Notice that the set of colors in $\Sigma(u)$ can be represented as contiguous range of numbers, and the task of computing $[x_u, y_u]$ for any given x, y and u can be reduced to two orthogonal range counting queries in two dimensions. We therefore maintain $O(n)$-word structure for executing this in $O(\log n / \log \log n)$ time [14].

2. An $\sqrt{n_u} \times \sqrt{n_u}$ boolean matrix M_u, for every node u in Δ. For this, we first partition E_u into blocks of size $\sqrt{n_u}$, where the ith block is given by $E_u[1 + (i-1)\sqrt{n_u}, i\sqrt{n_u}]$. Notice that the number of blocks is at most $\sqrt{n_u}$. Then, $M_u[i][j] = 1$, iff there is at least one color, which appear in both the ith block and the jth block of E_u. We also maintain a two-dimensional range maximum query structure (RMQ) with constant query time [2] over each M_u. The total space required is $O(\sum_{u \in \Delta} n_u) = O(n \sum_{u \in \Delta} 2^{-depth(u)}) = O(n \log n)$ bits.

3. Finding the leftmost/rightmost element in $E_u[x...y]$, which is from $\Sigma(u)$, for any given x, y, and u can be reduced to an orthogonal successor/ predecessor query. We therefore maintain $O(n)$-word structure for supporting this query in $O(\log^\epsilon n)$ time [18].

We use the following steps to answer if $\Sigma(u) \cap Out_{q,r,s,t}$ is empty.

1. Find $[q_u, r_u]$ and $[s_u, t_u]$ in $O(\log n / \log \log n)$ time.
2. Find $[q_u', r_u']$ and $[s_u', t_u']$, the ranges corresponds to the longest spans of blocks within $E_u[q_u, r_u]$ and $E_u[s_u, t_u]$ respectively. Notice that $q_u' - q_u, r_u - r_u', s_u' - s_u, t_u - t_u' \in [0, \sqrt{n_u})$. Check if there is at least one common in both $E_u[q_u', r_u']$ and $E_u[s_u', t_u']$ with the following steps.
   - Perform an RMQ on M_u with R as the input region, where
     $R = [1 + (q'-1)/\sqrt{n_u}, r'/\sqrt{n_u}] \times [1 + (s'-1)/\sqrt{n_u}, t'/\sqrt{n_u}]$.
   - If the maximum value within R is 1, then we infer that there is one color common in $E_u[q_u', r_u']$ and $E_u[s_u', t_u']$. Also we can return no as the answer to Subquery 2.

   The time spent so far is $O(\log n / \log \log n)$.
3. If the maximum value within R in the previous step is 0, we need to do some extra work. Notice that any color, which is an output must have an an occurrence in at least one of the following spans $E_u[q_u, q_u' - 1], E_u[r_u' + 1, r_u], E_u[s_u, s_u' - 1], E_u[t_u' + 1, t_u]$ of length at most $4\sqrt{n_u}$. Therefore, these colors can be retrieved using $O(\sqrt{n_u})$ successive orthogonal predecessor/successor queries and with Subquery 1, we can verify if a candidate belongs to the output. The total time required is $O(\sqrt{n_u} \log^\epsilon n)$.

This completes the proof of Lemma 4.

References

1. Bansal, N., Williams, R.: Regularity lemmas and combinatorial algorithms. Theory of Computing 8(1), 69–94 (2012)
2. Brodal, G.S., Davoodi, P., Rao, S.S.: On space efficient two dimensional range minimum data structures. Algorithmica 63(4), 815–830 (2012)
3. Chan, T.M., Durocher, S., Larsen, K.G., Morrison, J., Wilkinson, B.T.: Linear-space data structures for range mode query in arrays. In: STACS. LIPIcs, vol. 14, pp. 290–301. Schloss Dagstuhl - Leibniz-Zentrum fuer Informatik (2012)
4. Chan, T.M., Durocher, S., Skala, M., Wilkinson, B.T.: Linear-space data structures for range minority query in arrays. In: Fomin, F.V., Kaski, P. (eds.) SWAT 2012. LNCS, vol. 7357, pp. 295–306. Springer, Heidelberg (2012)
5. Chan, T.M., Larsen, K.G., Patrascu, M.: Orthogonal range searching on the ram, revisited. In: Symposium on Computational Geometry, pp. 1–10. ACM (2011)
6. Cohen, H., Porat, E.: Fast set intersection and two-patterns matching. Theor. Comput. Sci. 411(40-42), 3795–3800 (2010)
7. Ferragina, P., Koudas, N., Muthukrishnan, S., Srivastava, D.: Two-dimensional substring indexing. J. Comput. Syst. Sci. 66(4), 763–774 (2003)
8. Fischer, J., Gagie, T., Kopelowitz, T., Lewenstein, M., Mäkinen, V., Salmela, L., Välimäki, N.: Forbidden patterns. In: Fernández-Baca, D. (ed.) LATIN 2012. LNCS, vol. 7256, pp. 327–337. Springer, Heidelberg (2012)
9. Gall, F.L.: Powers of tensors and fast matrix multiplication. CoRR, abs/1401.7714 (2014)
10. Golynski, A., Munro, J.I., Rao, S.S.: Rankselect operations on large alphabets: A tool for text indexing. In: SODA, pp. 368–373. ACM Press (2006)
11. Hon, W.-K., Shah, R., Thankachan, S.V., Vitter, J.S.: String retrieval for multi-pattern queries. In: Chavez, E., Lonardi, S. (eds.) SPIRE 2010. LNCS, vol. 6393, pp. 55–66. Springer, Heidelberg (2010)
12. Hon, W.-K., Shah, R., Thankachan, S.V., Vitter, J.S.: Document listing for queries with excluded pattern. In: Kärkkäinen, J., Stoye, J. (eds.) CPM 2012. LNCS, vol. 7354, pp. 185–195. Springer, Heidelberg (2012)
13. Hon, W.-K., Shah, R., Thankachan, S.V., Vitter, J.S.: Space-efficient framework for top-k string retrieval. In: JACM (2014)
14. JáJá, J., Mortensen, C.W., Shi, Q.: Space-efficient and fast algorithms for multidimensional dominance reporting and counting. In: Fleischer, R., Trippen, G. (eds.) ISAAC 2004. LNCS, vol. 3341, pp. 558–568. Springer, Heidelberg (2004)
15. Matias, Y., Muthukrishnan, S.M., Şahinalp, S.C., Ziv, J.: Augmenting suffix trees, with applications. In: Bilardi, G., Pietracaprina, A., Italiano, G.F., Pucci, G. (eds.) ESA 1998. LNCS, vol. 1461, pp. 67–78. Springer, Heidelberg (1998)
16. Muthukrishnan, S.: Efficient algorithms for document retrieval problems. In: SODA, pp. 657–666. ACM/SIAM (2002)
17. Navarro, G.: Spaces, trees and colors: The algorithmic landscape of document retrieval on sequences. CoRR, abs/1304.6023 (2013)
18. Nekrich, Y., Navarro, G.: Sorted range reporting. In: Fomin, F.V., Kaski, P. (eds.) SWAT 2012. LNCS, vol. 7357, pp. 271–282. Springer, Heidelberg (2012)

Most Recent Match Queries
in On-Line Suffix Trees

N. Jesper Larsson

IT University of Copenhagen, Denmark
jesl@itu.dk

Abstract A suffix tree is able to efficiently locate a pattern in an indexed string, but not in general the most recent copy of the pattern in an
online stream, which is desirable in some applications. We study the most
general version of the problem of locating a most recent match: supporting queries for arbitrary patterns, at each step of processing an online
stream. We present augmentations to Ukkonen's suffix tree construction
algorithm for optimal-time queries, maintaining indexing time within a
logarithmic factor in the size of the indexed string. We show that the
algorithm is applicable to sliding-window indexing, and sketch a possible
optimization for use in the special case of Lempel-Ziv compression.

1 Introduction

The *suffix tree* is a well-known data structure which can be used for effectively
and efficiently capturing patterns of a string, with a variety of applications [1, 2,
3]. Introduced by Weiner [4], it reached wider use with the construction algorithm
of McCreight [5]. Ukkonen's algorithm [6] resembles McCreight's, but has the
advantage of being fully *online*, an important property in our work. Farach [7]
introduced recursive suffix tree construction, achieving the same asymptotic time
bound as sorting the characters of the string (an advantage for large alphabets),
but at the cost inherently off-line construction. The simpler *suffix array* data
structure [8,9] can replace a suffix tree in many applications, but cannot generally
provide the same time complexity, e.g., for online applications.

Arguably the most basic capability of the suffix tree is to efficiently locate a
string position matching an arbitrary given pattern. In this work, we are concerned with finding the *most recent* (rightmost) position of the match, which
is not supported by standard suffix trees. A number of authors have studied
special cases of this problem, showing applications in data compression and
surveillance [10, 11, 12], but to our knowledge, no efficient algorithm has previously been presented for the general case. One of the keys to our result is recent
advancement in online suffix tree construction by Breslauer and Italiano [13].

We give algorithms for online support of locating the most recent longest
match of an arbitrary pattern P in $O(|P|)$ time (by traversing $|P|$ nodes, one
of which identifies the most recent position). When a stream consisting of N
characters is subject to search, the data structure requires $O(N)$ space, and

A.S. Kulikov, S.O. Kuznetsov, and P. Pevzner (Eds.): CPM 2014, LNCS 8486, pp. 252–261, 2014.
© Springer International Publishing Switzerland 2014

maintaining the necessary position-updated properties takes at most $O(N \log N)$ total indexing time. If only the last W characters are subject to search (a *sliding window*), space can be reduced to $O(W)$ and time to $O(N \log W)$.

In related research, Amir, Landau and Ukkonen [10] gave an $O(N \log N)$ time algorithm to support queries for the most recent previous string matching a suffix of the (growing) indexed string. The pattern to be located is thus not arbitrary, and the data structure cannot support sliding window indexing.

A related problem is that of Lempel-Ziv factorization [14], where it is desirable to find the most recent occurrence of each factor, in order to reduce the number of bits necessary for subsequent encoding. For this special case, Ferragina et al. [11] gave a suffix tree based linear-time algorithm, but their algorithm is not online, and cannot index a sliding window. Crochemore et al. [12] gave an online algorithm for the rightmost *equal cost* problem, a further specialization for the same application. In section 5, we discuss a possible optimization of our algorithm for the special case of Lempel-Ziv factorization.

2   Definitions and Background

We study indexing a string $T = t_0 \cdots t_{N-1}$ of length $|T| = N$, characters $t_i \in \Sigma$ drawn from a given alphabet Σ. (We consistently denote strings with uppercase letters, and characters with lowercase letters.) T is made available as a *stream*, whose total length may not be known. The index is maintained online, meaning that after seeing i characters, it is functional for queries on the string $t_0 \cdots t_{i-1}$. Following the majority of previous work, we assume that $|\Sigma|$ is a constant.[1]

The data structure supports queries for the most recent longest match in T of arbitrary strings that we refer to as *patterns*. More specifically, given a pattern $P = p_0 \cdots p_{|P|-1}$, a *match* for a length-M prefix of P occurs in position i iff $p_j = t_{i+j}$ for all $0 \le j < M$. It is a *longest* match iff M is maximum, and the *most recent* longest match iff i is the maximum position of a longest match.

2.1   Suffix Tree Construction and Representation

By $\mathcal{ST}$, we denote the *suffix tree* [2,4,5,6] over the string $T = t_0 \cdots t_{N-1}$. This section defines $\mathcal{ST}$, and specifies our representation.

A string S is a nonempty *suffix* (of T, which is implied) iff $S = t_i \cdots t_{N-1}$ for $0 \le i < N$, and a nonempty *substring* (of T) iff $S = t_i \cdots t_j$ for $0 \le i \le j < N$. By convention, the empty string ϵ is both a suffix and a substring. Edges in $\mathcal{ST}$ are directed, and each labeled with a string. Each point in the tree, either coinciding with a node or be located between two characters in an edge label, corresponds to the string obtained by concatenating the edge labels on the path to that point from the root. $\mathcal{ST}$ represents, in this way, all substrings of T. We regard a point that coincides with a node as located at the end of the node's

---

[1] It should be noted, however, that ours and previous algorithms can provide the same *expected* time bounds for non-constant alphabets using hashing, and only a very small worst-case factor higher using efficient deterministic dictionary data structures.

edge from its parent, and can thus refer to the points of all represented strings as located on edges. An *external* edge is an edge whose endpoint is a leaf; other edges are *internal*. The endpoint of each external edge corresponds to a suffix of T, but some suffixes may be represented inside the tree. Note that the point corresponding to an arbitrary pattern can be located (or found non-existent) in time proportional to the length of the pattern, by scanning characters left to right, matching edge labels from the root down.

We do not require that T ends with a unique character, which would make each suffix correspond to some edge endpoint. Instead, we maintain points of implicit suffix nodes using the technique of Breslauer and Italiano [13] (section 3.5).

Following Ukkonen, we augment the tree with an auxiliary node $\perp$ above the root, with a single downward edge to the root. We denote this edge $\vdash$ and label it with ϵ. (Illustration in figure 1.) Although the root of a tree is usually taken to be the topmost node, we shall refer to the node below $\perp$ (the root of the unaugmented tree) as the root node of $\mathcal{ST}$.

Apart from $\vdash$, all edges are labeled with nonempty strings, and the tree represents exactly the substrings of T in the minimum number of nodes. This implies that each node is either $\perp$, the root, a leaf, or a non-root node with at least two downward edges. Since the number of leaves is at most N (one for each suffix), the total number of nodes never exceeds $2N + 1$.

We generalize the definition to $\mathcal{ST}_i$ over the string $T = t_0 \cdots t_{i-1}$, where $\mathcal{ST}_N = \mathcal{ST}$. In iteration i, we execute Ukkonen's *update* algorithm [6] to reshape $\mathcal{ST}_{i-1}$ into $\mathcal{ST}_i$, without looking ahead any further than t_{i-1}. When there is no risk of ambiguity, we refer to the current suffix tree simply as $\mathcal{ST}$, implying that N iterations have completed.

For downward tree navigation, we maintain $down(e, a) = f$ for constant-time access, where e and f are adjacent edges such that e's endpoint coincides with f's start node, and the first character in f's label is a. Note that a uniquely identifies f among its siblings. We define the string that *marks* f as the shortest string represented by f (corresponds to the point just after a). We also maintain $pred(f) = e$ for constant-time upward navigation.

For linear storage space in N, edge labels are represented indirectly, as references into T. Among the many representations available, we choose the following: For any edge e, we maintain $pos(e)$, a position in T of the string corresponding to e's endpoint, and for each *internal* edge e, we maintain $slen(e)$, the length of that same string. I.e., e is labeled with $t_i \cdots t_j$, where $i = pos(e) + slen(pred(e))$ and $j = pos(e) + slen(e)$. External edges need no explicit $slen$ representation, since their endpoints always correspond to suffixes of T, so $slen(e)$ for external e would always be $N - pos(e)$. Note that $pos(e)$ is not uniquely defined for internal e. Algorithms given in the following sections update pos values to allow efficiently finding the most recent occurrence of a pattern.

Ukkonen's algorithm operates around the *active point*, the point of the longest suffix that also appears earlier in T. This is the deepest point where $\mathcal{ST}$ may need updating in the next iteration, since longer suffixes are located on external edges, whose representations do not change. In iteration i, t_i is to be incorporated in

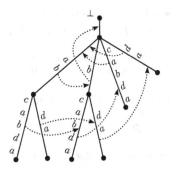

Fig. 1. Suffix tree over the string *abcabda*. Dotted lines show edge-oriented suffix links.

$\mathcal{ST}$. If t_i is already present just below the active point, the tree already contains all the suffixes ending at t_i, and the active point simply moves down past t_i. Otherwise, a leaf is added at the old active point, which is made into a new explicit node if necessary, and the active point moves to the point of the next shorter suffix. To make this move efficient, typically jumping to a different branch of the tree, the algorithm maintains a *suffix link* from any node corresponding to aA, for some character a and string A, directly to the node for A.

We choose a representation where suffix links are edge-oriented, rather than node-oriented as in McCreight's and Ukkonen's algorithms: for edges e and f, we let $suf(e) = f$ iff A marks f, and aA and is the shortest string represented by e such that A marks an edge. (Illustrated in figure 1.) Furthermore, we define *rsuf* to denote the *reverse suffix link*: $rsuf(f, a) = e$. We leave $suf(\vdash)$ undefined. Note that aA is the string that marks e, unless e is a downward edge of the root with an edge label longer than one character. We have $suf(e) = \vdash$ iff e's endpoint corresponds to a string of length one. This variant of suffix links facilitates the description of our *most recent match* scheme, but also has practical impact on runtime behavior, due to reduced branch lookup [15]. The change it implies in Ukkonen's algorithm is relatively straightforward, and has no impact on its asymptotic time complexity. We omit the details in this work.

We refer to the path from the active point to $\vdash$, via suffix links and (possibly) downward edges, as the *active path*. All suffixes that also appear as substrings elsewhere in T are represented along this path. We refer to those suffixes as *active suffixes*. A key to the $O(N)$ time complexity of Ukkonen's algorithm is that the active path is traversed only in the forward direction.

3   Algorithm and Analysis

To answer a most-recent longest-match query for a pattern P', we first locate the edge e in $\mathcal{ST}$ that represents the longest prefix P of P'. For an *exact-match* query, we report failure unless $P = P'$. The time required to locate e, by traversing edges from the root, while scanning edge labels, is $O(|P|)$ [2, 4, 5, 6]. In this section, we give suffix tree augmentations that allow computing the most recent match of P once its edge is located, while maintaining $O(|P|)$ query time.

Separation of Cases. The following identifies two cases in locating the most recent match of a pattern string P, which we treat separately.

Lemma 1. *Let e be the edge that represents P, and let the string corresponding to e's endpoint be PA, $|A| \geq 0$. Precisely one of the following holds:*

1. *The position of the most recent occurrence of P is also the position of the most recent occurrence of PA.*
2. *There exists a suffix PB, $|B| \geq 0$ such that $|B| < |A|$.*

(Proof in appendix [16].) Sections 3.1–3.4 show how to deal with case 1, and section 3.5 with case 2.

3.1 Naive Position Updating

We begin with considering a naive method, by which we update $pos(e)$ at any time when the string corresponding to e's endpoint reappears in the input.

Observe that any string that occurs later in $t_0 \cdots t_{N-1}$ than in $t_0 \cdots t_{N-2}$ must be a suffix $t_j \cdots t_{N-1}$, for some $0 \leq j \leq N - 1$. Hence, in each iteration, we need update $pos(e)$ only if e's endpoint corresponds to an active suffix. This immediately suggests the following: after update iteration i, traverse the active path, and for any edge e whose endpoint corresponds to a suffix, and *pos-update* e, which we define as setting $pos(e)$ to $i - slen(e)$. Thereby, we maintain $pos(e)$ as the most recent position for any non-suffix represented by e, and whenever case 1 of lemma 1 holds, we obtain the most recent position of P directly from the pos value of its edge.

The problem with this naive method is that traversing the whole active path in every iteration results in $\Omega(N^2)$ worst case time. The following sections describe how to reduce the number of pos-updates, and instead letting the query operation inspect $|P|$ edges in order to determine the most recent position.

3.2 Position Update Strategy

To facilitate our description, we define the *link tree* $\mathcal{LT}$ as the tree of $\mathcal{ST}$ edges incurred by the suffix links: edges in $\mathcal{ST}$ are nodes in $\mathcal{LT}$, and f is the parent of e in $\mathcal{LT}$ iff $suf(e) = f$. The root of $\mathcal{LT}$ is $\vdash$. In order to keep the relationship between $\mathcal{ST}$ edges and $\mathcal{LT}$ nodes clear, we use the letters e, f, g, and h to denote them in both contexts.

We define $depth_{\mathcal{LT}}(e)$ as the depth of e in $\mathcal{LT}$. Because of the correspondance between $\mathcal{LT}$ nodes and $\mathcal{ST}$ edges, we have $depth_{\mathcal{LT}}(e) = slen(pred(e))$.

By the current *update edge* in iteration i, we denote the edge e such that $depth_{\mathcal{LT}}(e)$ is maximum among the edges, if any, that would be updated by the naive update strategy (section 3.1) in that iteration: the maximum-$depth_{\mathcal{LT}}$ internal edge whose endpoint corresponds to an active suffix. Section 3.5 describes how the update edge can be located in constant time.

Our update strategy includes pos-updating *only* the update edge, leaving *pos* values corresponding to shorter active suffixes unchanged. When no update edge exists, we pos-update nothing. We introduce an additional value $repr(e)$ for each internal edge e, for which we uphold the following property:

Property 1. For every node g in the suffix link tree, let e be the most recently pos-updated node in the subtree rooted at g. Then an ancestor a of g exists such that $repr(a) = e$.

By convention, a tree node is both an ancestor and a descendent of itself. For new $\mathcal{LT}$ nodes e (without descendants), we set $repr(e)$ to $\vdash$. We proceed with first the algorithm that exploits property 1, then the algorithm to maintain it.

3.3 Most Recent Match Algorithm

Algorithm **mrm-find**(e) scans the $\mathcal{LT}$ path from node e to the root in search for any node g such that $f = repr(g)$ is a descendent of e. For each such f, it obtains the position $q = pos(f) + depth_{\mathcal{LT}}(f) - depth_{\mathcal{LT}}(e)$, and the value returned from the algorithm is the maximum among the q.

mrm-find(e):
1. Let $p = pos(e)$, and $g = e$.
2. If g is $\vdash$, we are done, and terminate returning the value p.
3. If $repr(g) = \vdash$ (i.e., it has not been set), go directly to step 6.
4. Let $f = repr(g)$. If e is not an ancestor of f in $\mathcal{LT}$, go directly to step 6.
5. Let $q = pos(f) + depth_{\mathcal{LT}}(f) - depth_{\mathcal{LT}}(e)$. If $q > p$, update p to q.
6. Set g to $suf(g)$, and repeat from step 2.

The following lemma establishes that when property 1 is maintained, the most recent occurrence of the string corresponding to e's endpoint is among the positions considered by **mrm-find**(e).

Lemma 2. *For an internal edge e, let A be the string corresponding to e's endpoint, and $t_{i-|A|} \cdots t_{i-1}$ the most recent occurrence of A in T. Then e has a descendent f in $\mathcal{LT}$ whose endpoint corresponds to BA for some string B, and $pos(f) = i - |B| - |A|$. (Proof in appendix [16].)*

Since **mrm-find**(e) returns the maximum among the considered positions, this establishes its validity for finding the most recent position of the string corresponding to e's endpoint. Under case 1 of lemma 1, this is the most recent position of *any* string represented by e. Hence, given that e represents pattern P, **mrm-find**(e) produces the most recent position of P in this case.

Lemma 3. *Execution time of **mrm-find**(e), where e represents a string P can be bounded by $O(|P|)$. (Proof in appendix [16].)*

3.4 Maintaining Property 1

Since $\vdash$ is an ancestor of all nodes in $\mathcal{LT}$, we can trivially uphold property 1 in relation to any updated node e simply by setting $repr(\vdash) = e$. But since this

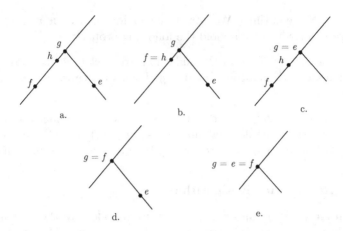

Fig. 2. Cases in **repr-update**: a, b, and c progress down the tree; d and e terminate

ruins the property in relation to other nodes (unless the *previous* value of $repr(\vdash)$ was an ancestor of e) we must recursively push the overwritten $repr$ value down $\mathcal{LT}$ to the root of the subtree containing those nodes.

More specifically, when $repr(r)$ is set to e, for some $\mathcal{LT}$ nodes r and e, let f be the previous value of $repr(r)$. Then find h, the minimum-depth node that is an ancestor of f but not of e, and recursively update $repr(h)$ to f. To find h, we first locate g, the lowest common ancestor of e and f. Figure 2 shows the five different ways in which e, f, g, and h can be located in relation to one another. In case a, h lies just under the path between e and the root, implying that we need to set $repr(h)$ to f. We find h via a reverse suffix link from g. Cases b (where $f = h$) and c ($g = e$) are merely special cases of the situation in a, and are handled in exactly the same way. In case d ($g = f$), the overwritten $repr$ value points to an ancestor of e, and the process can terminate immediately. Case e is the special case of d where the old and new $repr$ values are the same.

The following details the procedure. It is invoked as **repr-update**$(e,\vdash)$ in order to reestablish property 1, where e is the current update edge.

repr-update(e, r):

1. Let f be the old value of $repr(r)$, and set its new value to e.
2. If $f = \vdash$ (i.e. $repr(r)$ has not been previously set), then terminate.
3. Let g be the lowest common ancestor of e and f.
4. If $g = f$, terminate.
5. Let $h = rsuf(g, t_j)$, where $j = pos(f) + depth_{\mathcal{LT}}(f) - depth_{\mathcal{LT}}(g) - 1$.
6. Recursively invoke **repr-update**(f, h).

Correctness of **repr-update** in maintaining property 1, is established by the preceding discussion. We now turn to bounding the total number of recursive calls.

Lemma 4. *Given a sequence $V = e_1, \ldots, e_N$ of nodes to be updated in a tree $\mathcal{T}$ with M nodes, there exists a tree $\mathcal{T}'$ with at most $2N$ nodes, such that the depths of any two leaves in $\mathcal{T}'$ differ by at most one, and a sequence of $\mathcal{T}'$ nodes $V' = e_1', \ldots, e_N'$, such that invoking **repr-update**$(e', root(\mathcal{T}'))$ for each $e' \in V'$ results in at least as many recursive **repr-update** calls as invoking **repr-update**$(e, root(\mathcal{T}))$ for each $e \in V$.*

Proof (sketch). V can be replaced by a sequence V' containing only leaves, and $\mathcal{T}$ by a balanced binary tree $\mathcal{T}'$ with at most $2N$ nodes, without increasing the number of recursive **repr-update** calls. (Extended proof in appendix [16].)   □

3.5   Maintaining Implicit Suffix Nodes and Main Result

To conclude our treatment, we disucss handling case 2 in lemma 1: finding the most recent match of a pattern that corresponds to a point in $\mathcal{ST}$ with an implicitly represented suffix on the same edge. Once such an implicit suffix node is identified, the most recent pattern position is trivially obtained (the position of the corresponding suffix). Furthermore, identifying implicit suffix nodes has a known solution: Breslauer and Italiano [13] describe how Ukkonen's algorithm can be augmented with a stack of *band trees*, whose nodes maap top $\mathcal{ST}$ edges, by which implicit suffix nodes are maintained for amortized constant-time access, linear-time suffix tree online construction. (Further details in appendix [16].)

The band stack scheme has one additional use in our scheme: in each $\mathcal{ST}$ update operation, Breslauer and Italiano's algorithm pops a number of bands from the stack, and keeps the node that is the endpoint of the last popped edge. This node is the first explicit node on the active path, and, equivalently, the edge is the maximum-$depth_{\mathcal{LT}}$ internal edge whose endpoint corresponds an active suffix. This coincides with our definition of the *update edge* in section 3.2. Thus, we obtain the current update edge in constant time.

Theorem 1. *A suffix tree with support for locating, in an input stream, the most recent longest match of an arbitrary pattern P in $O(|P|)$ time, can be constructed online in time $O(N \log N)$ using $O(N)$ space, where N is the current number of processed characters.*

Proof (sketch). By lemma 4, the number of **repr-update** calls is $O(N \log N)$, each of which takes constant time, using a data structure for constant-time lowest common ancestor queries. [17], which bounds the time for maintenance under case 1 in lemma 1 to $O(N \log N)$. In case 2, we achieve $O(N)$ time by the data structure of Breslauer and Italiano. (Extended proof in appendix [16].)   □

We assert that an adversarial input exists that results in $\Omega(N \log N)$ recursive calls, and hence this worst-case bound is tight. (Further details in appendix [16].)

4   Sliding Window

A major advantage of online suffix tree construction is its applicability for a *sliding window*: indexing only the most recent part (usually a fixed length) of the input stream [18, 19]. We note that our augmentations of Ukkonen's algorithm can efficiently support most recent match queries in a sliding window of size W:

Corollary 1. *A suffix tree with support for locating, among the most recent W characters of an input stream, the most recent longest match of an arbitrary pattern P in $O(|P|)$ time, can be constructed online in time $O(N \log W)$ using $O(W)$ space, where N is the current number of processed characters.*

Proof (sketch). The suffix tree is augmented for indexing a sliding window using $O(W)$ space with maintained time bound [18, 19]. Deletion from the data structure for ancestor queries takes $O(1)$ time [20]. Node deletion from band trees takes $O(1)$ time using *pmerge* [21]. Hence, a $O(N \log W)$ term obtained analogously to lemma 4 dominates. (Extended proof in appendix [16].)   □

5   An Optimization for the Lempel-Ziv Case

While our data structure supports arbitrary most-recent-match queries, some related work has considered only the queries that arise in Lempel-Ziv factorization, i.e., querying $\mathcal{ST}_i$ only for the longest match of $t_i \cdots t_N$. The desire for finding the most recent occurrence of each factor is motivated by an improved compression rate in a subsequent entropy coding pass.

Ferragina, Nitto, and Venturini [11] gave an $O(N)$ time algorithm for this case, which is not online, and hence cannot be applied to a sliding window. Crochemore, Langiu, and Mignosi [12] presented an online $O(N)$ time suffix tree data structure that, under additional assumptions, circumvents the problem by replacing queries for most recent match with queries for matches with lowest possible entropy-code length. An interesting question is whether the time complexity of our method can be improved if we restrict queries to those necessary for Lempel-Ziv factorization. We now sketch an augmentation for this case.

As characters of one Lempel-Ziv factor are incorporated into $\mathcal{ST}$, we need not invoke **repr-update** for the update edge in each iteration. Instead, we push each update edge on a stack. After the whole factor has been incorporated, we pop edges and invoke **repr-update** for the reverse sequence, updating edge e only if it would have increased $pos(e)$. In other words, we ignore any updates superseded by later updates during the same sequence of edge pops. In experiments we noted drastic reduction in recursive calls, but whether worst case asymptotic time is reduced is an open question. (Extended discussion in appendix [16].)

6   Conclusion

We have presented an efficient online method of maintaining most recent match information in a suffix tree, to support optimal-time queries. The question whether the logarithmic factor in the time complexity of our method can be improved upon is, however, still open. Furthermore, precise characteristics of application to restricted inputs or applications (e.g. Lempel-Ziv factorization) is subject to future research, as is the practicality of the result for, e.g., data compression use.

References

1. Apostolico, A.: The myriad virtues of subword trees. In: Apostolico, A., Galil, Z. (eds.) Combinatorial Algorithms on Words. NATO ASI SERIES, vol. F12, pp. 85–96. Springer, Heidelberg (1985)
2. Gusfield, D.: Algorithms on Strings, Trees, and Sequences. Cambridge University Press (1997)
3. Larsson, N.J.: Structures of String Matching and Data Compression. Ph.D. thesis, Department of Computer Science, Lund University, Sweden (September 1999)
4. Weiner, P.: Linear pattern matching algorithms. In: Proc. 14th Ann. IEEE Symp. Switching and Automata Theory, pp. 1–11 (1973)
5. McCreight, E.M.: A space-economical suffix tree construction algorithm. J. ACM 23(2), 262–272 (1976)
6. Ukkonen, E.: On-line construction of suffix trees. Algorithmica 14(3), 249–260 (1995)
7. Farach, M.: Optimal suffix tree construction with large alphabets. In: Proc. 38th Ann. IEEE Symp. Foundations of Comput. Sci. pp. 137–143 (October 1997)
8. Manber, U., Myers, G.: Suffix arrays: A new method for on-line string searches. SIAM J. Comput. 22(5), 935–948 (1993)
9. Puglisi, S.J., Smyth, W.F., Turpin, A.H.: A taxonomy of suffix array construction algorithms. ACM Computing Surveys (CSUR) 39(2), 4 (2007)
10. Amir, A., Landau, G.M., Ukkonen, E.: Online timestamped text indexing. Information processing letters 82(5), 253–259 (2002)
11. Ferragina, P., Nitto, I., Venturini, R.: On the bit-complexity of Lempel-Ziv compression. In: Proc. Twentieth Ann. ACM – SIAM Symp. Discr. Alg. pp. 768–777 (2009)
12. Crochemore, M., Langiu, A., Mignosi, F.: The rightmost equal-cost position problem. In: Proc. IEEE Data Compression Conf. pp. 421–430 (March 2013)
13. Breslauer, D., Italiano, G.F.: On suffix extensions in suffix trees. Theoretical Computer Science 457, 27–34 (2012)
14. Ziv, J., Lempel, A.: A universal algorithm for sequential data compression. IEEE Trans. Inf. Theory IT 23(3), 337–343 (1977)
15. Larsson, N.J., Fuglsang, K., Karlsson, K.: Efficient representation for online suffix tree construction. Preprint, arXiv:1403.0457 [cs.DS], http://arxiv.org/abs/1403.0457
16. Larsson, N.J.: Most recent match queries in on-line suffix trees (with appendix), arXiv:1403.0800 [cs.DS], http://arxiv.org/abs/1403.0800
17. Cole, R., Hariharan, R.: Dynamic lca queries on trees. SIAM Journal on Computing 34(4), 894–923 (2005)
18. Fiala, E.R., Greene, D.H.: Data compression with finite windows. Commun. ACM 32(4), 490–505 (1989)
19. Larsson, N.J.: Extended application of suffix trees to data compression. In: Proc. ACM Data Compression Conf. pp. 190–199 (March-April 1996)
20. Dietz, P., Sleator, D.: Two algorithms for maintaining order in a list. In: Proc. 19th Ann. ACM Symp. Theory of Computing, pp. 365–372. ACM (1987)
21. Westbrook, J.: Fast incremental planarity testing. In: Kuich, W. (ed.) ICALP 1992. LNCS, vol. 623, pp. 342–353. Springer, Heidelberg (1992)

Encodings for Range Majority Queries[*]

Gonzalo Navarro[1] and Sharma V. Thankachan[2]

[1] Dept. of Computer Science, Univ. of Chile
gnavarro@dcc.uchile.cl
[2] Cheriton School of Computer Science, Univ. of Waterloo
thanks@uwaterloo.ca

Abstract. We face the problem of designing a data structure that can report the majority within any range of an array $A[1, n]$, without storing A. We show that $\Omega(n)$ bits are necessary for such a data structure, and design a structure using $O(n \log^* n)$ bits that answers majority queries in $O(\log n)$ time. We extend our results to τ-majorities.

1 Introduction

Given an array $A[1, n]$ of n numbers or arbitrary elements, an *array range query* problem asks to build a data structure over A, such that whenever an interval $[i, j]$ with $1 \le i \le j \le n$ comes as an input, we can efficiently answer queries on the elements in $A[i, j]$ [16]. Many array range queries arise naturally as subproblems of combinatorial problems, and are also of direct interest in data mining applications. Well-known examples are range maximum queries (RMQs, which seek the largest element in $A[i, j]$) [7] and top-k queries (which report the k largest elements in $A[i, j]$) [3].

An *encoding* for array range queries is a data structure that answers the queries without accessing A. This is useful when the values of A are not of interest themselves, and thus A may be deleted, potentially saving much space. It is also useful when array A does not fit in main memory, so it can be kept in secondary storage while a much smaller encoding can be maintained in main memory, speeding up queries. In this setting, instead of reporting an element in A, we only report a position in A where it occurs. Otherwise in many cases the encodings would be able to reconstruct A, and thus could not be small. As examples of encodings, RMQs can be solved in constant time using just $2n + o(n)$ bits [7], and top-k queries can be solved in $O(k)$ time using $O(n \log k)$ bits [10].

Frequency based array range queries, in particular variants of heavy-hitter-like problems, are very popular in data mining. Queries such as finding the most frequent element in a range (known as the range mode query) are known to be harder than problems like RMQs. For range mode queries, known data structures with constant query time require nearly quadratic space [14,13]. The best known linear space solution requires $O(\sqrt{n/\log n})$ query time [4], and conditional lower bounds in that paper show that a significant improvement is highly unlikely.

[*] Funded in part by Millennium Nucleus Information and Coordination in Networks ICM/FIC P10-024F, Chile.

A.S. Kulikov, S.O. Kuznetsov, and P. Pevzner (Eds.): CPM 2014, LNCS 8486, pp. 262–272, 2014.
© Springer International Publishing Switzerland 2014

Still, efficient solutions exist for some useful variations of the range mode problem. An example are approximate range mode queries, where we are required to output an element whose number of occurrences in $A[i,j]$ is at least $1/(1+\epsilon)$ times the number of occurrences of the mode in $A[i,j]$ [9,2].

In this paper we focus on a popular variation of range mode queries called *range majority queries*, which ask to report the range mode in $A[i,j]$ only if it occurs more than half of the times in $A[i,j]$. We also consider an extension where any element occurring a fraction larger than τ of the times in $A[i,j]$ can be reported. More formally, a majority is defined in the following way.

Definition 1. *A* majority *in an array $B[1,m]$, if it exists, is the element that occurs more than $m/2$ times in B. Given a real number $0 < \tau \leq 1/2$, a τ-*majority *in an array $B[1,m]$, if it exists, is any element that occurs more than τm times in B. Thus a majority is a τ-majority for $\tau = 1/2$.*

The problem we address in this paper can be stated as follows.

Definition 2. *Given an array $A[1,n]$, a* range majority query *gives an interval $[i,j]$ and must return whether $A[i,j]$ has a majority, and if it has, also return any position $i \leq k \leq j$ where the majority of $A[i,j]$ occurs. A* range τ-majority query *is defined analogously, returning any position of any τ-majority in $A[i,j]$.*

Range majority queries can be answered in constant time by maintaining a linear space (i.e., $O(n)$ words or $O(n \log n)$ bits) data structure [6]. Similarly, range τ-majority queries can be solved in time $O(1/\tau)$ and linear space if τ is fixed at construction time, or $O(n \log \log n)$ space (i.e., $O(n \log n \log \log n)$ bits) if τ is given at query time [1].

In this paper, we focus for the first time on *encodings for range majority (and τ-majority) queries*. In this scenario, a valid question is how much space is necessary for an encoding that correctly answers such queries (where we recall that A itself is not available at query time). We easily show in Section 2 that any such encoding needs $\Omega(n)$ bits, which reduces to $\Omega(\tau \log(1/\tau)n)$ bits for τ-majorities. Our main result is that it is possible to solve range majority queries within logarithmic time and almost linear-bit space. We achieve $O(n \log \log n)$ bits in Section 3, and our final result in Section 4:

Theorem 1. *There exists an encoding using $O(n \log^* n)$ bits answering range majority queries in time $O(\log n)$.*

In Section 5 we extend the results to τ-majorities, where the time and space obtained for range majority queries are divided by τ. Finally, in Section 6 we show how to build our structures in $O(n \log n)$ time.

Related Work. Range τ-majority queries were introduced by Karpinski and Nekrich [11], who presented an $O(n/\tau)$-words structure with $O((1/\tau)(\log \log n)^2)$ query time. Durocher et al. [6] improved their space and query time to $O(n \log(1/\tau))$ and $O(1/\tau)$, respectively. The currently best result is by Belazzougui et al. [1], where the space is $O(n)$ words and the query time is $O(1/\tau)$. All these results assume τ is fixed during the construction time. For the case where τ is also a part of the query input, a data structure of $O(n \log n)$ words was proposed by Chan et al. [5]. Very recently, Belazzougui et al. [1] brought down the space occupancy to $O(n \log \log n)$ words. The query time is $O(1/\tau)$ in all cases. All these solutions include a representation of A (sometimes aiming at compressing it [8,1]), thus they are not encodings. As far as we know, ours is the first encoding for this problem.

2   Lower Bounds

We first derive a couple of simple lower bounds on the minimum size our encodings may have. First, $\Omega(n)$ bits are needed to answer majority queries.

Lemma 1. *Any encoding for range majority queries requires $\lfloor n/2 \rfloor$ bits, even for an array with 2 distinct symbols.*

Proof. We can encode any bitmap $C[1, n]$ using an encoding for range majorities on an array $A[1, 2n]$, hence establishing the result. Set $A[2k+1] = 0$ for all valid k values, and $A[2k] = C[k]$. For example, let $C[1, 3] = \langle 0\ 1\ 1 \rangle$, then $A[1, 6] = \langle 0\ 0\ 0\ 1\ 0\ 1 \rangle$. Then, if $C[k] = 0$ then $A[2k - 1, 2k]$ has a majority, whereas if $C[k] = 1$ it does not.   □

Second, we show that τ-majority queries require $\Omega(\tau \log(1/\tau)n)$ bits.

Lemma 2. *Any encoding for range τ-majority queries requires $n \lg \lceil 1/\tau \rceil / (1 + \lceil 1/\tau \rceil) > (\tau \lg(1/\tau)/2)n$ bits.*

Proof. Let $c = \lceil 1/\tau \rceil$. We can encode any array $C[1, n]$ over alphabet $[1, c]$ using an encoding for range majorities on an array $A[1, (c+1)n]$. In each bucket $A[(c+1)k+1, (c+1)(k+1)]$ we write the values $\langle 1, 2, 3, \ldots, c \rangle$, except that the value $C[k + 1]$ is written twice. Therefore, $A[(c + 1)k + 1, (c + 1)(k + 1)]$ has only one τ-majority, precisely at offset $C[k + 1]$ within the bucket. Therefore, the encoding for τ-majorities in $A[1, (c+1)n]$ requires at least $n \lg c$ bits, as any possible array C can be reconstructed from it.   □

3   An $O(n \log \log n)$ Bits Encoding for Range Majorities

In this section we obtain an encoding using $O(n \log \log n)$ bits and solving majority queries in $O(\log n)$ time. In the next section we reduce the space.

Consider each distinct symbol x appearing in $A[1, n]$. Now consider the set of segments S_x within $[1, n]$ where x is a majority (this includes, in particular, all the segments $[k, k]$ where $A[k] = x$). Segments in S_x may overlap each other. For example, if $A[1, 7] = \langle 1\ 3\ 2\ 3\ 3\ 1\ 1 \rangle$, then

$$S_1 = \{[1, 1], [6, 6], [7, 7], [6, 7], [5, 7]\},$$
$$S_2 = \{[3, 3]\},$$
$$S_3 = \{[2, 2], [4, 4], [5, 5], [4, 5], [2, 4], [3, 5], [4, 6], [2, 5], [1, 5], [2, 6]\}.$$

Now let $A_x[1, n]$ be a bitmap such that $A_x[k] = 1$ iff position k belongs to some segment in S_x. In our example, $A_1 = \langle 1\ 0\ 0\ 0\ 1\ 1\ 1 \rangle$, $A_2 = \langle 0\ 0\ 1\ 0\ 0\ 0\ 0 \rangle$, and $A_3 = \langle 1\ 1\ 1\ 1\ 1\ 1\ 0 \rangle$.

We recall operation $rank(B, i)$ in bitmaps $B[1, m]$, which returns the number of 1s in $B[1, i]$. Operation $rank$ can be implemented using $o(m)$ bits on top of B and in constant time [12].

We define a second bitmap related to x, M_x, so that if $A_x[k] = 1$, then $M_x[rank(A_x, k)] = 1$ iff $A[k] = x$. In our example, $M_1 = \langle 1\ 0\ 1\ 1 \rangle$, $M_2 = \langle 1 \rangle$, and $M_3 = \langle 0\ 1\ 0\ 1\ 1\ 0 \rangle$. Then the following result is not difficult to prove.

Lemma 3. *An element x is a majority in $A[i, j]$ iff $A_x[k] = 1$ for all $i \leq k \leq j$, and 1 is a majority in $M_x[rank(A_x, i), rank(A_x, j)]$.*

Proof. If x is a majority in $A[i, j]$, then by definition $[i, j] \in S_x$, and therefore all the positions $k \in [i, j]$ are set to 1 in A_x. Therefore, the whole segment $A_x[i, j]$ is mapped bijectively to $M_x[rank(A_x, i), rank(A_x, j)]$, which is of the same length. Finally, the number of occurrences of x in $A[i, j]$ is the number of occurrences of 1 in $M_x[rank(A_x, i), rank(A_x, j)]$, which establishes the result.

Conversely, if $A_x[k] = 1$ for all $i \leq k \leq j$, then $A[i, j]$ is bijectively mapped to $M_x[rank(A_x, i), rank(A_x, j)]$, and the 1s in this range correspond one to one with occurrences of x in $A[i, j]$. Thus, if 1 is a majority in $M_x[rank(A_x, i), rank(A_x, j)]$, then x is a majority in $A[i, j]$. □

In our example, 1 is a majority in $A[5, 7]$, and it holds $A_1[5, 7] = \langle 1\ 1\ 1 \rangle$ and $M_1[rank(A_1, 5), rank(A_1, 7)] = M_1[2, 4] = \langle 0\ 1\ 1 \rangle$, where 1 is a majority. Thus, with A_x and M_x we can determine whether x is a majority in any range.

Lemma 4. *It is sufficient to have rank-enabled bitmaps A_x and M_x to determine, in constant time, whether x is a majority in any $A[i, j]$.*

Proof. We use Lemma 3. We compute $i' = rank(A_x, i)$ and $j' = rank(A_x, j)$. If $j' - i' \neq j - i$, then $A_x[k] = 0$ for some $i \leq k \leq j$ and thus x is not a majority in $A[i, j]$. Otherwise, we find out whether 1 is a majority in $M_x[i', j']$, by checking whether $rank(M_x, j') - rank(M_x, i' - 1) > (j' - i' + 1)/2$. □

To find out any position $i \leq k \leq j$ where $A[k] = x$, we need operation $select(B, j)$, which gives the position of the jth 1 in a bitmap $B[1, m]$. This operation can also be solved in constant time with $o(m)$ bits on top of B [12]. Then, for example, if x is a majority in $A[i, j]$, its first occurrence in $A[i, j]$ is $i - i' + select(M_x, rank(M_x, i' - 1) + 1)$. With a similar formula we can retrieve any of the positions of x in $A[i, j]$.

We cannot afford to store all the bitmaps A_x and M_x for all x, however. The next lemma is the first step to reduce the total space to slightly superlinear.

Lemma 5. *Any position $A[k] = x$ induces at most five 1s in A_x.*

Proof. Consider a process where we start with $A[k] = \perp$ for all k, and set the values $A[k] = x$ for increasing positions k (left to right). Setting $A[k] = x$ induces a segment $[k, k] \in S_x$, which may induce a new 1 in A_x. It might also induce some segments of the form $[i, k] \in S_x$, for $i < k$, depending on previous values. If x is a majority in $[i, k]$ with $A[k] = x$ and it was not a majority in $[i, k]$ with $A[k] = \perp$, then x occurs $\lfloor (k - i + 1)/2 \rfloor$ times in $A[i, k - 1]$. If $A[k - 1] \neq x$, then x also occurs $\lfloor (k - i + 1)/2 \rfloor > (k - i - 1)/2$ times in $A[i, k - 2]$, and thus it is a majority in $A[i, k - 2]$. Thus all the range $A_x[i, k - 2]$ was already 1s and setting $A[k] = x$ has only induced two new 1s, $A_x[k - 1] = A_x[k] = 1$. If, on the other hand, $A[k - 1] = x$, let l be the smallest value such that $[l, k - 1] \in S_x$. Setting $A[k] = x$ will add new 1s to A_x only if $i < l$. By the definition of l, it must hold that $A[l - 1] \neq x$ and $A[l - 2] \neq x$, that x occurs $\lfloor (k - l + 1)/2 \rfloor$ times in $A[l, k - 1]$, and that $\lfloor (k - l + 1)/2 \rfloor > (k - l)/2$. That is, $k - l$ must be odd and therefore $\lfloor (k - l + 1)/2 \rfloor = (k - l + 1)/2$. Now, this implies that x occurs $\lfloor (k - l + 1)/2 \rfloor + 1 = (k - l + 3)/2$ times in $A[l - 1, k]$, so x is a majority in $A[l - 1, k]$ and setting $A[k] = x$ could induce a new 1 in $A_x[l - 1] = 1$. On the other hand, x is not a majority in $A[l - 2, k]$. To be a majority in $A[i, k]$ with $i < l - 2$, x has to be a majority in $A[i, l - 3]$, and therefore only positions $A_x[l - 2] = A_x[l - 1] = 1$ could be new 1s induced by $A[k] = x$.

The consideration of the new induced segments of the form $[i, k + 1] \in S_x$ is simpler, because we know that at this point $A[k + 1] = \perp$. Therefore, if x is a majority in $A[i, k + 1]$, it occurs more than $(k - i + 2)/2$ times in $A[i, k + 1]$, and thus it occurs more than $(k - i)/2$ times in $A[i, k - 1]$, thus it is also a majority in $A[i, k - 1]$. Therefore the only new 1 that can be added is $A_x[k + 1] = 1$.

Finally, we consider the new induced segments of the form $[i, j] \in S_x$, with $i < k$ and $j > k + 1$. We know that at this point $A[k + 1, j] = \perp$. Therefore, if x is a majority in $A[i, j]$, it occurs more than $(j - i + 1)/2$ times in $A[i, j]$, and thus it occurred more than $(j - i - 1)/2$ times in $A[i, j]$ before setting $A[k] = x$. Thus x occurred more than $(j - i - 1)/2$ times in $A[i, j - 2]$ and thus it was already a majority in $A[i, j - 2]$. Therefore the only new 1s that can be added by setting $A[k] = x$ are $A_x[j - 1] = A_x[j] = 1$.

Overall, each new value $A[k] = x$ may induce up to five new 1s in A_x. $\square$

The lemma shows that all the A_x bitmaps add up to $O(n)$ 1s, and the lengths of the M_x bitmaps adds up to $O(n)$ as well (recall that M_x has one position per 1 in A_x). Therefore, we can store all the M_x bitmaps within $O(n)$ bits of space. We cannot, however, store all the A_x bitmaps, as they may add up to $O(n^2)$ 0s (note there can be $O(n)$ distinct symbols x).

Instead, we will *coalesce* different bitmaps A_x into one, as long as their areas of contiguous 1s do not overlap or touch (that is, there must be at least one 0 between any two areas of 1s of two coalesced bitmaps). The bitmaps M_x are merged accordingly, in the same order of the areas. In our example, we can coalesce A_1 and A_2 into $A_{12} = \langle 1\ 0\ 1\ 0\ 1\ 1\ 1 \rangle$, with the corresponding $M_{12} = \langle 1\ 1\ 0\ 1\ 1 \rangle$.

Then, at query time, we check for the area $[i, j]$ of each coalesced bitmap using Lemma 4. We cannot confuse the areas of different symbols x because we force that there is at least one 0 between any two areas. If we find one majority in one coalesced bitmap, we know that there is a majority and can spot all of its occurrences (or one, as the problem is defined), even if we cannot tell which particular symbol x is the majority.

This scheme will work well if we obtain just a few coalesced bitmaps overall. Next we show how to obtain only $O(\log n)$ coalesced bitmaps.

Lemma 6. *At most* $2 \lg n$ *distinct values of* x *can have* $A_x[k] = 1$ *for a given* k.

Proof. First, $A[k] = x$ is a majority in $A[k, k]$, thus $A_x[k] = 1$. Now consider any other element $x' \neq x$ such that $A_{x'}[k] = 1$. This means that x' is a majority in some $[i, j]$ that contains k. Since $A[k] \neq x'$, it must be that x' is a majority in $[i, k]$ or in $[k, j]$ (or in both). We say x' is a left-majority in the first case and a right-majority in the second. Let us call $y_1, y_2, \ldots$ the x' values that are left-majorities, and $i_1, i_2, \ldots$ the left endpoints of their segments (if they are majorities in several segments covering k, we choose one arbitrarily). Similarly, let $z_1, z_2, \ldots$ be the x' values that are right-majorities, and $j_1, j_2, \ldots$ the right endpoints of their segments. Assume the left-majorities are sorted by decreasing values of i_r and the right-majorities are sorted by increasing values of j_r. If a same value x' appears in both lists, we arbitrarily remove one of them. As an exception, we will start both lists with $y_0 = z_0 = x$, with $i_0 = j_0 = k$.

It is easy to see by induction that y_r must appear at least 2^r times in the interval $[i_r, k]$. This clearly holds for $y_0 = x$. Now, by the inductive hypothesis, values $y_0, y_1, \ldots, y_{r-1}$ appear at least $2^0, 2^1, \ldots, 2^{r-1}$ times within $[i_{r-1}, k]$ (which contains all the intervals), adding up to $2^r - 1$ occurrences. In order to be a left-majority, element y_r must appear at least 2^r times in $[i_r, k]$, to outweigh all the $2^r - 1$ occurrences of the previous symbols. The case of right-majorities is analogous. This shows that there cannot be more than $\lg n$ left-majorities and $\lg n$ right-majorities. □

In the following it will be useful to define C_x as the set of maximal contiguous areas of 1s in A_x. That is, C_x is obtained by merging all the segments of S_x that touch or overlap. In our example, $C_1 = \{[1, 1], [5, 7]\}$, $C_2 = \{[3, 3]\}$, and $C_3 = \{[1, 6]\}$. Note that segments of C_x do not overlap, unlike those of S_x. Since a segment of C_x covers a position k iff some segment of S_x covers position k (and iff $A_x[k] = 1$), it follows by Lemma 6 that any position is covered by at most $2 \lg n$ segments of C_x of distinct symbols x. Clearly, a pair of consecutive positions is covered by at most $4 \lg n$ such segments (this is a crude upper bound).

We obtain $O(\log n)$ coalesced bitmaps as follows. We take the union of all the sets C_x of all the symbols x and sort the segments by their starting points. Then we start filling coalesced bitmaps. We check if the current segment can be added to an existing bitmap without producing overlaps (and leaving a 0 in between). If we can, we choose any appropriate bitmap, otherwise we start a new bitmap. If at some point we need more than $4 \lg n$ bitmaps, it is because all the last segments of the current $4 \lg n$ bitmaps overlap the starting point of the current segment or the previous position, a contradiction.

In our example, we take $C_1 \cup C_2 \cup C_3 = \{[1,1], [1,6], [3,3], [5,7]\}$, and the process produces precisely the coalesced bitmaps A_{12}, corresponding to the set $\{[1,1], [3,3], [5,7]\}$ and A_3, corresponding to $\{[1,6]\}$. Note that in general the coalesced bitmaps may not correspond to the union of complete original bitmaps A_x, but areas of a bitmap A_x may end up in different coalesced bitmaps.

Therefore, the coalescing process produces $O(\log n)$ bitmaps. Consequently, we obtain $O(\log n)$ query time by simply checking the coalesced bitmaps one by one using Lemma 4.

Finally, representing the $O(\log n)$ coalesced bitvectors, which contain $O(n)$ 1s and have total length $O(n \log n)$, requires $O(n \log \log n)$ bits if we use a compressed bitmap representation [15] that still offers constant-time *rank* and *select* queries. This concludes the first part of our result.

4   An $O(n \log^* n)$ Bits Encoding for Range Majorities

We introduce a different representation of the coalesced bitmaps that allows us storing them in $O(n \log^* n)$ bits, while retaining all the mechanism and query time complexity. We will distinguish segments of C_x by their lengths, separating lengths by ranges between 2^ℓ and $2^{\ell+1} - 1$, for any ℓ. In the process of creating the coalesced bitmaps described in the previous section, we will have separate coalesced bitmaps for inserting segments within each range of lengths; these will be called bitmaps of level ℓ. There may be several bitmaps of the same level. It is important that, even with this restriction, our coalescing process will still generate $O(\log n)$ bitmaps, because only $O(1)$ coalesced bitmaps of each level ℓ will be generated.

Lemma 7. *There can be at most 8 segments of any C_x, of length between 2^ℓ and $2^{\ell+1} - 1$, covering a given position k, for any ℓ.*

Proof. Any such segment must be contained in the area $A[k - 2^{\ell+1}, k + 2^{\ell+1}]$, and if x is a majority in it, it must appear more than $2^{\ell-1}$ times. There can be only 8 different values of x appearing $2^{\ell-1}$ times in an area of length $2^{\ell+2}$.   □

To represent any coalesced bitmap $B[1, n]$, we cut the universe $[1, n]$ into chunks of length $b = \lg n$. We store a string K of length $n / \lg n$, where for each position a 0 indicates that the chunk is all 0s, a 1 that the chunk is all 1s, and a 2 indicates that there are 0s and 1s in the chunk. We store explicitly only the chunks with value 2, concatenated one after the other. Let B_1 be a bitmap such that $B_1[k] = 1$ iff $K[k] = 1$, B_2 such that $B_2[k] = 1$ iff $K[k] = 2$, and C the bitmap where the explicit chunks are concatenated. Then it holds

$$rank(B, i) = b \cdot rank(B_1, \lfloor (i-1)/b \rfloor) +$$
$$rank(C, b \cdot rank(B_2, \lfloor i/b \rfloor) + [\textbf{if } B_2[1 + \lfloor i/b \rfloor] = 1 \textbf{ then } i \bmod b \textbf{ else } 0]),$$

which takes constant time. Operation $select(B, j)$ can be done by binary search on $rank$, which takes $O(\log n)$ time but has to be done once per query, hence retaining the $O(\log n)$ query time. Note that K is not explicitly stored, but it is represented with B_1 and B_2.

In our example, we would have three coalesced bitmaps: $B^0 = \langle 1\ 0\ 1\ 0\ 0\ 0\ 0 \rangle$, of level $\ell = 0$, for the segments $[1, 1]$ and $[3, 3]$; $B^1 = \langle 0\ 0\ 0\ 0\ 1\ 1\ 1 \rangle$, of level $\ell = 1$, for the segment $[5, 7]$; and $B^2 = \langle 1\ 1\ 1\ 1\ 1\ 1\ 0 \rangle$, of level $\ell = 2$, for the segment $[1, 6]$. Assume $b = 2$. Then, for B^0 we would have $K^0 = \langle 2\ 2\ 0\ 0 \rangle$, $B_1^0 = \langle 0\ 0\ 0\ 0 \rangle$, $B_2^0 = \langle 1\ 1\ 0\ 0 \rangle$, and $C^0 = \langle 1\ 0\ 1\ 0 \rangle$. For B^1 we would have $K^1 = \langle 0\ 0\ 1\ 1 \rangle$, $B_1^1 = \langle 0\ 0\ 1\ 1 \rangle$, $B_2^1 = \langle 0\ 0\ 0\ 0 \rangle$, and $C^1 = \langle \rangle$. Finally, for B^2 we would have $K^2 = \langle 1\ 1\ 1\ 0 \rangle$, $B_1^2 = \langle 1\ 1\ 1\ 0 \rangle$, $B_2^2 = \langle 0\ 0\ 0\ 0 \rangle$, and $C^2 = \langle \rangle$.

Consider a fixed bitmap B of some level ℓ, which has been formed by adding n' segments. We store at most $2n' \lg n$ bits in the explicit chunks of C, as there are only n' transitions from 0 to 1 and n' from 1 to 0 in B. For any level $\ell \geq \lg \lg n$, there are at least $n' \lg n$ 1s, because the segments have length at least $2^\ell \geq \lg n$. Therefore, in those levels, the number of bits stored in C bitmaps is of the same order of the total number of 1s in the corresponding bitmaps B. Thus we store only $O(n)$ bits over all the chunks of all coalesced bitmaps of levels $\ell \geq \lg \lg n$. As for the sequences B_1 and B_2 describing the chunks, they are of length $n/\lg n$, so they add up to $O(n)$ bits over all the possible $O(\log n)$ levels.

Now, for the levels up to $\lg \lg n$, we use chunk size $b = \lg \lg n$, storing a sequence of length $n/\lg \lg n$. The explicitly stored chunks C add up to $n' \lg \lg n$ bits, and for any level $\ell \geq \lg \lg \lg n$, the total number of 1s is over $n' \lg \lg n$, thus the total number of stored bits is of the same order of the 1s. The sequences B_1 and B_2 describing the chunks add up to $O(n)$, because there are only $O(\log \log n)$ levels where this is applied.

We continue with the remaining (lowest) $\lg \lg \lg n$ levels, and so on. Then the total number of stored bits is $O(n \log^* n)$, dominated by the sequences B_1 and B_2. This proves Theorem 1.

5   Extension to τ-Majorities

We first consider the case where τ is fixed at the time the data structure is built, and then move on to the case of τ given at query time. For lack of space we only sketch the results, which follow relatively easily from our results on majorities. First, Lemmas 3 and 4 hold verbatim if we define S_x as the segments where x is a τ-majority. Lemma 5 can be extended to this case, so that any position $A[k] = x$ induces $O(1/\tau)$ 1s in A_x. As a consequence, there are $O(n/\tau)$ 1s in all the A_x bitmaps. Lemma 6 can also be extended, so that $O(\log_{1/(1-\tau)} n) = O((1/\tau) \log n)$ distinct values of x can have $A_x[k] = 1$ for a given k. Therefore, the coalescing process produces $O((1/\tau) \log n)$ bitmaps, and this is the query time. Lemma 7 can be extended similarly, so that there can be only $O(1/\tau)$ coalesced bitmaps of any given level, and there are $\lg n$ levels. Thus the mechanism of Section 4 can be applied verbatim, so that the total number of bits used is $O((n/\tau) \log^* n)$. Therefore we obtain the following result.

Theorem 2. *For a fixed threshold* $0 < \tau \leq 1/2$, *there exists an encoding using* $O((n/\tau) \log^* n)$ *bits answering range* τ-*majority queries in time* $O((1/\tau) \log n)$.

In order to allow τ to be specified at query time, we build the encoding of Theorem 2 for values $\tau = 1/2, 1/4, 1/8, \ldots, 1/2^{\lceil \lg 1/\mu \rceil}$, where μ is the minimum τ value to support. Then, given a τ-majority query, we run the query on the structure built for $\tau' = 1/2^{\lceil \lg 1/\tau \rceil}$. Note that $\tau/2 < \tau' \leq \tau$, therefore the query time is $O((1/\tau') \log n) = O((1/\tau) \log n)$. For each possible answer to the τ'-majority query, we use *rank* on the coalesced M_x bitmaps to find out whether the answer is actually a τ-majority. This verification does not change the worst-case time complexity. As for the space, the factor multiplying $O(n \log^* n)$ is $2 + 4 + 8 + \ldots + 2^{\lceil \lg 1/\mu \rceil} = O(1/\mu)$. Therefore we obtain the following result.

Theorem 3. *For a fixed threshold* $0 < \mu \leq 1/2$, *there exists an encoding using* $O((n/\mu) \log^* n)$ *bits answering range* τ-*majority queries, for any* $\mu \leq \tau \leq 1/2$ *given at query time, in* $O((1/\tau) \log n)$ *time.*

6 Construction

The most complex part of the construction of our encoding is to build the sets C_x; once these are built, the construction of the structure of Section 4 can be easily carried out in $O(n \log^* n)$ additional time.

We separate the set of increasing positions P_x where x appears in A, for each x. The P_x sets are easily built in $O(n \log n)$ time. Now we build C_x from each P_x using a divide and conquer approach, in $O(|P_x| \log |P_x|)$ time, for a total construction time of $O(n \log n)$.

We pick the middle element $k \in P_x$ and compute in linear time the segment $[l, r] \in C_x$ that contains k. To compute l, we find the leftmost element $p_l \in P_x$ such that x is a majority in $[p_l, k_r]$, for some $k_r \in P_x$ with $k_r \geq k$.

To find p_l, we note that it must hold $(w(p_l, k-1) + w(k, k_r))/(k_r - p_l + 1) > 1/2$, where $w(i, j)$ is the number of occurrences of x in $A[i, j]$. The condition is equivalent to $2w(p_l, k-1) + p_l - 1 > k_r - 2w(k, k_r)$. Thus we compute in linear time the minimum value v of $k_r - 2w(k, k_r)$ over all those $k_r \in P_x$ to the right of k, and then traverse all those $p_l \in P_x$ to the left of k, left to right, to find the first one that satisfies $2w(p_l, k-1) + p_l + 1 > v$, also in linear time. Once we find the proper p_l and its corresponding k_r, the starting position of the segment is slightly adjusted to the left of p_l, to be the smallest value that satisfies $w(p_l, k_r)/(k_r - l + 1) > 1/2$, that is, l satisfies $l > -2w(p_l, k_r) + k_r + 1$, that is, $l = k_r - 2w(p_l, k_r) + 2$.

Once p_r and then r are computed analogously, we insert $[l, r]$ into C_x and continue recursively with the elements of P_x to the left of p_l and to the right of p_r. Upon return, it might be necessary to join $[l, r]$ with the rightmost segment of the left part and/or with the leftmost segment of the right part, in constant time. The total construction time is $T(n) = O(n) + 2T(n/2) = O(n \log n)$. The construction for τ-majorities is similar, although for τ given at query time we must build $O(\log(1/\mu))$ similar structures.

7   Final Remarks

We have obtained the first result about encodings for answering range majority queries, that is, data structures that use less space than the data and do not need to access it. We have proved that $\Omega(n)$ bits are necessary for any such encoding, and have presented a particular encoding that uses $O(n \log^* n)$ bits and $O(\log n)$ time. It can be built in $O(n \log n)$ time. An open question is whether it is possible to reach $O(n)$ bits of space and/or constant query time.

We have also extended our result to range τ-majorities, where we have proved a lower bound of $O(\tau \log(1/\tau)n)$ bits and presented an encoding using $O((n/\tau) \log^* n)$ bits and $O((1/\tau) \log n)$ query time. An intriguing aspect of this result is that our lower bound suggests that τ-majorities require less space for smaller τ, whereas our upper bound uses more space (and time) for smaller τ, in line with previous work on data structures that are not encodings. It is an interesting problem to determine which is the case.

References

1. Belazzougui, D., Gagie, T., Navarro, G.: Better space bounds for parameterized range majority and minority. In: Dehne, F., Solis-Oba, R., Sack, J.-R. (eds.) WADS 2013. LNCS, vol. 8037, pp. 121–132. Springer, Heidelberg (2013)
2. Bose, P., An, H.-C., Morin, P., Tang, Y.: Approximate range mode and range median queries. In: Diekert, V., Durand, B. (eds.) STACS 2005. LNCS, vol. 3404, pp. 377–388. Springer, Heidelberg (2005)
3. Brodal, G.S., Fagerberg, R., Greve, M., López-Ortiz, A.: Online sorted range reporting. In: Dong, Y., Du, D.-Z., Ibarra, O. (eds.) ISAAC 2009. LNCS, vol. 5878, pp. 173–182. Springer, Heidelberg (2009)
4. Chan, T., Durocher, S., Larsen, K., Morrison, J., Wilkinson, B.: Linear-space data structures for range mode query in arrays. In: STACS, pp. 290–301 (2012)
5. Chan, T.M., Durocher, S., Skala, M., Wilkinson, B.T.: Linear-space data structures for range minority query in arrays. In: Fomin, F.V., Kaski, P. (eds.) SWAT 2012. LNCS, vol. 7357, pp. 295–306. Springer, Heidelberg (2012)
6. Durocher, S., He, M., Munro, I., Nicholson, P., Skala, M.: Range majority in constant time and linear space. Inf. Comput. 222, 169–179 (2013)
7. Fischer, J., Heun, V.: Space-efficient preprocessing schemes for range minimum queries on static arrays. SIAM J. Comput. 40(2), 465–492 (2011)
8. Gagie, T., He, M., Munro, J.I., Nicholson, P.: Finding frequent elements in compressed 2d arrays and strings. In: Grossi, R., Sebastiani, F., Silvestri, F. (eds.) SPIRE 2011. LNCS, vol. 7024, pp. 295–300. Springer, Heidelberg (2011)
9. Greve, M., Jørgensen, A.G., Larsen, K.D., Truelsen, J.: Cell probe lower bounds and approximations for range mode. In: Abramsky, S., Gavoille, C., Kirchner, C., Meyer auf der Heide, F., Spirakis, P.G. (eds.) ICALP 2010. Part I, LNCS, vol. 6198, pp. 605–616. Springer, Heidelberg (2010)
10. Grossi, R., Iacono, J., Navarro, G., Raman, R., Rao Satti, S.: Encodings for range selection and top-k queries. In: Bodlaender, H.L., Italiano, G.F. (eds.) ESA 2013. LNCS, vol. 8125, pp. 553–564. Springer, Heidelberg (2013)

11. Karpinski, M., Nekrich, Y.: Searching for frequent colors in rectangles. In: CCCG (2008)
12. Ian Munro, J.: Tables. In: Chandru, V., Vinay, V. (eds.) FSTTCS 1996. LNCS, vol. 1180, pp. 37–42. Springer, Heidelberg (1996)
13. Petersen, H.: Improved bounds for range mode and range median queries. In: Geffert, V., Karhumäki, J., Bertoni, A., Preneel, B., Návrat, P., Bieliková, M. (eds.) SOFSEM 2008. LNCS, vol. 4910, pp. 418–423. Springer, Heidelberg (2008)
14. Petersen, H., Grabowski, S.: Range mode and range median queries in constant time and sub-quadratic space. Inf. Process. Lett. 109(4), 225–228 (2009)
15. Raman, R., Raman, V., Srinivasa Rao, S.: Succinct indexable dictionaries with applications to encoding k-ary trees, prefix sums and multisets. ACM Trans. Alg. 3(4), 43 (2007)
16. Skala, M.: Array range queries. In: Brodnik, A., López-Ortiz, A., Raman, V., Viola, A. (eds.) Munro Festschrift. LNCS, vol. 8066, pp. 333–350. Springer, Heidelberg (2013)

On the DCJ Median Problem

Mingfu Shao and Bernard M.E. Moret

Laboratory for Computational Biology and Bioinformatics
EPFL, Switzerland
{mingfu.shao,bernard.moret}@epfl.ch

Abstract. As many whole genomes are sequenced, comparative genomics is moving from pairwise comparisons to multiway comparisons framed within a phylogenetic tree. A central problem in this process is the inference of data for internal nodes of the tree from data given at the leaves. When phrased as an optimization problem, this problem reduces to computing a median of three genomes under the operations (evolutionary changes) of interest. We focus on the universal rearrangement operation known as double-cut-and join (DCJ) and present three contributions to the DCJ median problem. First, we describe a new strategy to find so-called adequate subgraphs in the multiple breakpoint graph, using a seed genome. We show how to compute adequate subgraphs w.r.t. this seed genome using a network flow formulation. Second, we prove that the upper bound of the median distance computed from the triangle inequality is tight. Finally, we study the question of whether the median distance can reach its lower and upper bounds. We derive a necessary and sufficient condition for the median distance to reach its lower bound and a necessary condition for it to reach its upper bound and design algorithms to test for these conditions.

Keywords: genomic rearrangement, network flow, dynamic programming.

1 Introduction

The combinatorics and algorithmics of genomic rearrangements have seen much work since the problem was formulated in the 1990s [1]. Genomic rearrangements include inversions, transpositions, circularizations, and linearizations, all of which act on a single chromosome, and translocations, fusions, and fissions, which act on two chromosomes. These operations can all be described in terms of the single double-cut-and-join (DCJ) operation [2, 3], which has formed the basis for most algorithmic research on rearrangements since its publication [4–9]. A DCJ operation makes two cuts in the genome, either in the same chromosome or in two different chromosomes, producing four cut ends that it then rejoins, giving rise to three possible outcomes.

A basic problem in genome rearrangements is to compute the edit distance between two genomes, i.e., the minimum number of operations that are needed to transform one genome into another. Under the inversion model, Hannenhalli and Pevzner gave the first polynomial-time algorithm to compute the edit distance

A.S. Kulikov, S.O. Kuznetsov, and P. Pevzner (Eds.): CPM 2014, LNCS 8486, pp. 273–282, 2014.
© Springer International Publishing Switzerland 2014

between two unichromosomal genomes [10]; a linear-time algorithm for the same problem was later designed [11]. Under the DCJ model, the edit distance can also be computed in linear time, this time for two multichromosomal genomes [2]. The median problem is a generalization of the edit distance: given three genomes, we want to construct a fourth genome, the *median*, that minimizes the sum of the edit distances between itself and each of the three given genomes. The median problem is NP-hard for almost all formulations [12, 13]. Under the inversion model, several exact algorithms [14, 15] and heuristics [16, 17] have been proposed. Under the DCJ model, Zhang *et al.* presented an exact solver using a branch-and-bound framework [18]. In [19], Xu *et al.* proposed a decomposition scheme that preserves optimality by using *adequate subgraphs*, particular substructures of the multiple breakpoint graph [20]. Later, Xu produced the ASMedian software to implement a median search based on adequate subgraphs using an optimistic branch-and-bound search [21]. ASMedian uses a precomputed set of small adequate subgraphs; at each step, it tests whether the current multiple breakpoint graph contains a subgraph from that set.

We propose a new strategy to find adequate subgraphs in the multiple breakpoint graph, based on a *seed genome*. We give a polynomial-time algorithm to decide whether there exists an adequate subgraph w.r.t. this seed genome (and to identify such a subgraph if one exists) using a network flow formulation.

The DCJ *median distance* (the sum of the distances of the given genomes to their median) can be lower- and upper-bounded using the sum of the three pairwise DCJ edit distances among the three given genomes. The lower bound was recently proved to be tight [22]. We show that the upper bound is also tight. Moreover, we give testable characterizations of the equality problem: for a given instance, is the median distance equal to its upper or lower bound? We give a necessary and sufficient condition for equality with the lower bound—the necessary condition can be tested using a dynamic programming formulation—and we give a necessary condition for equality with the upper bound, a condition that can also be tested effectively.

2   Preliminaries

We assume that each genome consists of the same set of n distinct genes and that those genes form one or more circular chromosomes in each genome. The head and tail of a gene g, represented by g_h and g_t, are called *extremities*. Two consecutive genes form one *adjacency*, represented as the set of its two extremities. Since all genes are distinct, each genome is uniquely determined by its n adjacencies. We build a graph (V, E), where V has $2 \cdot n$ vertices representing the extremities and E has n edges representing the adjacencies. Note that a genome thus corresponds to a perfect matching on V (see Fig. 1).

Given genomes G_1 and G_2 represented by perfect matchings M_1 and M_2 on V, the corresponding *breakpoint graph* is defined as the multigraph (V, M_1, M_2). In the multigraph, two vertices may be connected by two edges, one from M_1 and the other from M_2. These edges are distinguished by their provenance.

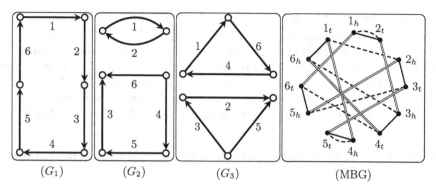

Fig. 1. Three genomes and the corresponding complete MBG. Genes, adjacencies, and extremities are represented by arrows, circles, and solid circles, respectively. Adjacencies in G_1, G_2, and G_3 are represented by solid, dashed and double lines respectively.

Each vertex in this breakpoint graph has degree 2, so that the graph consists of vertex-disjoint cycles; let $c(M_1, M_2)$ denote the number of these cycles. The *DCJ distance* between G_1 and G_2, denoted as $d(M_1, M_2)$, can be expressed as $d(M_1, M_2) = n - c(M_1, M_2)$ [2]. We can extend this concept to three given genomes, M_1, M_2 and M_3, yielding a *multiple breakpoint graph* (MBG for short, see an example in Fig. 1), denoted by (V, M_1, M_2, M_3). Given a MBG (V, M_1, M_2, M_3), the DCJ median problem asks for a perfect matching M_0 on V (another genome) that minimizes $\sum_{k=1}^{3} d(M_0, M_k)$.

We generalize the definition of MBG by allowing nonperfect matchings, distinguishing MBGs with three perfect matchings as *complete* MBGs. If M_1' and M_2' are not perfect matchings on V', then the breakpoint graph (V', M_1', M_2') consists of isolated vertices, simple paths, and vertex-disjoint cycles; we continue to use $c(M_1', M_2')$ to denote the number of cycles.

Let B' be a MBG and B a complete MBG; B' is a *subgraph* of B if we have $V' \subseteq V$ and $M_k' \subseteq M_k$, $k = 1, 2, 3$. A matching M_0' on V' is a *median* of B' if it maximizes $\sum_{k=1}^{3} c(M_0', M_k')$ over all possible matchings on V'. B' is *adequate* if, for any median M_0' of B', we have $\sum_{k=1}^{3} c(M_0', M_k') \geq 3 \cdot |V'|/4$.

Theorem 1. *[19] If B' is an adequate subgraph of B, then for any median M_0' of B', there exists one median M_0 of B such that $M_0' \subset M_0$.*

This result leads to a decomposition scheme to compute the median by iteratively finding adequate subgraphs and resolving each separately; ASMedian uses a precomputed set containing all adequate subgraphs with size less than 10.

3   Adequate Subgraphs w.r.t. a given Matching

We describe a new algorithm to compute adequate subgraphs in a complete MBG, based on the use of a "seed" genome—a perfect matching on V. In practice, this seed genome can be one of the three given matchings. Let M be a perfect matching on V. An MBG $B' = (V', M'_1, M'_2, M'_3)$ is adequate w.r.t. M if there exists a matching M' on V' satisfying $\sum_{k=1}^{3} c(M', M'_k) \geq 3 \cdot |V'|/4$ and $M' \subseteq M$. If B' is adequate w.r.t. M, then clearly it is adequate. Given a complete MBG $B = (V, M_1, M_2, M_3)$ and a perfect matching M on V, let $\mathcal{C}_k$ be the set of cycles in the breakpoint graph (V, M_k, M), $k = 1, 2, 3$, and write $\mathcal{C} = \cup_{k=1}^{3} \mathcal{C}_k$. For a cycle $C \in \mathcal{C}$, let $V(C)$ be the set of vertices covered by C and $E(C)$ be the set of edges covered by C. For a subset $\mathcal{S} \subseteq \mathcal{C}$, set $V(\mathcal{S}) = \cup_{C \in \mathcal{S}} V(C)$ and $E(\mathcal{S}) = \cup_{C \in \mathcal{S}} E(C)$.

Lemma 1. *There exist adequate subgraphs of B w.r.t. M iff there exists a subset $\mathcal{S} \subseteq \mathcal{C}$ obeying $|\mathcal{S}| \geq 3 \cdot |V(\mathcal{S})|/4$.*

Proof. If such $\mathcal{S}$ exists, we can define the subgraph as $(V(\mathcal{S}), M_1 \cap E(\mathcal{S}), M_2 \cap E(\mathcal{S}), M_3 \cap E(\mathcal{S}))$. Let $M' = M \cap E(\mathcal{S})$; then the sum $\sum_{k=1}^{3} c(M', M_k \cap E(\mathcal{S}))$ is exactly equal to $|\mathcal{S}|$, which is larger than or equal to $3 \cdot |V(\mathcal{S})|/4$. Thus, our subgraph is adequate w.r.t. M. Conversely, suppose that there exists one adequate subgraph (V', M'_1, M'_2, M'_3) of B w.r.t. M and let $M' \subseteq M$ be a matching on V' satisfying $\sum_{k=1}^{3} c(M', M'_k) \geq 3 \cdot |V'|/4$. Let $\mathcal{S}$ be the set of all cycles in the three breakpoint graphs (V', M', M'_k), $k = 1, 2, 3$. We can write $|\mathcal{S}| = \sum_{k=1}^{3} c(M', M'_k)$. Since M'_1, M'_2 and M'_3 are all matchings on V', we have that $|V'| \geq |V(\mathcal{S})|$. Combining these formulas yields $|\mathcal{S}| \geq 3 \cdot |V(\mathcal{S})|/4$.   □

We use a network flow formulation to compute such $\mathcal{S}$. Fig. 2 illustrates the construction. We add to N one vertex for each extremity in V, one vertex for each cycle in $\mathcal{C}$, plus a source s and sink t. We add to N directed edges of capacity $3/4$ from s to each extremity in V and directed edges of capacity 1 from each cycle in $\mathcal{C}$ to t. For each pair of $v \in V$ and $C \in \mathcal{C}$ with $v \in V(C)$, we add one directed edge of infinite (very large) capacity from v to C. Let f be a maximum s-t flow of N, N_f the residual network w.r.t. f, S the set of vertices reachable from s in N_f, and T the set of all other vertices.

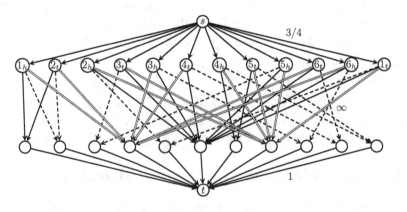

Fig. 2. The network for the complete MBG of Fig. 1 with the seed $M = M_2$

Lemma 2. *A subset $S \subseteq C$ with $|S| \geq 3 \cdot |V(S)|/4$ exists iff we have $\{t\} \subsetneq T$.*

Proof. By construction of S and T, we must have $s \in S$ and $t \in T$; moreover, (S, T) is a minimum s-t cut of N. For any other minimum s-t cut (S', T'), we have $|S| \leq |S'|$. The total capacity of cut (S, T) is at most $|C|$, since it is a minimum s-t cut and there is a trivial s-t cut (containing just the sink t on one side) whose total capacity is $|C|$.

Assume we have $\{t\} \subsetneq T$. Let $S \subseteq C$ be the set of cycles in T and let $V' \subseteq V$ be the set of extremities that are in T. The edges of infinite capacity cannot belong to the (S, T) cut, so that the total capacity of the (S, T) cut is exactly $3 \cdot |V'|/4 + |C| - |S|$. Since the total capacity of any minimum s-t cut is at most $|C|$, we must have $|S| \geq 3 \cdot |V'|/4$. Because the edges of infinite capacity are not in the (S, T) cut, we also have $V(S) \subseteq V'$. Thus, we can conclude $|S| \geq 3 \cdot |V(S)|/4$.

Now assume there exists a subset S satisfying $|S| \geq 3 \cdot |V(S)|/4$. We prove $\{t\} \subsetneq T$ by contradiction. Assume $T = \{t\}$; then the total capacity of the cut (S, T) is $|C|$. Now we construct another s-t cut (S', T'), where T' consists of the extremities in $V(S)$ and the cycles in S and sink t. The capacity of this cut (S', T') is $3 \cdot |V(S)|/4 + |C| - |S|$, less than or equal to $|C|$ since we have $|S| \geq 3 \cdot |V(S)|/4$. Thus (S', T') is also a minimum s-t cut, but clearly we have $|S'| < |S|$, the desired contradiction. $\qquad\square$

Thus, if there exist adequate subgraphs w.r.t. a perfect matching, one such subgraph can be found from the residual network.

4   The Upper Bound is Tight

Let M_0 be a median of a complete MBG $B = (V, M_1, M_2, M_3)$. We denote by $d_m = \sum_{k=1}^{3} d(M_0, M_k)$ the median distance of B and by $d_t = d(M_1, M_2) + d(M_1, M_3) + d(M_2, M_3)$ the *triangle distance* of B. According to the triangle inequality (the DCJ distance is a metric), we have $d(M_0, M_i) + d(M_0, M_j) \geq d(M_i, M_j)$, $1 \leq i < j \leq 3$, which yields a lower bound for the median distance, $d_m \geq d_t/2$. However, by using any of M_1, M_2, and M_3 as a possible median, we get $d_m \leq d(M_1, M_2) + d(M_1, M_3)$, $d_m \leq d(M_2, M_1) + d(M_2, M_3)$, and $d_m \leq d(M_3, M_1) + d(M_3, M_2)$, which yields an upper bound for the median distance, $d_m \leq 2 \cdot d_t/3$. Fig. 3 shows a subgraph where the upper bound is reached. Notice that this subgraph is also adequate. Thus, the combination of any number of copies of this subgraph yields a graph that also reaches the upper bound.

5   Deciding Equality to the Bounds

We now study whether the median distance of a complete MBG reaches its lower or upper bound. Let $u, v \in V$ be two distinct vertices. A DCJ operation *induced* by (u, v) on M_1 removes (u, u_1) and (v, v_1) from M_1 and adds (u, v) and (u_1, v_1) to M_1, where u_1 and v_1 are the neighbors of u and v in M_1. (If u is matched to v in M_1, then the DCJ operation induced by (u, v) on M_1 is an identity.)

Fig. 3. Tightness of the upper bound. M_1, M_2, M_3 are represented by solid, dashed and double edges. We have $d(M_1, M_2) = d(M_1, M_3) = d(M_2, M_3) = 4$ and thus $d_t = 12$. Any M_k is a median with $d_m = \sum_{k=1}^{3} d(M_1, M_k) = 8$. Thus we have $3 \cdot d_m = 2 \cdot d_t$.

Property 1. Let M and M_1 be two perfect matchings on V and $u, v \in V$ two distinct vertices with $(u, v) \in M$ and $(u, v) \notin M_1$. Then we can write $d(M, M_1') = d(M, M_1) - 1$, where M_1' is the perfect matching obtained from M_1 after performing the DCJ operation induced by (u, v).

Definition 1. (u, v) *is strong w.r.t.* M_1 *and* M_2 *if* u *and* v *are in the same cycle of* (V, M_1, M_2) *and the distance between them is odd—see Fig. 4. Otherwise,* (u, v) *is weak w.r.t.* M_1 *and* M_2.

Property 2. Let M_1' and M_2' be the two perfect matchings after performing the two DCJ operations induced by (u, v) on M_1 and M_2 respectively. Then (u, v) is strong w.r.t. M_1 and M_2 iff we have

$$d(M_1', M_2') = \begin{cases} d(M_1, M_2) & \text{if } (u, v) \in M_1 \cap M_2; \\ d(M_1, M_2) - 1 & \text{if } (u, v) \in (M_1 - M_2) \cup (M_2 - M_1); \\ d(M_1, M_2) - 2 & \text{if } (u, v) \notin M_1 \cup M_2. \end{cases}$$

Two strong edges (u, v) and (u', v') w.r.t. M_1 and M_2 are *independent* w.r.t. M_1 and M_2 if (i) they are in different cycles of (V, M_1, M_2) or (ii) they do not "intersect" in the same cycle—where an intersection would mean that u' and v' are on the different paths from u to v.

Property 3. Let (u, v) be a strong edge w.r.t. M_1 and M_2, and M_1' and M_2' be the matchings after performing two DCJ operations induced by (u, v) on M_1 and M_2 respectively.

(a) If (u', v') is weak w.r.t. M_1 and M_2, then (u', v') is weak w.r.t. M_1' and M_2'.
(b) If (u', v') is strong w.r.t. M_1 and M_2 and (u, v) and (u', v') are independent w.r.t. M_1 and M_2, then (u', v') is strong w.r.t. M_1' and M_2'.
(c) If (u', v') is strong w.r.t. M_1 and M_2 and (u, v) and (u', v') are not independent w.r.t. M_1 and M_2, then (u', v') is weak w.r.t. M_1' and M_2'.
(d) If, w.r.t. M_1 and M_2, (u', v') and (u'', v'') are strong, (u, v) and (u', v') are independent, (u, v) and (u'', v'') are independent, but (u', v') and (u'', v'') are not independent, then (u', v') and (u'', v'') are (strong but) not independent w.r.t. M_1' and M_2'.

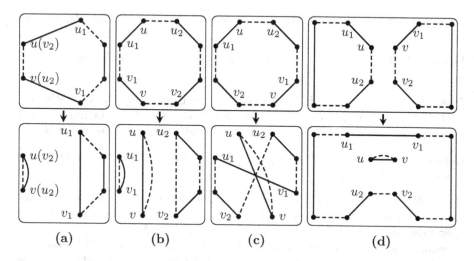

Fig. 4. The four cases for two DCJ operations induced by edge (u, v) on M_1 and M_2 (represented by solid and dashed edges respectively). u_1 and v_1 (u_2 and v_2) are the neighbors of u and v in M_1 (M_2). **(a)** u and v are neighbors in M_2; **(b)** u and v are in the same cycle at odd distance; **(c)** u and v are in the same cycle at even distance; and **(d)** u and v are in different cycles. In (a) ad (b), (u, v) is strong w.r.t. M_1 and M_2.

Lemma 3. *Let M, M_1 and M_2 be three perfect matchings on V. We have $d(M, M_1) + d(M, M_2) = d(M_1, M_2)$ iff M consists of n mutually independent strong edges w.r.t. M_1 and M_2.*

Proof. Choose one edge from M that is not in $M_1 \cup M_2$ and perform the DCJ operations induced by this edge on M_1 and M_2, and repeat until no more such operations can be performed. Let $2 \cdot o$ be the number of DCJ operations performed in this process and let M_1^* and M_2^* be the final matchings thus obtained. We must have $M = M_1^*$ or $M = M_2^*$ since at the final state we cannot find any edge in M that is not in $M_1 \cup M_2$. Without loss of generality, assume $M = M_1^*$. Using Property 1, we have $d(M, M_1) = d(M, M_1^*) + o = d(M_1^*, M_1^*) + o = o$ and $d(M, M_2) = d(M, M_2^*) + o = d(M_1^*, M_2^*) + o$.

Assume M consists of n mutually independent strong edges w.r.t. M_1 and M_2. By Property 3(b), all edges used to perform DCJ operations must be strong w.r.t. their current states. Using Property 2, we get $d(M_1^*, M_2^*) = d(M_1, M_2) - 2 \cdot o$ and thus also $d(M, M_1) + d(M, M_2) = d(M_1, M_2)$. Now suppose that there exists an edge in M that is weak w.r.t. M_1 and M_2 or that there exist two edges in M that are not independent. By the end of the iterative process, all edges in M are mutually strong w.r.t. M_1^* and M_2^*. By Property 3, there exists a weak edge that is used to perform DCJ operations. Thus, by Property 2, we have $d(M_1^*, M_2^*) > d(M_1, M_2) - 2 \cdot o$, which implies $d(M, M_1) + d(M, M_2) > d(M_1, M_2)$. $\square$

(u, v) is strong w.r.t. to $B = (V, M_1, M_2, M_3)$ if (u, v) is strong w.r.t. M_1 and M_2, M_1 and M_3, and M_2 and M_3. Two strong edges (u_1, v_1) and (u_2, v_2) w.r.t.

B are independent w.r.t. B if they are independent w.r.t. M_1 and M_2, M_1 and M_3, and M_2 and M_3.

Lemma 4. *We have $d_m = d_t/2$ iff there are n mutually independent strong edges w.r.t. B.*

Proof. We have $d_m = d_t/2$ iff there exists a perfect matching M_0 of V satisfying $d(M, M_i) + d(M, M_j) = d(M_i, M_j)$ for all $1 \leq i < j \leq 3$. By Lemma 3, such matching consists exactly of n mutually independent strong edges w.r.t. B. □

Lemma 5. *Assume $M_1 \cap M_2 \cap M_3 = \emptyset$; then we have $d_m = 2 \cdot d_t/3$ only if there is no strong edge w.r.t. B.*

Proof. Assume edge (u, v) is strong w.r.t. B and let M_0 be a median of B. We have three cases. First, assume $(u, v) \notin M_1 \cup M_2 \cup M_3$. We perform the DCJ operations induced by (u, v) on M_1, M_2 and M_3. Let M'_k, $k = 1, 2, 3$, be the corresponding new matchings and denote by $B' = (V, M'_1, M'_2, M'_3)$ be the new complete MBG. We have $(u, v) \in \cap_{k=1}^3 M'_k$ and the subgraph induced by $\{u, v\}$ is clearly adequate, so that there exists a median of B', call it M'_0, with $(u, v) \in M'_0$. Set $d'_m = \sum_{k=1}^3 d(M'_0, M'_k)$ and $d'_t = d(M'_1, M'_2) + d(M'_1, M'_3) + d(M'_2, M'_3)$. Since each DCJ operation can increase the DCJ distance by at most one, we have

$$d_m \leq \sum_{k=1}^3 d(M'_0, M_k) \leq \sum_{k=1}^3 (d(M'_0, M'_k) + 1) = d'_m + 3.$$

However, since (u, v) is strong w.r.t. B, by Property 2, we have $d(M'_1, M'_2) = d(M_1, M_2) - 2$, $d(M'_1, M'_3) = d(M_1, M_3) - 2$, and $d(M'_2, M'_3) = d(M_2, M_3) - 2$, which gives us $d'_t = d_t - 6$. Applying the upper bound on B', we get $d'_m \leq 2 \cdot d_t/3$. By combining these formulas, we finally get $d_m \leq d'_m + 3 \leq 2 \cdot (d_t - 6)/3 + 3 = 2 \cdot d_t/3 - 1$, implying that the upper bound cannot be achieved.

Second, assume $(u, v) \in M_1 - (M_2 \cup M_3)$. Now we perform the DCJ operations induced by (u, v) on just M_2 and M_3. We can thus write $d_m \leq d'_m + 2$. By Property 2 and using the fact that M'_1 is just M_1, we can write $d(M'_1, M'_2) = d(M_1, M_2) - 1$, $d(M'_1, M'_3) = d(M_1, M_3) - 1$, and $d(M'_2, M'_3) = d(M_2, M_3) - 2$, which gives us $d'_t = d_t - 4$. We thus get $d_m \leq d'_m + 2 \leq 2 \cdot (d_t - 4)/3 + 2 = 2 \cdot d_t/3 - 2/3$, implying that the upper bound cannot be achieved.

Third, assume $(u, v) \in (M_1 \cap M_2) - M_3$. Now we perform the DCJ operation induced by (u, v) on M_3 only. By similar reasoning, we get $d_m \leq d'_m + 1$, $d(M'_1, M'_2) = d(M_1, M_2)$, $d(M'_1, M'_3) = d(M_1, M_3) - 1$, and $d(M'_2, M'_3) = d(M_2, M_3) - 1$. Thus, we have $d_m \leq d'_m + 1 \leq 2 \cdot (d_t - 2)/3 + 1 = 2 \cdot d_t/3 - 1/3$, implying again that the upper bound cannot be achieved. □

The necessary condition of Lemma 5 is not sufficient, as illustrated in Fig. 5: the subgraph shown has no strong edge, but the median distance is not equal to its upper bound. Since the subgraph is adequate, we can build a general example by combining an arbitrary number of copies of this subgraph.

By Lemma 4, we can decide whether the median distance reaches its lower bound by checking whether there exist n mutually independent strong edges

Fig. 5. A subgraph with no strong edge where the median distance does not reach its upper bound. Matchings M_1, M_2, M_3, and M_0 are represented by solid, dashed, double and dotted edges respectively. We have $d(M_1, M_2) = d(M_1, M_3) = d(M_2, M_3) = 5$, yet $d(M_0, M_1) = 2$, $d(M_0, M_2) = d(M_0, M_3) = 3$.

w.r.t. B. We can reduce this question to a maximum independent set problem by setting a vertex for each strong edge and linking two strong edges if they are not independent. Clearly, there exist n mutually independent strong edges iff the size of the maximum independent set is n. The independent set problem is NP-hard, but we can test in polynomial-time whether there exist n mutually independent strong edges w.r.t. M_1 and M_2, a necessary condition. The algorithm enumerates all possible strong edges w.r.t. M_1 and M_2; this can be done in $O(n^3)$ time. Let $C_1, C_2, \ldots, C_m$ be the cycles in the breakpoint graph (V, M_1, M_2). Because each strong edge must have both endpoints on the same cycle, we can handle each cycle separately. For cycle C_i with $V(C_i)$ vertices, we use dynamic programming to compute the maximum number of non-crossing edges, taking time in $O(|V(C_i)|^3)$. If this maximum number is less than $V(C_i)/2$, then we cannot find enough independent strong edges and thus the algorithm returns false. If the algorithm terminates after examining all cycles, it returns true. The total running time is $O(n^3)$. The necessary condition of Lemma 5 can be tested in $O(n^3)$ time as well, by checking each pair of vertices to see whether it is strong w.r.t. B.

Acknowledgements. We thank Yu Lin for helpful discussions.

References

1. Fertin, G., Labarre, A., Rusu, I., Tannier, E., Vialette, S.: Combinatorics of Genome Rearrangements. MIT Press (2009)
2. Bergeron, A., Mixtacki, J., Stoye, J.: A unifying view of genome rearrangements. In: Bücher, P., Moret, B.M.E. (eds.) WABI 2006. LNCS (LNBI), vol. 4175, pp. 163–173. Springer, Heidelberg (2006)
3. Yancopoulos, S., Attie, O., Friedberg, R.: Efficient sorting of genomic permutations by translocation, inversion and block interchange. Bioinformatics 21(16), 3340–3346 (2005)
4. Alekseyev, M.A., Pevzner, P.A.: Whole genome duplications, multi-break rearrangements, and genome halving problem. In: Proc. 18th ACM-SIAM Symp. Discrete Algs. SODA 2007, pp. 665–679. SIAM Press (2007)

5. Braga, M.D., Willing, E., Stoye, J.: Double cut and join with insertions and deletions. J. Comput. Biol. 18(9), 1167–1184 (2011)
6. Chen, X., Sun, R., Yu, J.: Approximating the double-cut-and-join distance between unsigned genomes. In: Proc. 9th RECOMB Workshop Compar. Genomics RECOMB-CG 2011, BMC Bioinformatics 12(S.9), S17 (2011)
7. Shao, M., Lin, Y.: Approximating the edit distance for genomes with duplicate genes under DCJ, insertion and deletion. In: Proc. 10th RECOMB Workshop Compar. Genomics RECOMB-CG 2012, BMC Bioinformatics 13(S. 19), S13 (2012)
8. Shao, M., Lin, Y., Moret, B.M.E.: Sorting genomes with rearrangements and segmental duplications through trajectory graphs. In: Proc. 11th RECOMB Workshop Compar. Genomics RECOMB-CG 2013, BMC Bioinformatics 14(S. 15), S9 (2013)
9. Moret, B.M.E., Lin, Y., Tang, J.: Rearrangements in phylogenetic inference: Compare, model, or encode? In: Chauve, C., et al. (eds.) Models and Algorithms for Genome Evolution, Computational Biology, vol. 19, pp. 147–172. Springer (2013)
10. Hannenhalli, S., Pevzner, P.A.: Transforming cabbage into turnip (polynomial algorithm for sorting signed permutations by reversals). In: Proc. 27th ACM Symp. Theory of Computing STOC 1995, pp. 178–189. ACM Press (1995)
11. Bader, D.A., Moret, B.M.E., Yan, M.: A fast linear-time algorithm for inversion distance with an experimental comparison. J. Comput. Biol. 8(5), 483–491 (2001)
12. Caprara, A.: The reversal median problem. INFORMS J. Comput. 15, 93–113 (2003)
13. Tannier, E., Zheng, C., Sankoff, D.: Multichromosomal genome median and halving problems. In: Crandall, K.A., Lagergren, J. (eds.) WABI 2008. LNCS (LNBI), vol. 5251, pp. 1–13. Springer, Heidelberg (2008)
14. Siepel, A.C., Moret, B.M.E.: Finding an optimal inversion median: experimental results. In: Gascuel, O., Moret, B.M.E. (eds.) WABI 2001. LNCS, vol. 2149, pp. 189–203. Springer, Heidelberg (2001)
15. Moret, B.M.E., Siepel, A.C., Tang, J., Liu, T.: Inversion medians outperform breakpoint medians in phylogeny reconstruction from gene-order data. In: Guigó, R., Gusfield, D. (eds.) WABI 2002. LNCS, vol. 2452, pp. 521–536. Springer, Heidelberg (2002)
16. Arndt, W., Tang, J.: Improving reversal median computation using commuting reversals and cycle information. J. Comput. Biol. 15(8), 1079–1092 (2008)
17. Rajan, V., Xu, A.W., Lin, Y., Swenson, K.M., Moret, B.M.E.: Heuristics for the inversion median problem. In: Proc. 8th Asia-Pacific Bioinf. Conf. APBC 2010, BMC Bioinformatics 11(S. 1), S30 (2010)
18. Zhang, M., Arndt, W., Tang, J.: An exact solver for the DCJ median problem. In: Proc. 14th Pacific Symp. Biocomputing PSB 2009, pp. 138–149 (2009)
19. Xu, A.W., Sankoff, D.: Decompositions of multiple breakpoint graphs and rapid exact solutions to the median problem. In: Crandall, K.A., Lagergren, J. (eds.) WABI 2008. LNCS (LNBI), vol. 5251, pp. 25–37. Springer, Heidelberg (2008)
20. Alekseyev, M.A., Pevzner, P.A.: Breakpoint graphs and ancestral genome reconstructions. Genome Research 19(5), 943–957 (2009)
21. Xu, A.W.: A fast and exact algorithm for the median of three problem: A graph decomposition approach. J. Comput. Biol. 16(10), 1369–1381 (2009)
22. Aganezov, S., Alekseyev, M.A.: On pairwise distances and median score of three genomes under DCJ. In: Proc. 10th RECOMB Workshop Compar. Genomics RECOMB-CG 2012, BMC Bioinformatics 13(S.19), S1 (2012)

Author Index